Listening for the Light

A New Perspective on Integration Disorder
in
Dyslexic Syndrome, Schizophrenia,
Bipolarity, Chronic Fatigue
Syndrome, and Substance Abuse

Laurna Tallman

Northern Light Books
2010

Place names and the names of most of the individuals in this true story have been changed to protect persons' privacy.
Cover photograph by Richard Tallman.

Library and Archives Canada Cataloguing in Publication

Tallman, Laurna
Listening for the light : a new perspective on integration disorder in dyslexic syndrome, schizophrenia, bipolarity, chronic fatigue syndrome, and substance abuse / by Laurna Tallman.

Includes bibliographical references and index.
ISBN 978-0-9811672-0-6

1. Dyslexia--Treatment. 2. Mental illness--Treatment.
3. Cerebral dominance. 4. Tomatis method.
5. Sound--Therapeutic use. 6. Schizophrenia--Etiology.
7. Chronic fatigue syndrome--Etiology. 8. Substance abuse.
9. Music therapy. I. Title.

RJ496.A5T34 2009 616.85'5306 C2009-903667-3

1 2 3 4 13 12 11 10

Published in Canada by
Northern Light Books
P.O. Box 494, Marmora, Ontario, Canada K0K2M0
www.northernlightbooks.ca

Printed in Canada by KingsWeb Inc.

Contents

Acknowledgements

Integral to this story, as to my life, are the exceptional love and patience of my husband Richard, our children and their partners, my sister, and our extended families. The accolades they deserve will be evident as the story unfolds.

For their inspiration and encouragement I owe special thanks to Ruth Welham-Umphrey and Pat Gall. Among those who listened to me for years on end and were unfailing in grace and wisdom are not only Ruth and Pat, but Margaret Anne Scott Hunt, Mary Jane (Minka) Case, Charis Wahl, Nancy Gall, Faith Holwyn, and Leona and Paul Brady. People from many different religious and denominational traditions have lightened our burdens. Heartfelt thanks to each and every one of you.

Perpetual gratitude is due to the late Dr. Alfred Tomatis, and to Paul Madaule and Morana Petrofski of The Listening Centre in Toronto and their former staff, including Edna Druxerman and Sophie Garceau.

Daniel's long road to recovery was supported by the compassion and expertise of our family physician and all-too-briefly by the reliable counsel and enduring influence of Lauder Smith and his associate Steven Spicer.

I appreciate the kindness of the following people in granting permission to quote from their works: Professor Doctor Lars Heslet; Barbara Lewis on behalf of her late husband Nicholas Regush; Dr. Michael D. Mann, *The Nervous System in Action* (online) published earlier as *The Nervous System and Behavior* (Harper & Row, 1981). Thanks, also, to the following publishers for permission to quote from authors' works: Eastland Press for Dr. John Upledger and Jan Vredevoogd, *Craniosacral Therapy*; Lucky Press for Thelma Wheatley, *My Sad Is All Gone*; Dr. Jay A. Goldstein for his *Checklist of Symptoms for ME/CIFIDS*; Oxford University Press, for Dr. Terry Wotherspoon, "Education," in Lorne Tepperman, Patrizia Albanese, and James Curtis, eds., *Sociology*; Scarecrow Press for Alfred Tomatis, *The Ear and the Voice.*

My mother immersed her family in music, through her own accomplished musicianship on the piano and church organ, in teaching and singing, and in her conducting of several choirs and musical groups. My father, a research physicist and engineering chemist, taught me to reason, to observe, and to be uncompromising in the pursuit of truth. Their legacy endures here, also.

"Those who have ears to listen, let them listen."
Jesus of Nazareth

"As we move towards effective listening uncontrolled behaviour occurs less often."
Alfred Tomatis, *The Ear and the Voice* (2005 [1987]), p. 65

Introduction

This book is not about the good times in our family or about the achievements and gifts of its members. This book is about the problems—and at that, only some of them. Its point of view is sociological and psychological. Its discoveries impact our understanding of neurology and the function of the brain. It illuminates aspects of two of those "20 greatest unsolved problems,"[1] namely, free will and consciousness, and investigates little-understood factors that control various so-called mental states. As a study of the microcosmic social unit, the family, it does not attempt to be comprehensive, but restricts its considerations primarily to the issues of brain function mentioned in the title. Our son Daniel's role is central, because it was through trying to find solutions to his severe problems that I learned more about how the brain affects certain kinds of behaviour and states of consciousness that all of us experience to some extent. While his escalating needs tended to write the agenda for the family still at home after his diagnosis of schizophrenia—two of the children were adults and living away from home and the third soon would be—such was not the case to the degree that seems so in this telling. Other family members, too, are presented primarily in terms of the central problems they were dealing with or that bear on my discoveries. Please keep in mind, when you glimpse other dimensions of relationships, that remarkable accomplishments characterized each family member and resonated positively throughout the family during the troubled times recounted here.

This story draws together many strands of my experience, but focuses on a period of about a dozen years during which my husband Richard and I tried to balance our love and concern for every stressed member of an impoverished family with the special needs of one child who started to slide out of control when he was about 11 years old. The counterpoint to Daniel's story is my journey of learning and discovery, which did not proceed on schedule like academic study and was multi-layered—one problem partially solved was topped by another less easily solved, topped with another more intractable—a pyramid that eventually explained itself.

When family crises arise that are not adequately solved within existing social institutions, parents may have tough choices to make. A sociological tenet permeating this story is that the lack of social resources in rural settings is detrimental to the health of those living there. Another sociological fact, less often expressed by those actually caught in the talons of poverty, is that a subsistence-level struggle to survive also is detrimental to physical and mental health. Sociologists know this, but the individuals who might alter this situation with their votes, their wealth, and their choices continue to accept privileges without regard to responsibility—as formerly I did. And this problem is urban as well as rural, although the superficial features differ according to the differences in settings. The unjust distribution of wealth in our society leads not simply to inconvenience and a simpler lifestyle for an increasing sector of the population, but to intellectual deprivation, poor nutrition, inadequate protection from the climate, illnesses, exhaustion from excessive work, and the sense of hopelessness that perpetuates addictions and leads to untimely death.

Because Richard and I were committed to one another and to our children we could cover one another for some of the risks involved in the decisions we took in ways that most people, perhaps, would not undertake and that certainly would not be possible, for example, in an urban setting for a single parent who is employed. We took unusual risks. Because our employment was based in our home and in a rural location, we had space and relative isolation for Daniel's periods of mental illness. My sister's generosity with her house nearby created a more comfortable situation for the family members living in our house during Daniel's first efforts to survive on his own and during his most recent episode of illness. The deaths of at least 50 people recently who were attempting to care for mentally ill relatives at home in Quebec was the subject of a provincial inquiry. Daniel's story is not intended to be prescriptive in the dimension of home care.

Chronic Fatigue Syndrome Chapters 1 and 2 introduce the family in its social context and delineate some of the trials and other challenges we had already faced while the serious problems were developing. They are not intended to be a depressing litany of grief and stress to win pity, scorn, sympathy, or to invite moral and social judgement on those whose behaviour may seem to some readers undesirable. They provide a showcase of catastrophes, consequences, and other difficulties that will be seen, further down the road, as dramatic instances of greater and lesser integration of the hemispheres of the brain. A test of one's grasp of the new knowledge gained by the end of this story is to re-read those chapters in the light of that new understanding. My view of myself and of others changes profoundly in some ways from the attitudes and reactions I held when I began this journey of discovery.

The narrative is primarily from my point of view, with obvious contributions from Daniel. In any family with a disabled child, siblings are likely to feel they have been treated unfairly as the disabled one receives special treatment. When the disability is "invisible" and poorly understood even by the parents, inequities and inconsistencies loom large to siblings trying to grow up in a situation they consider far from ideal. All of this narrative takes place under the shadow of the illness of at least one of the parents, often of both, and of the same disabling CFS in one of the children. Maintaining family relationships—the microcosm of the society—includes coming to terms with human diversity and inequities and seeing what one can do to level the playing field—not for competition, but for the sharing according to individual abilities that produces genuine community. I hope that *Listening for the Light* will bring to other families the knowledge they need to see themselves on a continuum that is inclusive and that helps to define the human family.

Dyslexic Syndrome and a Strategy for Healing Chapters 3, 4, and 5 tell a story of learning, discovery, and healing. I brought considerable academic and professional experience to Daniel's foundational problem, which was dyslexic syndrome. Having been raised by a research scientist, I was made aware early in my life that observation need not be restricted to the laboratory. Just as Jean Piaget turned his powers of observation upon his own children and observed patterns in the developmental stages of their cognitive maturation, many of us may bring our learning to our observations of our children and others around us. My academic and professional experience proved to be germane to the problems I was observing in our family and to the discoveries I made.

To those who raise children, who care about children, and who want to understand better the world of the child whose listening skills are developing in ways that lead to confusion, learning problems, and anti-social behaviour, Daniel's story will be of interest. I have met a great many educators whose interest in the children they teach and whose schools they administer made them knowledgeable and insightful about the learning problems children bring with them to the classroom. The fact that research has not brought answers to important questions in pedagogy does not diminish the value of the determined efforts of educators to do the best they can for learning disabled children. These are the educators who will be prepared to lend an ear to a story that may change their perceptions of their students with learning disabilities and lead them to make important changes in their teaching strategies, especially in regard to children's listening activities.

My first experience of chronic fatigue syndrome healing through the method of middle- and inner-ear stimulation led me to related experiences, including the LiFT® Program through The Listening Centre in Toronto. The Listening Centre serves a range of clients' needs with "listening retraining" developed by the late Dr. Alfred Tomatis, which also helped to keep my chronic fatigue syndrome in remission. For those suffering the debilitating symptoms of chronic fatigue syndrome, the Tomatis Method may offer an avenue of hope for improved well-being, as it did for me. My husband was relieved of chronic fatigue syndrome and remained in remission for over a year; he continues to use focused listening to improve his health.

Our son Daniel was cured of all of his symptoms of dyslexic syndrome through sound stimulation at The Listening Centre. Daniel was an unusual client for The Listening Centre, although none of us realized he was at the time of his exposure to the Electronic Ear® and consequent healing from the symptoms of dyslexic syndrome. He had become addicted to alcohol and drugs, a disclosure that began to emerge just before the listening intervention. I was not sufficiently informed about addictions and about the drugs he was using to realize he had come to a precipice. In retrospect, it is well I did not; otherwise Daniel might have been deprived of a healing that provided a foundation for recovery from his psychotic episodes and from his addictions, and for a more rewarding life—or for life at all.

Psychosis/Schizophrenia Chapters 6 and 7 tell of unnecessary suffering. Some readers will live in places and in financial circumstances where they can pick and choose among extensive psychiatric and counselling resources. We do not. In some ways we were very fortunate, in others, less so.

I brought ignorance and naiveté to Daniel's addictions, but could educate myself once we discovered what he was doing to his brain. My strong interest and significant academic and professional experience in the way people think and express themselves at first brought objectivity into my observations of how Daniel's medications affected him. I soon found myself at odds with the two communities providing the drugs Daniel craved: both the local subculture and those in psychiatric practice.

A great many people in jails and mental hospitals reside there as a tribute to the failure of society to diagnose problems early and to remediate them effectively. How can we as a society—as educators, as doctors, and as parents—alter the circumstances that bring a dyslexic syndrome child to the verge of self-extinction? Answers to these questions may be surprisingly simple; the prevention of despair in children is one of my purposes here.

Substance Abuse Chapters 8, 9, and 10 are about addictions and about psychosis in addictions. The level of success I attained in observing Daniel's reactions to medication helped me to be aware of real changes taking place in his addictive behaviours that others missed or dismissed. For those fighting addictions or supporting those who are, the details and patterns of behaviour reported here will not seem merely tedious, messy, and unpleasant. They provide familiar points of reference. They illustrate the impact of illicit and prescription drugs on the ear and brain and hence on behaviour, but in in a new light, namely the differing functions of the two halves of the brain and the innate inability of some people with audio-processing deficits to control hemispheric shifting. They also explain the hemispheric shifts that take place in anyone under the influence of alcohol and drugs. The patterns of ascent out of psychosis that Dan became active in facilitating demonstrate the learning process that was possible because of his listening intervention. His experience at The Listening Centre underlay Daniel's ability and willingness to pick himself up and keep trying, despite his losses.

Many young people are doing damage to themselves with drugs and other harmful substances and behaviours. A great deal is said about why they should not. Much less is said about why they do—about the factors that work upon a young person to tempt him or her into experiences from which he or she may never recover. Even less is said about how they may not. The fundamental new knowledge in *Listening for the Ligh*t posits an underlying physiological condition that may be treatable in some addicts. It will help those already reaching out to addicts to better understand where some addicts "are coming from," why their brains make them behave in particular ways, and how existing methods might be modified to be more helpful.

Those near the struggling addict want the whole process over directly and may add to the suffering addict's burden with their impatience, criticism, anger, and scorn. At least, I did. Gerald May's observation[2] that the human tendency to addictions is built into human neurology and defines the human condition has helped me to put my own reactions to others' efforts into a more empathetic perspective. People conditioned to their addictions may also be innovative, creative, and compassionate. Now, I try to find ways to affirm, not criticize.

Another important lesson was learning to forgive myself for making the same mistake twice. As our counsellor Lauder Smith pointed out, we usually pay insufficient attention to the anomalous error; only when we make the mistake twice or more often do we notice the cause-and-effect pattern resulting in our lack of success. Directing this wisdom towards our children helped me to be patient as they made the same mistakes repeatedly. Recovery from addiction is usually a "two steps forward and one back" sort of climb; it is a learning process that may seem to the addict to offer little in the way of rewards.

The particular psychological and neurological dangers of overmedication to substance abusers—drug addicts and alcoholics—are sometimes overlooked or poorly understood. Neuroleptic medications for psychosis can perpetuate drug abuse by providing precisely the "escape" for addicts that they find on the street. To some degree, masking symptoms also removes from the addict the uncomfortable symptoms that may provide negative reinforcement against the addiction; but reduction or withdrawal of medication is a manoeuvre that can be overdone. Because the medications currently in use can cause psychosis and, possibly by damaging the ear, make that condition permanent, they can be as much a part of the problem as other factors impinging on psychosis. I think the learning presented here will show that a much finer line of mercy and wisdom could be attained in regard to medications for addicts in psychosis.

As has been dramatized in reference to the beginning of the AIDS crisis,[3] the sharing of information about medical discoveries across the Atlantic and around the world can be a slow process detrimental to the well-being of the millions of people at risk. I found this to be true when I was seeking answers for myself and the members of my family suffering from chronic fatigue syndrome. The same situation prevails in reference to dyslexic syndrome. Daniel's drug-induced experiences of psychosis that were diagnosed as schizophrenia suggest that parallels may exist for at least some of those individuals who are spontaneously schizophrenic. In my view, and in the view of some psychiatrists, other psycho-affective disorders—bipolarity (with and without psychosis), depression, autism, Tourette's, epilepsy, and likely other problems currently considered "mental"—apparently fall along the same continuum, suggesting those problems also may originate in ear dysfunction. If the reader finds many questions raised for which answers are not yet available, other questions are being answered here in perhaps unexpected ways that merit, at the very least, consideration and further testing.

Integration Disorder Chapters 11, 12, 13, 14, and 15 present a new perspective on all of the principal mental health problems experienced not only by Daniel, but by some other members of the family. The final stage of my learning proceeded relatively swiftly under the intense pressure of Daniel's recent episode of psychosis when the observations I had made earlier came into focus as the pieces of an intellectual puzzle that, with some research, fell into place. That perspective has implications for all of us because the fundamental neurological relationships, between what enters the ears and how the two hemispheres respond, define a universal spectrum of brain function.

I have had the peculiar advantages of not being steeped in the expertise of neurology or of medicine, but of having particular expertise in the critical analysis of language and training in objectivity (which involves self-knowledge), so that I could observe—watch and listen—"outside of the box" and keep records. Indeed, I had been doing so for more than a decade in particular reference to Daniel and for a lifetime more fortuitously. Hence, Chapter

13 is somewhat technical both in terms of language analysis and of neurology. Chapter 14, also, brings a close analysis of language in psychosis to psychology and the microcosmic social setting, the family. Research studies that inform or support aspects of my observations and my critiques and commentary have been referred to primarily in the Notes in order to maintain the narrative flow of Daniel's story.

As my husband and I had to learn how to let go of prejudices and erroneous information as we travelled towards new ways of understanding Daniel's disabilities and addictions, those who work within strict protocols for dealing with such disabled and mentally ill persons, such as social workers, nurses, and doctors, may need to take a step back from the assumptions and misconceptions built into their protocols in order to assess fairly the paradigm introduced here. Daniel's story also is for those who treat and medicate children, young people, and adults suffering from the range of neurological maladies that they may not realize could be related to ear function.

The widely differing philosophical platforms from which individual psychiatrists operate ensure bias in the way different cohorts of patients are treated. Bias issues, such as researchers' personal investment in their careers and the pharmaceutical industry's vast financial commitments along particular lines of research, can continue to influence the thinking of practitioners and their protocols for their patients. I crave the patience of those in positions of absolute power over their patients to listen to one story that may have enormous implications for their understanding differently the function of the brains of the spectrum of mental patients under their care and for considering revision in their modes of treatment.

The studies of schizophrenic individuals frequently are confounded by the drug treatments to which they are subjected amid confusion and disagreement about the causes of the range of mental problems subsumed under the term "schizophrenic." For reasons of their frequently extreme physical symptoms and their anguish—and the consequent distress in their loved ones and communities—most schizophrenics today are heavily drugged and encouraged very strongly to remain on these high dosages of medication. This procedure may be equivalent to telling a right-handed person who has broken the left arm that henceforth both arms must be kept permanently in casts and slings. Furthermore, from the standpoint of research it is difficult to know what you are learning about the underlying mental condition of individuals whose physiology has been severely impacted by pharmaceuticals. Even when medication cannot be avoided, like the crutch that is an aid to mobility given to someone in physical therapy following a foot injury, it can also hinder or prevent healing and healthy maintenance.

Daniel's story suggests new avenues of research. For those who recognize my extension of Tomatis's ear-brain paradigm as the Gestalt shift they have been seeking, Daniel's experiences also will suggest new modes of instruction and new therapies to be explored. While I have asked more questions than I can answer, I hope those questions reach the medical, psychiatric, psychological, and educational communities as a clarion call. The methods we used to support Daniel's recovery have drawn exclusively on the discoveries and learning of allopathic medical practitioners. There was nothing "alternative" about Daniel's treatments (which is not to denigrate therapies that still come under that rubric). Yet, the fact that some of Daniel's treatments derived from medical doctors who worked in languages and cultures different from one another has made the dissemination of their findings awkward. In North America the discoveries of Dr. Tomatis have not yet been integrated into the mainstream of medical and educational practice, and the pharmaceutical companies already have a grip on ADD, ADHD, and the other symptoms associated with dyslexic syndrome that may effectively mask certain problems, but without curing the underlying deficits, as appears to be the case also in regard to psychosis and psychotropic drugs. Children and mental patients are being medicated to make it easier for those charged with their care, their supervision, and their teaching, but these medications do not necessarily help in any substantive way the children and patients being medicated; and they may be doing irreparable harm.

To those who minister valiantly and compassionately to the mentally ill and who may be seeking inspiration and direction for exploring alternatives to present therapies, I beg forbearance as I tell the story as I lived it. As had also been the case in the educational environment, not all of our experiences with medical personnel seemed helpful, yet in retrospect I am sure that in most instances kindness was intended. If you will stay with me to the end of the story, you will share with me the 20/20 vision of hindsight in which my own mistakes glare forth as dreadfully as those of most others, once the etiologies and the useful strategies for healing arise clear and obvious above the miasma of ignorance and confusion.

The hidden world of the subconscious is made overt in the decisions people make, and those decisions affect the society in which one lives. I explore briefly in Chapter 15 some of the implications of the function of the brain that I have observed and how that function affects the microcosmic social unit—the family—to influence the macrocosmic social unit, the human family. For sociology and for the study of other cultures ancient, primitive, and modern, my findings raise issues and suggest new directions for research.

As I have been inspired by the histories of others pressing through challenging circumstances to find alleviation for their families' suffering, perhaps parents dealing with entirely different challenges from the ones we faced will find hope from our pursuit of answers to our ill health and from Daniel's climb from mental darkness into the light of sanity and sobriety. Sometimes, the solutions others proposed did not fit our circumstances or Daniel's needs at a particular phase of his illness but became useful in another phase. Be wary of generalizations and do not assume that the answers we found necessarily apply simplistically to people you know. With that caution, we hope that in the sharing of our stories you will find inspiration for solving the unique challenges in your life.

Prologue

On the Cutting Edge

> *Two roads leading west through Toronto cross like scissors. I once lived not far from the thumb-hold, but in literal déjà vu I discover that following the lower blade has brought me two blocks south of my friends' house, so I must backtrack with left turns among one-way streets to find the cutting edge—to approach their one-way street from the north. I used to drive right past the house, too, although its dark bricks rise three floors to command an intersection. Perhaps others also overlooked its austere presence: ten-inch brass numerals have been mounted on its façade at clerestory level. It had been a small art gallery before C. and her partner gradually choreographed their collection of exquisitely rendered drawings, watercolours, mixed media, and sculptural art against its expansive, well-lit walls. I park my muddy van and gain the deck. A hot pink notice protected in plastic and taped to a mailbox sized to accept book manuscripts warns: No junk/flyers/missionaries/ conservatives. Nonetheless, I am welcomed with hugs, hot tea, nut cookies, and a profusion of oranges. C. structures manuscripts for authors. She observes that no one reading my book will guess who I am that she knows. I need a Prologue.*

Decades ago, as a precocious Senior Editor of a major publisher, I hired this unpretentious woman, who was a little younger and better-educated than I, into her first editing job. I remembered her slightly from when she was a university student and I was struggling to remain one at summer- and night-school while immersed more than full-time in semi-professional theatre in production and design. Part of my job at Hart House Theatre was organizing the volunteer help of a couple of hundred students. C. had managed props for several productions—a perfectionist in the wings. During her job interview, she carefully explained all the reasons I ought not to hire her, including the fact that she was left-handed.

My publishing concerns employed my B.A. in English and ideals for educating children and young people and had largely supplanted my passion for theatre. I had drawn away even from the Stratford Shakespeare Festival where I had worked for a season, and from the plays and operas and ballets that had dominated my free time for years and enhanced my short career in theatre. I was transferring my expertise in dramatic art to people in real-life drama; at the same time, I was becoming a painter. C., on the other hand, was an arts patron.

Imprinting my childhood was the summer community of teachers, all from the same west-Toronto church, who had built cottages on the east shore of a small, fox-shaped lake in Muskoka. Dad was the pioneering Protestant who had earned his lot by helping to cut roads through the bush to its shores. I learned to walk at the cottage. Sunday services were held without fail at "Uncle" Jack and "Aunt" Laura's (I was named for her) "Loon Echo Lodge," where eventually I played their pump organ for the hymns. In our short holidays at "our" lake, my sister and I learned to swim, paddle a canoe, build a campfire, and revere the wilderness. The only other presence on Fox, at its south paw, was the Jewish Camp Ogama, which I knew only from the canoes on the beach. It also could be reached on the left fork of the access roads, but faster by water if you could paddle. We knew this because communication between these communities occurred (rarely), such as when an axe missed its mark creating a medical emergency: because the camp had a telephone and a motorboat. C. and I discovered during a coffee break that we had learned more or less the same lessons during the same summers in, on, or by the same small lake.

I missed the camaraderie among the editors when I was promoted; but I was a perfectionist, too—fiercely career-oriented without knowing quite where or how I wanted to make a difference. Weekends, I painted in oils, designed books, wrote poetry, entertained friends at dinner parties, and became intensely and widely involved in the Charismatic renewal. I fell in love with one of my editors. I joined the Catholic Church. I fretted to C. when I was to be interviewed for a career change. She asked, "Do they know you could have organized the Invasion of Normandy?"

I made a right-turn into educational research that sent me into dozens of schools across the multicultural city; I hired staff, supervised projects, designed testing instruments, advised curriculum consultants, joined researchers from school boards in the public system and from other cities in founding an association. I enjoyed every encounter; I tried to be useful. My apartment was tiny but centrally located. C.'s dinner parties and mine kept us in the eye of a Venn diagram of fascinating people.

I lived a protected life moving in business, church, and intellectual milieux in the safest modern city on the planet. The only threats I perceived were from the men who wanted to go to bed with me; or from careless drivers on the highway. I knew no one, I thought, who could deal death and destruction, or frighten me out of my wits or capacity for charity. I vacationed in England and Scotland with Presbyterians and changed my mind about justifiable warfare at the plain of Culloden. In scenes reminiscent of the ostriches in *Fantasia*, C. and I took ballet classes together with her sister, and went to the opera. I took lessons in conversational French. The day the man I had thought I would marry died suddenly, C. was the first of my friends who simply arrived at my door.

Michael's death answered profound questions. I still had my executive job, church, family, and friends. I sang in two choirs. I moved to a Rosedale apartment with my friend Margaret. We vacationed in Liège with a friend of her physician brother. But the rhapsody of openness and seeking had ended. Minka, my flower-child friend from a prayer meeting, wisely drew me into the total-body concentration of wheel-thrown ceramics. In a couple of years I resumed painting my landscapes and visions in oils. At night school I completed honours certification and the coursework for my master's degree, intent on teaching in the university. I negotiated a thesis topic crossing the departments of English, Fine Art, and Psychology. My dinner parties gathered other young professionals settling into their careers—a smaller group of cherished friends. Then, the sister of one of them was murdered.

At her wake, I met the man she was to have married. Marie had died after his estranged wife shot her while he was teaching a university class in folklore. We talked and walked; he wept; I listened. Two months later when he returned to the city, we knew we would marry. I would be immigrating to the US to live among the folk in Cajun Louisiana. Church friends met at our apartment for Richard's baptism and for Holy Communion before the farewell dinner. Demurely agnostic, C. declined to attend the early formalities; she brought the gifts of loaves from the Spadina Bakery, and the wine for the party.

1

Desert Wine

Our Family

After we moved from Aurora to rural southern Ontario, I rarely painted. Nowhere in the house would an easel and oil paints be safe from the tumble of children and pets, guests, projects, and renovations. But one day I came across my watercolours, found some paper, and sketched a landscape. What grew under my brush was sere, desolate, like the desert in Nevada or a wasteland. Over it I drew a pair of champagne glasses, knowing they symbolized the always effervescent love between Richard and me. I do not know what became of the sketch; the image floats still in memory symbolizing unimaginable suffering and incomprehensible love held in fragile vessels.

A Family Divided

Richard and I had expected that April's mother would go to jail for a long, long time and that he would be given custody of his five-year-old daughter April. In any other state than Louisiana that most likely would have happened, but Louisiana (and Florida) law is based on the French civil code in which crimes of passion by a spouse against a rival are viewed paternalistically as somewhat justifiable homicide, unlike the precedent-based English common law that is more prevalent and familiar. Everyone, including Richard's attorneys, assumed that she would be found guilty, even in Louisiana. What actually happened at her trial for Marie's murder was that his wife was acquitted on the grounds of insanity from before puberty and she was remitted to outpatient treatment. Thus, six weeks before we were to marry, I learned that someone deemed at the murder trial to be a grave danger to me and possibly to any children I might have with my soon-to-be husband was at large and would remain so.

Four days before our wedding, April's custody suddenly became a matter of debate rather than a foregone conclusion. During her parents' divorce hearing, a bomb scare cleared the courthouse. The judge, who had said he would not separate the issues of divorce and custody, reversed his inclination with the lawyers on the sidewalk under the broiling Louisiana sun. If the parties would reassemble after the building had been checked he would reconvene the hearing, grant the divorce (enabling our wedding fully prepared for a week hence), and continue on custody. April was placed temporarily in her mother's custody; and her mother would not allow her to come back to Canada with us for our wedding.

In Toronto, I hugged my welcoming mother and burst into tears as I tried to explain the situation to her; both our families were as aghast as we were at the fate, even temporarily, of Richard's little girl.

After the wedding I traversed the rituals of immigration and, once we had settled in Louisiana, a slew of social workers' investigations (two of them quit in rapid succession before a third one completed the process). Then, Richard lost his claim for the custody of April. The judge, who refused to consider the information about April's mother read into the murder trial—because she had been acquitted, and without consideration for the reason for acquittal—made his decision on my testimony that we had an unlisted telephone number. Although April's mother had instant access to us and her daughter through Richard's attorney, apparently that situation represented some technical illegal separation of mother from child. His attorney had not seen the phone arrangement as a problem until the judge admonished her during the custody hearing. April's father was accorded "generous" visitation rights.

My Beloved was shattered. I felt responsible. His parents came to the little cement block house that we were renting and the four of us cried together. The first theme of our marriage, which had begun with my husband dealing with his shock and grief from Marie's death, modulated rapidly into his grief and dismay over his loss of April. And by then I was experiencing physiological depression as a side effect of my first pregnancy and cultural shock from being miles from any city and living in isolation more complete than a hermitage. It was a huge adjustment for me to learn to live without my friends and the cultural activities of my city, although Margaret came to visit, my mother was a faithful letter-writer, and a few friends also wrote to me. My faith, my executive skills, and the love growing between Richard and me as we worked in tandem through one legal or relational problem after another kept me calm and clear-headed (most of the time) and focused on April's fragile mental and physical health. During the weekends she was with us, we usually drove to Richard's parents' condo near the Gulf. His mother had heavy drapes installed that she drew each evening so no one could see in to take a bead on her.

In the Ozark Mountains

Believing that we might not be safe in Louisiana, Richard left his career track at the end of the semester for a teaching position at a small college in the

Ozarks, so we moved from the bayous to the mountains, where I learned to survive in still deeper isolation. Essentially, there were few places to eat outside of our own kitchen and no money to spend even if there had been. The nearest city was Little Rock, a two-hour drive away. The life of modest comfort I'd enjoyed before I'd married, partly the outcome of sharing accommodation with a well-paid teacher, was replaced by penury—U.S. college teachers were being paid about half of what was earned by their Canadian counterparts. The spheres of social behaviour in which I had defined myself by my choices, whether picking out food in a supermarket or sorting through racks in a department store for clothing I needed or even window shopping for the fun of it, had vanished. Richard had his work at the college and at the new Folk Center and had papers to mark at home in the evenings. I was at home all day without the means of travel or opportunities for socializing. The one young couple living nearby were at work during the day, too, but were neighbourly. Charlotte and Bub had a new baby girl and helped me talk Richard into Lamaze birthing classes. Even occasional get-togethers over coffee with Charlotte eased my loneliness. We had minimal food, few possessions, no alteration from year to year in our plain clothing except for a few maternity clothes for me, and visitors infrequently. We eked out some money for a few canvasses and some paints so I could continue painting.

Richard loved his work, and after Luke was born I was drawn into it, by helping him with the technicalities of book publishing as he produced a monograph series. Developing an archive and overseeing aspects of the burgeoning Folk Center at Mountain View were aspects of his job that taught me the ways, arts, and music of the mountain people, too. We collaborated on folklore curricula for schools in both Arkansas and Kentucky, and had fun producing a book for use by folklore students that included my folksy illustrations. Although I worked at home, these collaborations got me out of the house occasionally. Of course, there were few people where we lived in the mountains. The college was nested in a small town. The Folk Center drew people out of the woods and "hollows," the steep, winding valleys through the mountains, who lived as ascetically as we. The balladeer of fame, Almeda (Granny) Riddle, for example, had things to teach me about raising children in impoverishment as we shared a simple meal with her under her quilting frame that was suspended from her low parlour ceiling.

Because April would become upset by questions about her life at home, we learned not to ask them; but the fragments she spontaneously conveyed suggested a feast-or-famine quality to her emotional environment. Extravaganzas of her mom's attention, such as the dazzling birthday party with April's entire class attending, alternated with emptiness, as when she left six-year-old April with her partner for several months (except weekends) soon after the custody judgement while she attended school in a distant city. Richard was furious at such an arrangement for April's care, especially when he could have parented April during that interval.

On our side, maintaining the relationship of father and daughter and family seemed to require heroic measures. The child support payments took everything we had above subsistence living. Richard drove through blizzards three years in a row (once, when even the police would not use the roads) to fetch April from Illinois, where she had moved with her mother, to Arkansas. I would spend half the night listening to the interstate reports for truckers, anxious for their safety in unprecedented storms and dangerous road conditions.

We had warm reunions. We celebrated the arrival of a baby brother the first year; and the next, the advent of a baby sister. The longer summer visits mingled small adventures such as trips to Ozark natural sites and folk festival venues with the routines of a happy family and outings to swim and picnic. April would ask to go to church with us and then become too upset to remain. We discovered she was being taught contrary to our own beliefs, so we soon learned to skip church when April was with us.

At the Townhouse

We had moved to the Toronto area when summer again brought April to us; she was extremely thin and much too tired for a ten-year-old. One night I awakened to find her sobbing, hunched over the stair railing outside our room. She had dreamed that her mother had come and shot everyone in the family except her. It had been five years since Marie's death. I cuddled her, comforted her, and reassured her. We were aware from her attitudes that she was moving beyond our control and influence on our watch and perhaps beyond her mother's at her home base as well. Not yet 11, she was wearing makeup, mascara, metallic-red nail polish. The reading material she brought with her was *Petals on the Wind*, a sequel to *Flowers in the Attic*, which she said was a terrible book but that her mom had given it to her. I hesitated to take from her the reading her mother provided, although I considered it inappropriate to her age despite her psychologist-assessed genius intellect. I wondered if it had been given to her precisely because it provided a damaging perspective on sibling relationships. Or perhaps her mother had nothing to do with it. Between the disparate realities of April's two lives, a great gulf of silence, superficial responses, and deceptions created an unbridgeable barrier.

She was moody, sarcastic, and showed a rough sort of independence tinged with the justice of the street. She could play enthusiastically with her young siblings, or tune them out completely. The gradations of interaction normal to sibling relationships were lacking: she performed, or she withdrew. Years later, she would admit that she and some of the neighbour children had amused themselves that summer by making detours out of the common playground to enter other people's townhouses, apparently without taking or damaging property, but taking other liberties, such as playing records.

Many or most divorced parents make an effort to stay in communication in ways that will benefit their children, but to say relations between April's parents were not cordial was an understatement: whenever Richard tried to communicate with his ex-wife she was hostile, angry, or abrupt and usually conveyed only the most minimal information necessary for arranging April's transfers. Her unpleasant manner with Richard kept us anxious about April and uncertain about our own safety. And Richard, for his part, hesitated in attempting to bridge the gap. He could not know, as April would claim in her teens, that the manner he fielded in phone calls was reserved for him and that her mother treated others with greater civility. Thus, two sorts of "sound barriers" prevented her parents from co-operating in any way regarding April's upbringing. Some of April's behaviour showed her increasing stress from her feelings of conflict as she tried to encompass or alternately adjust to the contrasting conditions and values of her two families.

Letting go of April at the end of the summer was indescribably difficult for us. Richard and I discussed privately the possibility of kidnapping her and forcing a retrial. April didn't want to leave, but made some terrible I-don't-care-it-doesn't-really-matter shift in her thinking and retreated into another persona. After he left with her for the airport I lay down, feeling pain through my whole body. Several years later, I read that kidnapping, even with the best of intentions, has had poor outcomes for the children.

Learning How to Survive

The weaving of our lives seemed to be making a smoother fabric. Our second move, from the Ozarks—via Richard's summer teaching job in the Maritimes—to the Toronto area, had put us again within visiting distance of my family and friends. Through the generosity of Richard's parents and a small pension windfall from the college we bought a townhouse in a new development. We were past the new-parent anxious stages with our first child, Luke, and the colicky digestive upsets of baby Sarah's early months; their learning and antics as toddlers brought us continuous joy and much delighted laughter. We enjoyed friendly neighbours; we attended church; we joined an association of Christian artists. A modestly successful year financially as freelance editors encouraged my expectations of renewing a busy social life, although the prospect of Richard's getting back into academe remained uncertain. Our son Daniel was welcomed into the family in 1980.

Then, the recession of 1982 reset our financial clock, leaving us without sufficient editorial work to survive. We had a desperate year. I miscarried our fourth baby. We sold our townhouse and found a 93-acre marginal farm property with a dilapidated century farmhouse. C. phoned and suggested editorial work might be shipped to our new location by bus parcel express. Neither Richard nor I had imagined we might maintain that meagre stream of employment in rural Ontario. The managing editor at McClelland & Stewart was willing to take the trouble to send out work and the rivulet that expanded or shrank with the fortunes and seasons of the publisher remained vital to our survival.

Once again, we had become surprisingly poor for professional people with multiple degrees. We were falling behind in child support payments and scraped to hoard nickels and dimes to pay April's travel expenses. If at times we should have been receiving welfare support, we were too blindly middle-class to know that. Being so sub/urban, and not only inexperienced but uninformed about life on the land, we made ridiculous mistakes as we tried to make use and to make sense of our rural property. Guided by Department of Agriculture publications, we planted a garden big enough to feed three families and then nearly did ourselves in trying to look after it. We did learn how much to grow and how to preserve most of our own food, picking berries from the fields and apples from the trees around the house; collecting eggs from our laying hens and turkeys penned in the old barn; slaughtering a flock of chickens for meat in the fall; cutting our own wood supply for the stoves. I made almost all our food from scratch: bread and all other baked goods, corned beef, preserves, spaghetti sauce, and pickles. My husband's hilarious first efforts at wood-cutting—with a hand saw and sliding the laboriously cut logs down the icy slope of the west field—quickly sent him after a chainsaw, which even the mechanically challenged Richard learned to use and care for. After the frightening deprivation we had struggled with in the city, we felt blessed.

Part of our intent in choosing a rural location with a large house was to offer our place for counselling and retreat, an extension of our work with the Association of Christian Artists in Toronto and in our church. Once we were settled, some artists (and some not) came to relax and return refreshed to the challenges of their lives in the city. Visitors could easily walk over the hill and out of sight of domestic preoccupations. Camping out was an attraction for a few. People with a variety of problems sought our counselling. Some stayed for weeks or months. Occasionally, people saw the sign at the gate and came to talk. Not all who came wanted what we offered; one or two left deep wounds.

We watched our own children adapt to the rural landscape, learning a relationship to the fields and streams that for us had been brief cottage or holiday experiences. While the peace was not unbroken, a general harmony prevailed among the children, between us, with our neighbours, and for our family in relationship to the church and school in the village a few miles away. That was epitomized for me when Richard's annulment was granted, and with our children present—our fourth baby Eric in the arms of our close neighbour Linda—our marriage was blessed by the Catholic Church.

Waging the Peace

During our first summer in the Ontario countryside, at the end of April's visit, her six-year-old brother Luke recognized for the first time his loss when she had to return to the Midwest. He cried for a long time in terrible distress while his father held him and tried to console him. I had my first glimpse of what our decision to keep April integral to our family was going to do to the other children. At the same time, I learned far better what happened to April that she kept inside and how she felt after she had left us that we didn't see or hear about. What we saw in Luke her mother may have had to face tenfold in April.

For people accustomed to inches of space between houses and the semi-permeable membranes of apartment walls, the spaciousness of our country estate was a munificent gift. Richard and I were thrilled with our land, its woods, fields, and valleys; spring, creek, and wildlife. Even after many walks over and around the richly treed great hill we could not encompass the magnitude of its loveliness. The century farmhouse was another matter: snow drifted through the broken lath of the plaster walls onto our pillows the first night we slept there—an enormous task of renovation lay ahead. Within our first months on the land I was introduced to a nitty-gritty version of domesticity far from sentimentality and far from the life I had been living. I had to figure out where I stood on a number of unfamiliar issues. I had held a naive notion that rural peace would mean living undisturbed.

During the first year I kept very careful accounts to learn what life in the country cost and yielded—for example, the price to the penny of making our own daily bread down to the last quarter cup of flour. The next year April's mother sued Richard for back child support. In our effort to keep up payments to her (who was comfortably employed) we had already cashed in all our life insurance, sent her every dollar from the sale of the townhouse over and above the cost of the move (while we carried a charge account debt for the minimum of food we had consumed), and always gave her a portion of our small earnings, even though that deprived our other children of essentials. We would have continued in this fashion, as he explained to her, until she had been paid in full. She would not be patient. Richard walked through low doors of humility seeking legal aid as he wept for all of his children caught in different webs of deprivation. A lien was put on the property. When the Ontario court asked for information in regard to our finances, my accounts were as detailed as could be imagined. After months under this cloud, the suit was decided in Richard's favour; the amount of child support set years before by the Louisiana court was reduced to the amount of the Ontario "Mother's Allowance."

Growing in Community

Despite our straitened finances and the constant ache regarding April, I felt we were a comfortable and reasonably normal family. Luke and Sarah did well in school; the children usually got on together creatively, especially when April was on hand with suggestions for new entertainments. Friends and family came to check out our "farm." In the fall, I was enjoying the company of my sister-in-law, who characteristically had pitched into the current task—we were stacking logs for winter fuel into the driveshed. She expressed her admiration at my cheerful patience as I fielded the children's endless questions while they ran about and got in the way. She surprised me; I was simply doing what I had learned from the patience of my own mum and dad, both of them gifted teachers.

Although remote from the urban lifestyle I enjoyed, rural Ontario was a social experience compared to my isolation in Arkansas. We attended the Catholic Church and its activities and joined in community events: Victoria Day fireworks, summer baseball games in the village, hayrides with neighbours in the fall. With a loan from C. and the astonishing generosity of neighbours who hardly knew us but let us borrow their truck, we drove to Bezek Centre west of Toronto, the place of retreat and ministry where we had been married, to obtain their potter's wheel. More amazing, when our neighbours' truck broke down, they drove the four hours to Milton to retrieve us and it with remarkable good humour. Later, because of the generosity of all of those people, I was prepared to teach wheel-thrown ceramics at a community college and to conduct clay art classes from my home studio and in the schools. Those activities helped to keep flour in my grandmother's big, old lard tin and milk in the children's mugs.

My sister Lainey, a teacher, bought a house on the same road as ours that brought her some rental income; then she sold it to our parents when they moved from Toronto in their mid-seventies to be near us. Mother was very generous and helpful with her limited resources, although our parents on both sides held us responsible for our poverty—as though the economic disaster right across Canada wasn't real to them. However, we were no worse off than many of our neighbours and new friends, and better off than others. Then, some of our peers were back-to-the-landers, choosing for their own ethical reasons what had befallen us through the recession. A friend who shared her Mennonite cookbooks told me stories of her childhood that I understood better in our minimalist economy. The Mennonites, Amish, and other groups attached to the land had been living according to such ideals for centuries. Wine-quaffing gourmet cook though I had been, I admitted ruefully their ways were becoming important to me. As a constant re-evaluation process was taking place in my head, I was dimly aware that our children were going to have to solve problems different from the ones we had faced as suburban children of a more optimistic era.

Some of our guests viewed our life in the country as idyllic. At first, I thought it would be, too. The churches seemed quaint and charming, for they were conservative and a comfortable half-century out-of-sync with their urban confreres. The old-fashioned ways and friendliness of neighbours we could wave to across the fields reminded me of the community Lucy Maude Montgomery had portrayed in *Anne of Green Gables*. I thought my own world of imagination, like Anne's, would flourish in such a community; I would paint and sculpt and form a community of artists. The need for hard labour pushed most of my creative plans to the backs of filing cabinets and desk drawers, although I continued to write when I could. The accounts given by our neighbour, Marion, of her adventures on the farm across the road also tempered that romantic view. From her I learned that maternity care during a winter storm might be whatever one could provide for oneself with help from a neighbour. More often than not, care of sick and weak farm animals and pets came down to some very harsh alternatives. We sheltered a few stray dogs and cats that found their way to our place. Marion and her husband Barney instructed us in the killing and dressing of chickens. Linda and Brian helped us in the slaughter of four of our huge turkeys and shared a drink with us as we toasted the sacrificed birds that would top up our freezers with holiday meat. Bernadette and Morris let me help them wrap the beef they had butchered as they taught me about cutting and trimming and cooking. Country life was not idyllic for most of us living on the land, but camaraderie in the work did lighten the burdens.

Assault by Children

As the children matured and took larger roles in the community, and the community sometimes came to us, we found that relative isolation did not protect us from the same sorts of problems that occur in the larger society—however, this community was not supplied with urban resources. Gradually, distressing incidents accumulated: a child played sexually with one of the children, but the parents apparently were very abusive and complaining to them would have helped no one. Some of the neighbourhood children were very rough with our children. This happened at a time when zero tolerance for violence was becoming standard policy in provincial schools, but certainly was not a prevalent standard in many communities. Luke became silently defensive and aggressive; Danny and little Eric were treated violently a few times and then Luke saw to it that they stopped playing with those children. Our children passed off as indifference their aversion to their former playmates so that we missed what lay behind it.

Eventually, I saw that I had assumed that the churches were providing teaching and a kind of moral security in the community, as had been my own childhood experience. My naiveté was shared by the greater part of Canadian society, which found its confidence in the institutional churches shaken by revelations of sexual abuse and violence in the Native residential schools and at such orphanages as Mount Cashel in Newfoundland. I would learn such abuses were by no means restricted to the particular institution we attended.

One Child Was Different

Danny became extremely lonely once Sarah started school. He was not interested in preschool learning about letters and numbers as the other children had been. It was difficult for me to engage him in activities in quite the same way I had done with his older siblings. I knew there was something different about the way he was learning and behaving. He was sweet-natured and intelligent, with a prodigious memory, but he could become very distracted. I suspected I was seeing the early manifestations of dyslexic syndrome.

My development work in educational publishing included reading programs designed to help all children to learn to read, including those with dyslexic syndrome, and as a result of my subsequent career in educational research I knew much more about the problem than most educators. But I could see that even my expertise was not enough to bridge all of Danny's difficulties. Once Danny started school, he had increasing trouble with some kinds of learning, but remained a happy child at school and home generally.

Only around Grade 5, when he was picked on by an especially violent schoolmate, did he begin to show behaviour problems at home, too.

Assault in the Community

The community environment was hardly idyllic, either. The bank in the village was held up by robbers. And the convenience store. Regularly. Among the people who came from the city to seek our counsel were abused women, sexually abused children, people who had been assaulted when they were children. Among the people who never would have considered seeking help from us or anyone else (as I would learn from friends and neighbours) were victims of precisely the same abuses scattered through our community. Rural people go to great lengths to maintain their family ties. They may also be willing to keep serious problems under a cover of silence for the same reason, even when they perceive them as problems. The churches and charitable organizations retain influence, but they are poorly resourced.

The relatively fewer people in a rural social setting more obviously impact one another. Put another way, the dozen young people who are of the same age may be scattered over a 10-mile radius and they will make extraordinary efforts to connect that city young people can manage by walking a few blocks. Families, as a social unit, in the rural setting more easily come under duress. For example, a farm teen who "drops out" through drugs can mean the loss of the needed additional help to keep the small farm running, which is the source of livelihood for the rest of the family. Farm families are no longer as large as formerly; the margin for loss has shrunk.

Loved ones, even those living in other places, assume a larger place in consciousness because there is no hubbub of city crowds, no passing parade. As the circle of friends and acquaintances shrinks, it also intensifies. I noticed this in my reliance on the few friends who stayed in touch with me even though my life had become far removed from the lifestyle we had shared.

The labour of farming—especially in an age of small families and fewer hands to share the work—allows for relatively less leisure than it used to and far less than the spare time enjoyed by regularly employed townsfolk, even those in the nearby village. Such differences were evident between our farming neighbours, the retirees who still kept a cow and a few chickens along with their gardens, and even those who commuted 90 miles to the General Motors plant in Oshawa. The employed and under-employed made up the civic organizations, the politically involved; the farm folk kept agricultural organizations lively.

While few had much leisure, in an emergency of any kind, neighbours could count on one another for willing help—whether with chicks that were dying at 3 a.m. ("Barney 'n' me 'll be right over on the four-wheeler"), a chimney fire ("our boys 'll be over in two minutes if you need help"), an impending birth ("Now you call me if you go into labour"), or a car that slid off an icy road into a ditch after a meeting ("Brian's on his way with the tractor"). Folks would do what they could for one another. A special kind of peace grew from those responses, in the feeling that, despite hard work, some loneliness, and the pain inherent in impoverishment—and despite our differences of occupation, education, and our ignorance as newcomers—we were accepted and contributing members of a caring community.

The Family as Community

After a few years, we began to feel that our own children were developing problems that seemed unusual to us. The first issue was that the back-and-forth circumstances of April's relationships with her siblings were not only hard on her, but became deeply distressing for her brothers and sister as well, especially during the weeks after she had left us to return to her other home. The second was an intractable unpleasantness that had grown up between two of the children; their reconciliations became perfunctory. Then, Danny's difficulties at school increased despite testing that showed he could function in some ways near genius level. The forms of remediation offered by the school showed minimal effect and an open-area classroom had disastrous consequences for his learning. His differences were in some ways fostered, in other ways rejected, by his siblings. Lines of opposition formed gradually within the family and we spent many sessions trying to teach the children to be fair, forgiving, and loving despite their differences.

Our youngest son, Eric, like the tail on a kite, was pulled in many different directions as his older brothers and sisters struggled with the pressures both inside and outside the family. Eric, like Sarah, tended to shyness; it was difficult to get him to express his feelings about the powerful currents swirling about him.

The pressures of poverty were felt by all of the children in the clothing we could and could not provide, the after-school activities available, and the furnishings and material resources of their home. Their birthday gifts compared unfavourably with those of their school friends; the only holiday gift-giving was made possible by the generosity of grandparents and my sister.

Assault of a Teen by a Teen

During one of April's visits when was in her mid-teens, she said she wanted to remain with us, even though she already had a boyfriend at home. All the arrangements were made for her transferring to high school here. Then, her mother apparently found out that April's boyfriend was seeing someone else, and phoned to tell her about it. This tactic had not only the effect of fracturing April's nascent heterosexual friendship, but of reminding April of the lengths her mother had gone to in order to try to get what she wanted. She phoned every couple of days and put a lot of pressure on April to return.

Her father and I refused to ramp up the pressure on April to stay. She knew we wanted her to; we had tried to smooth the way for that transition. Essentially, we felt the way we had over the kidnapping possibility, as though the paradox of King Solomon had been set squarely before us. Should Richard declare "possession" of April and tear her into two halves? April awakened distressed one morning; later in the day she shared her troubling dream that suggested she feared that choosing her father over her mother would destroy her. It had been almost 10 years since Marie's death. Having lingered for a few weeks, April acceded to her mother's wishes. Against my pleading, Richard risked both their lives by driving through a major snowstorm to try to get her to her flight from Toronto. Partway there, the roads became impassable; they had to retrace the dangerous journey and rebook the flight, which had been cancelled. April received a clear lesson in the sort of risks her father had been willing to undertake for her, driving through snowstorms being the least of them. But she stuck with her plans.

The summer before she turned 16, April refused to comply with her father's visitation rights. Her mother had found a chance for April to take a trip with other teens to South America. Legally, April was constrained under the court decision that gave her father access that summer, but the alternative, understandably, was very enticing. For some inexplicable reason, I had a very, very bad feeling about the situation. We both tried by phone to persuade her not to go, but to come back to Ontario as she usually did. We did not succeed, and April's alternative summer itinerary resulted in serious illness later compounded by shingles, sexual assault, the move of her mother to Nevada, and April's being shuttled from South America, to Illinois, to Nevada, to a psychiatrist.

We could only grieve. Richard did not have the means to travel to her; even if he had, her location kept shifting unpredictably. Phone calls seemed an inadequate means of support, even as he reached out that way to comfort her.

Assault of Children by a Family Friend

Three years after the fact, two of the children disclosed that a Christian friend who had come to us for counselling about his failing marriage had behaved inappropriately with them. We located him in Vancouver, and confronted him in writing and by phone, which was awkward and somewhat unsatisfactory. He wrote to apologize and apologized in person a few years later. We stopped inviting people into our home for retreat counselling.

Our location offered few choices compared to a city, where organizations, neighbourhoods, schools, churches, other resources—even trustworthy friends—provide a range of alternatives. We hadn't the means to access the mentors and counsellors who at one time had been so important in my life in the Toronto area. There were no priests or ministers or other counsellors we trusted that we could turn to. While we accepted the responsibility for dealing as carefully and sensitively as we could with this difficulty, that is not to say we dealt brilliantly with all of its ramifications. For example, the distance in time of the disclosures from the events made it difficult for Richard to access more emotion. The children felt that their father showed too little outrage at the violation of their innocence. Richard tends to be quiet, laid back, and contemplative about any serious problem that presents itself before he tries to deal with it in a rational manner. In this instance, along with his sense of failure in not having adequately protected his own children, he was dealing with the betrayal of someone he had considered a friend, the failure of his own discernment in counselling, his awareness (which he could not discuss with his children) of the horrific abuses the friend had been subjected to in his childhood that had given rise to aberrant behaviour as an adult. I, on the other hand, was furious at the man's actions and what that revealed further of his lies, deceit, and betrayal of our trust, and my anger spilled over unfairly onto other family members.

Dreaming of Golf

Richard surprised me after we had been here a few years by commenting that our property was suitable for a 9-hole golf course. An excellent golfer, he began to explore an idea that evidently had occurred to him at first sight of the contoured landscape. In some ways the property was limited—by the hydro lines running across the land on the far side of the hill, for example—but as he started to map out the tees and greens, it looked like an idea with real possibilities. Working in his spare time over the next couple of years he brought the dream to the brink of realization. Enthusiasm in the community was high and some brave souls turned up one spring day for a game of golf across the roughed-out course that had been prepared with mown tees and greens equipped with holes and flags over the proposed course layout. Marie's mother, generous as always, made a contribution to our start-up expenses for a driving range, and Luke was old enough by then to greet customers and hand out baskets of balls, and at dusk to gather them by hand for the next day's practising golfers. He managed the little business—and much of the mowing—very responsibly during two summers.

We drew up a business plan that won considerable local respect and we consulted with a course architect in Toronto who came out and enthused about Richard's design while making some useful suggestions for refinements. My mother and Richard's mother had come into inheritances that would have made the business plan work. Neither was inclined to invest in our project. As we sought other investors we discovered that pursuing the golf course would jeopardize our outright ownership of the land, a risk we would not take. While it was a great learning experience for us and for the children old enough to grasp what was in the works—and Luke certainly benefited from the experiences associated with it—the outcome was not just a disappointment for us, it reverberated through the immediate and extended families as our failure.

"Asian Flu"

After I had regained my health from severe tonsillitis in my teens, it was very unusual for me to feel so ill that I would remain in bed, even for a day. I had considered myself of low average strength or endurance, but fit enough for my managerial office work of a demanding kind. After we settled in the countryside, the physical work of extensive gardening and house renovations gradually made me stronger in my forties than I'd ever been when younger. I managed rather physical eighteen-hour days with enough energy still for joyful nights. Then, in the spring of our fifth year in the country (1988), four members of the family caught a virus that was being identified as "Asian flu." For three weeks, among other symptoms, I suffered such extreme vertigo that I remained in bed, and sometimes needed assistance navigating the stairs on the way to the bathroom.

Luke, who missed little school over the illness, appeared to have recovered fully after about three months. Sarah seemed to recover, too, but found she was less able to perform physical exercise or sports. As she rarely complained (and I'd never been sports-minded), we did not pay close attention to her occasional remarks on her fatigue. Richard and I did not recover. We continued to feel as though we had the flu, sometimes with and sometimes without the debilitating vertigo. We continued to struggle to maintain our self-employment as college-text editors, to care for the children, and to look after the house, garden, and property, but we knew we were functioning with sharply reduced abilities. Frequently, we drew on all of the discipline of our academic training to force our concentration on our work. Some days we felt so foggy mentally that we had to stop working. Eventually, I found myself turning down or avoiding new editing assignments to be able to take care of family responsibilities. The expansion of our editing business we had planned was dropped and my art projects had to be abandoned. Most of our involvement in the children's schools and in the community diminished, except for our minimal involvement in our church. Richard had very much enjoyed coaching a baseball team of junior boys but could not continue doing it. Both of us would strain muscles in some mild activity, such as tossing a Frisbee or moving a bag of garbage, causing permanent damage. We consulted two doctors, who told us some forms of flu were hard to recover from. Two or three years? Oh, yes, they said. A chiropractor concurred. The garden became a chore we could barely manage, even when we neglected the flowers for the essential vegetables.

Turmoil in the School

In September 1989, for the first time all of our children were in school. I would not discover until many years had passed that on his first day of kindergarten, when four-year-old Eric tried to find Sarah's classroom—because we had explained that his brothers and sister were in the school and could help him if he got lost—the kindergarten teacher stopped him in the hall and pulled him by his ear into the classroom as he was crying.

In the dark days of November, I met a very upset parent at the village intersection. As the wind whipped us where we tried to talk in front of the bank, Paulette told me she was withdrawing her son from the class Luke was in. She said there were many children in the class showing symptoms of severe stress: several had skin rashes; two had gastrointestinal upsets so intense they had undergone hospital tests; boys who had not wet the bed for many years were waking up in soaking sheets; many children were refusing to go to school, using every excuse they could muster to be allowed to stay home. Luke, who had been dry at night from the age of three, was among the boys who were awakening drenched. I went home to share this parent's information with my husband and we took a fresh look at some of the things Luke had been grumbling about that fall.

We decided to contact some of these other parents and compare what their children had been reporting with what Luke had told us. We began to realize we should have put more credence in his observations. A large group of concerned parents met in the basement meeting room at the local library to discuss and learn about the school "problem." Richard attended while I stayed home with our brood. From that meeting, a smaller group gathered at our place a few days later; our priest attended the meeting. We decided to talk to the principal and the school board trustee. The priest felt he should inform his bishop.

As our concerns about the teacher's methods escalated, we discovered that the trustee was the mother of the principal, a conflict of interest that never should have been allowed to occur. The principal would prevaricate, telling us, for example, that he had not heard complaints from other parents, when in fact we had spent time outside his office minutes earlier with other parents from the two meetings who were consulting him to ask the same questions we were. The trustee did not return phone calls. We sent letters. She did not reply. At the next trustees' meeting, which several of the parents attended, she claimed no one had brought complaints to her. After a while, the bishop responded to the concerns of the parents. The letter he sent contained two sentences, each equally unhelpful.

Richard and I accepted a secretarial function as letter-writers, but we were in no sense the leaders of the group—the coalition was just that, a group of parents whose children were suffering for unknown reasons and who wanted to learn why. Two of the parents were teachers at other schools, one was an attorney, several were business people. Although no more participant than other parents, we were perceived as instigators of the investigation, perhaps because we were relative newcomers without roots or many relatives in the community. The allegations, of course, were that children were somehow being emotionally traumatized in the classroom. There was no hint whatsoever of overt sexual inappropriateness. Some children described experiences that shamed and humiliated them and said they felt punished in their efforts to learn. For example, Luke described working hard on a math assignment that the teacher crumpled into a ball and threw along the floor, insisting that he crawl on the floor to retrieve it, throw it out, and redo the assignment; nor was this an isolated incident.

Then, when the custodian (the father of two children in the school) heard Paulette's son in the washroom in obvious distress and suggested he go to the office and call his mother (who he knew as a neighbour as well as from the church), the staff was up in arms because the custodian had no rights *in loco parentis*. Paulette, one of the teachers, withdrew her son from school and kept him at home, hoping for a resolution to the issues in the class. The custodian lost his job. The crisis precipitated in the school, community, and church was unprecedented in this district. Some parents removed their children to the public school. Others kept their children at home. Intergenerational, interfamily, and interpersonal ties of many kinds produced criss-crossing loyalties among school staff, students in various classrooms, parishioners who had relatives in the school as staff or students, and other members of the community. One of the teachers stood her ground against the opposition of most of the school staff to the parents' coalition and refused to become part of what she called a smear campaign against those of us who had expressed concern. She not only offered her friendship to me, but found ways of sheltering children in her classroom when they felt harassed in other locations. Without betraying her professional standards or anyone's confidence, she offered support and solidarity with the parents. Her courage and dignity were outstanding.

Through C., I spoke with the wife of an Ontario Supreme Court justice who had wrestled with alleged incidents of sexual harassment by a teacher in her child's school. Even the formidable legal resources in her family were insufficient to find a way to sue the teacher. After he moved to another school board, she received calls from parents who wanted to know what his reputation had been in his previous school because they had developed similar concerns about their children.

A Resolution of Sorts

The bonds of openness and trust that had been forged in small groups during a parish program the priest had conducted earlier in the year became a significant force as the parish at large was drawn into what was happening at their school. Months of worry, gossip, genuine communication, anger, and upset ensued. Lines were drawn, choices made, sides chosen. Those who risked their jobs, their positions in the community, their friendships, and their roles in the church trod "a dark valley of decision." In general, the people of the church knew who they could trust.

In the end, the trustees met the parish at the school. When we arrived at the school gym we were stunned by the turnout—the entire parish appeared to have assembled, whether or not they had children in the school. Also present and representing the teachers' federation was a lawyer who stood up as soon as the meeting was called to order and stated bluntly that he was there to keep an eye on another couple and on us—we four apparently perceived as ringleaders. Ordinary men and women became warriors: the impoverished parents who stood to ask questions did so courageously under legitimate fear of being sued and losing everything material they owned. We knew that one mother suffered from a neurological disease and that the stress in the circumstances made her health more precarious; yet she sat by her husband in the meeting and supported his presentation. At the end of the evening, the trustees apologized to the parents and to the parish, the teacher in question was moved to another assignment, and the principal (after finishing the school term) accepted a position in another school.

A warehouse of dissatisfaction with teachers' privileges *in loco parentis* (as they act in place of the parents in their classrooms)—which most teachers and parents of schoolchildren everywhere care about deeply—was left unresolved. The disturbing truth that came to light through this experience was that parents have little voice in the education of their children, and that what transpires behind the closed door of the classroom that affects those children can neither be easily brought to light nor fairly examined under the law. For Luke and his remaining classmates, however, a change was ensured. The school year of "crisis" left Richard and me, with our constant flu-like symptoms, exhausted.

The Aftermath . . .

My health became more seriously compromised. Three surgeries that should have been minor had involved escalating reactions to the anaesthetic —which the anaesthetists failed to diagnose, in spite of my efforts to inform them before surgery of my difficulty recovering post-operatively. The third and worst of these anaesthetics came after the school crisis, caused central nervous system damage, and kept me dysfunctional for six months recovering from it. I had the unending "flu," the approach of a grim menopause, and incipient stomach ulcers. Our doctor provided no useful input. Somehow, with the help of my increasingly ill husband, I kept on keeping on.

The temporary teacher brought in to replace the transferred teacher had a knack for helping the children to get over their fears and she did not attempt to resume a normal curriculum. Nevertheless, it had been a very challenging year for Luke, who was still struggling with feelings of insecurity. The year was a total write-off for him academically. And his closest friend had been moved to the public school.

The male teacher for Grade 8 and the woman principal who came the following term were not only splendid educators, but skilled at building bridges between the concerned parents and the school. For two years under their leadership the wounded senior students and the entire school flourished. I was invited to teach ceramics classes to the older grades. In Luke's graduating year his teacher brought the class to our property for a gala day of orienteering and a barbecue. It seemed that the breach with the school, at least, had been healed.

. . . And Fallout

When the excellent principal retired she was replaced by someone our younger children feared. This time, when they brought home complaints about what was happening to them at school, we listened. "Harassment" seems an appropriate term for what they described. They were stopped in the halls while on the way to class and interrogated for no apparent reason. Danny's teacher gave him a broken pencil sharpener with a loose blade, which a few minutes later the principal confiscated as a "potential weapon" for which he issued an expulsion. It looked to us like a set-up. Eleven-year-old Danny was upset and confused.

A few days later, I waited in the hallway for seven-year-old Eric to be dismissed from his class. I stood a couple of feet behind and to the left of the teacher as she stood at the classroom door and dismissed the students; I could see him approaching, gathering up his jacket and backpack. He said a couple of quiet words to the boy beside him, which I could not hear and to which the other child did not noticeably respond. His teacher exploded. She grabbed Eric from among the other children and shook him, yelling at him that he had spoken rudely to her, blaming him for saying something she considered inappropriate. He twisted from her grasp and rushed past me sobbing to collapse on the row of outdoor boots in the hallway. I bent to pick him up and hugged him. She followed us and did not stop for breath in her tirade. Eric shrank down onto the boots. I turned and spoke to her, telling her she was mistaken, that I had been watching and he had not even looked at her and certainly had said neither anything rude to her nor about her. She kept up her vitriolic shouting at me. Danny came and stood by his brother and did say something quite rude and descriptive of her behaviour, but she was fulminating so that she could not hear him. Next, Sarah, who had been visiting with a former teacher, came upon the scene and began comforting her little brother. I kept my body between the children and the woman whose raging continued out of control and spoke over my shoulder to Sarah, "Take the boys to the car. I will be right there."

At last, the teacher was running out of steam. I tried to reason with her, tried to explain that I had been present for the entire time and that Eric—as he had been protesting between sobs—had said nothing about her. I was thinking that in all my years as a student, as a professional working with teachers in publishing and in research, I never had seen or heard about a teacher so brimming with malice in unleashed fury. And my experience includes my Grade 1 teacher who strapped children she considered slow, and who was admitted at the end of term to a mental hospital. As Eric's teacher merely continued to berate me and as I gave up trying to reason with her, I thought of all that Luke's year of unhappiness had taught us about the school system. I could have this woman arrested for assaulting Eric; I could have her in court for her abusive behaviour towards him and for her slanderous language towards me. But where would this leave our shy, sensitive, polite little boy? When finally she paused, I simply told her she would be hearing more from me and I left.

Our former ally in the school had retired. That night I phoned the one remaining teacher I knew and trusted and—knowing that my inquiry was putting that person's career in jeopardy—I asked: "Are you aware of what is happening to our children at the school?"

"Yes."

"If this happened to your own children, would you remove them from the school?"

The reply was prompt, "Yes." That is a very grave assertion for a Roman Catholic educator to make. I was grateful for such courage and frankness.

Richard and I talked about the situation that evening, consulting Sarah, who knew the teachers and students better than we. Despite our commitment to the ideals of parochial education we could not return our children to an environment that was toxic. We removed the boys from the Roman Catholic school, where they had learned both to love and to fear, to respect and to distrust teachers and school authorities—and authority and authority figures more generally. The next day we registered them in the public school, where there was more violence on the playground, but for the rest of that year a better quality of teaching.

A newspaper reporter told us that the earlier crisis at our Catholic school had reached ears at the provincial legislature, so that a new directive for teachers' treatment of children and parents had been distributed province-wide in the public schools. The efforts of a few parents at the Catholic school apparently had had a wide-ranging, positive effect and had smoothed the path for our children in the neighbouring public school.

A few months later we started attending the Anglican Church. The Anglican priest and his wife had four children of ages that more or less matched the ages of our children. Danny quickly became a friend of their son Len.

April

At the point where our concern had been mounting in regard to Luke's school environment, April had returned in real mental distress. For several days she barely spoke to me, which was bizarre. When at last I thought she might be willing to talk, I asked her what specifically had prompted her to decide to come. She said that she was "either going to check into a psychiatric ward or come here." She then explained that she was having panic attacks. She had become politically and socially active in her high school on behalf of gays and lesbians. Threats were made; bricks (quite literally) had been thrown at her and other protesters. With the move to Nevada, she had been befriended by a group of lesbians and recruited to their cause, immersing herself in lesbian literature, songs, lectures, and experiences. Although she continued to avoid speaking to me, she spent time with her father, often simply in silence.

April stayed for almost a year. As she improved, she took a volunteer job at a daycare and did well at it, which brightened her outlook. Her father devoted a great deal of time to her, especially during the summer following the school crisis, trying to reassure her. April's presence in the family during a time when another of the children also was feeling unusual stress had been important for her and for each of us. She was not neglected; at the same time and despite her own angst, she was able to be supportive of Luke in some ways at a time when he was in need. By August she had recovered her stability to the point of being able to find a university and fly back to Nevada to pack her belongings before the fall term.

2

Starflower Tree

Chronic Fatigue Syndrome

The first painting I attempted after I was married distressed me. The background purple and dark blues are sombre. The trunk of the apple tree in the foreground—symbolizing Richard and me—that might have grown straight is bent into an S from supporting its burden of foliage and fruit. The foliage spreads awkwardly in some directions according to an obscure logic, neither symmetrical nor, I thought as I worked, attractive; but I had learned to obey that other, truer source of suggestion in my mind. Its blossoms—our children actual and symbolic—are stars, huge and shining. The foliage kept spreading under my brush to provide places for them and would have filled the entire canvas, except I had other vital components to include; whether it pleased me or not, I would have to accept it. No one but my husband liked the look of it, either; but I framed it and let it speak from the otherwise empty walls.

When the Rope Runs Out

In 1991 I had the beginnings of a stomach ulcer as well as the mysteriously unending flu. For the first time in my adult life I no longer thought of myself as well—then menopause hit me. I was taken to the hospital twice by ambulance with symptoms similar to the onset of heart attack and again diagnosed with "the flu." When I nearly collapsed taking my mother to her dentist in a neighbouring town, I found in the same medical centre a doctor who would see me, although he was not accepting new patients. He prescribed estrogen and Zantac; I stopped having blood pressure drops and became functional again, but not well.

Two years later a couple we knew in Virginia invited us to help them establish a theatre. It seemed like the right opportunity at the right time and we devoted a year to their project, a year in which many things appeared to go badly wrong. Our children, also, seemed marooned; they had looked forward with anticipation to life in the great American land of promise—promise that had proven illusory. On a hot day the following spring, we were cramming a rental truck full with the help of Luke, who suddenly was on the verge of tears and saying good-bye. I had not known until this moment of our leaving Virginia, with the heat radiating from the asphalt and our disappointment as sulphurous as its fumes, that our son had already spoken to his father about remaining in the household of his girlfriend Kathleen's mother. I stood in the parking lot feeling a landslide of "nevers" hurtle over me. We would not be seeing him graduate from high school. We would not be present as a resource of any kind as he embarked on the first serious romantic relationship of his life. His awareness of his many abilities was just beginning to flower; his self-esteem was growing, his confidence improving. I felt it was too soon to have to let go; in Canada he would have another full year to finish high school. I had assumed he planned to do that. Why had Richard said nothing to me of Luke's decision? I was in shock as we drove away, yet so accustomed to emotional stress by then that my husband and younger sons did not recognize my distress.

I struggled to put this event in context. Luke's relationship with Dan was deeply problematical, which was clear to me, but opaque to the other members of the family, who also found Dan's behaviour difficult to cope with sometimes. Dan's inability to conform to our expectations was interpreted by all of the other children and usually by Richard, too, as deliberate disobedience due to some inadequacy and failure in our parenting or recalcitrance on his part. Following on the upheaval at their former school, the year in Virginia that had seemed so full of promise and yet proved so disappointing meant the children were losing confidence in our values, standards, and choices. Our efforts at intervention and guidance with the older children seemed to echo from the walls of an empty valley now that our first son had moved far down another road. I tried to look on the bright side: separation of the brothers from one another could be important for their maturing, especially for Dan. American teens of Luke's grade all would be ending their high school careers that spring. Nevertheless, the suddenness of the decision and the unbridgeable distances were tender sorrows. For our youngest son, Eric, who adored his oldest brother, the loss surely would be traumatic. No one had prepared him for this turn of events. I focused my attention towards him in the car, knowing he would be the person in greatest need for awhile.

One of the best things about that desperately difficult year had been the visit of my parents and sister, the touristy things we'd done with them, the sharing of anxieties and hopes and dreams over cups of tea. As we drove north, how much I wished that I had been able to throw off the anxiety, spend more time on the good memories with her—for Mum had died just five weeks ago. I supposed I would become some sort of mainstay for my loved, but surely difficult, father. I wondered what it would be like to live near Dad now that Mum, his angel and martyr and buffer to the family, was gone.

Neither Richard nor I was improving in health. We had talked to several doctors who had various ideas of what might be the cause of our ailing, including the Australian Dr. Barry Marshall we had met socially and who was responsible for discovering the bacterium (helicobacter pylorus) that causes stomach ulcers. His groundbreaking solution was instrumental in healing that aspect of our imperfect health. Certainly, stress could not be discounted as a factor; however, something beyond stress and the normal aging process was affecting us. Our Virginian physician wondered if I had lupus, but we were on the verge of leaving and had no money to undertake tests. At least, I thought, as I followed the U-haul in our aging van, we're going home, back to the familiar house and garden, the fields and woods and friendly neighbours, a place of peace.

"The Longest Way 'Round Is the Shortest Way Home"

The next day we were parked in our own barnyard. The U-haul needed to be unpacked and promptly returned to a depot; the children were anxious to reclaim their rooms—and our tenants, who had been notified of our date of return six weeks prior, flatly refused to leave our house. With some doubt as to how safe our belongings would be under its leaky shelter, we unloaded the contents of the U-haul into the driveshed. Dad, now 80, decided very kindly that he would forgo his use of the cottage after the end of June to give us a place to stay while we contended with the renters to reclaim our home. In the meantime, we could use his house and he would go to the cottage, a five-hour drive away. (Almost a year later we would find it odd that Dad, her executor, had not mentioned to us that on Mother's death my sister and I had inherited the house he lived in.) The boys felt this to be a happy turn of events. For my father and Richard and me, considerable additional effort ensued. The tenants now had a lawyer's letter requesting their removal and we had a couple of months in which to reorganize our lives.

Being at the Muskoka cottage was an enforced vacation we needed very much. Sarah returned from Virginia, where she had stayed at a friend's in order to finish the school year, via a babysitting assignment on Cape Cod, and while she and I made summer clothes from cheap fabric and old curtains on my Great Aunt Gerty's ancient sewing machine, we spent sunny hours discussing her time away, the ups and downs of her babysitting job, and topics of interest on the radio. Neither of us had the strength for canoeing, which seemed peculiar, but we swam a little and raked weeds off the sandy beach and cooked for the family.

Richard and I kept the symptoms of our illness to ourselves, for the most part, trying to bear one another's burdens but protecting the children from our distress. Near the end of July, however, Richard developed heart palpitations. This went on until he had enough non-stop pain one night to scare us and we explained to Sarah why we were leaving her with her sleeping brothers as we set out for the Huntsville hospital. Tests there indicated that a heart attack did not seem to be imminent; the palpitations appeared to be related to neurology rather than to Richard's vascular system. We learned from the young doctor in emergency that former classmates of his were running a new clinic in our village. At first we could hardly believe he was serious; but we had been away. The encounter set us on a new path. Two days later as we drove home to reclaim our house (although the tenants' horses remained in the fields munching our hay for another year) we decided to try the new clinic. The new doctor immediately began aggressive testing programs for Richard and for me.

A Descending Arpeggio

As Sarah contemplated the new school year she was far from happy and feeling less than healthy. She first tried the 30-mile bus commute to the RC high school she had attended before the sojourn in the U.S. The long bus ride added to the school day was more than she could tolerate. While in the U.S., Sarah had been diagnosed as asthmatic, but her puffer made no difference whatever in the sharp fatigue and breathlessness that assailed her in gym classes. We began to connect her present problems with her occasional complaints over a couple of years that she had been having trouble with endurance and breathing when physically active. She hated the idea of leaving school friends and going to the nearer public high school in the next village, a 13-mile trip, but felt she had no alternative. The contrasts that faced her there, by comparison with her recent experiences in Virginia, were bitterly disappointing. Although she knew some students, few were friends. She had left a boyfriend who was moving to another state to go to art school; she had left the closest girlfriend she'd ever had. The brother nearest to her in age was gone. Furthermore, her younger brothers were not merely at the annoying age: one of them was developing a damaged self-image in reference to his learning problems and her youngest brother was struggling under his own emotional burden of losses and frustrations.

On an overnight visit—that I had strongly encouraged—to her closest friend from the school she had quit in the city, Sarah apparently caught a virus and became severely ill. She did not recover. After a few weeks of struggling to attend school and feeling worse each day, she turned to the new doctor, submitting to the sort of extensive testing already underway with Richard and me. It never occurred to us that there could be any connection between her problems and ours; her symptoms were so different. However, it became clear to the doctor, who had seen a lot of CFS in a Toronto clinic before he moved to our village, that, as all other possibilities were ruled out, each of us had chronic fatigue syndrome.

Learning about CFS

At first I thought of "chronic" as "frequent" or "intermittent," as, say, with some migraine headaches. If just Richard and I were ailing, I suppose we would have continued to soldier on indefinitely; but Sarah was even sicker than we were and becoming sicker, week by week. The doctor had suggested I contact another of his CFS patients, Yolanda, to learn about what support there might be in the community for people afflicted in this way. From pamphlets that Yolanda had collected I learned about fibromyalgia and "pressure points"—nerve junctions on the body that, when touched, are particularly painful for those with CFS and/or fibromyalgia. We checked our pressure points; the information was accurate. I read that mild exercise might help, but that vigorous exercise could cause serious and permanent setbacks. That explained what had happened to my arm when I tossed a grocery bag of scraps into the compost. Yolanada told me she exercised by skipping through her trailer home while singing, and that helped her. It sounded refreshing, if a bit unusual. She told me her CFS had started following an automobile accident.

Among the printed materials she generously passed on to me was the "Checklist" of Dr. Jay Goldstein of California (Appendix A). His pioneering work with this disabling condition had produced a list of symptoms and preliminary statistics on the frequency of the symptoms among his patients. In our family, his list helped our understanding of one another because we could see how this ailment affected individuals differently as well as similarly. I learned that the condition was first called "fibromyalgia" in Britain; that the terms were being used interchangeably by many physicians in North America. CFS was also called "myalgic encephalitis" and such unhelpful terms as "the yuppie flu."

Our new doctor was brilliant (special awards hung among his degrees on the wall), attentive, concerned about each individual in the context of the family, and saw the family as a group as well as individually. His approach, as much as any other intervention, conveyed to the children that they were part of a family that cared and was being cared for. It was a refreshing change and our values as parents were to some extent reinforced in that setting. The doctor explained that it might be possible to find some supportive therapies. I still was thinking that "chronic" meant there would be some respite.

I was determined to find out more about chronic fatigue syndrome and what could be done about it. Although we could not afford to drive the 60-mile round trip to the nearest city regularly, I went, spent $30 on a library card (while thinking about the groceries it might have purchased), and brought home the 10 or so books available on CFS. There was no doubt that Sarah's and Richard's symptoms were more severe than mine, so I decided I had more leeway to make myself "the guinea pig" of the family. The doctor showed amazing patience in sharing his expertise, even explaining to me on one occasion how information I picked up from a veterinary column in the newspaper might or might not apply to humans in regard to our disease. From my library reading, I discussed with Dr. A. the options in diet, rest, and exercise some authors recommended and made some lifestyle changes accordingly. These changes made no difference whatever in my level of energy or health or joint pain or in the strength of my wasting muscles. While the range of symptoms by now familiar to me was fairly reported by these authors, most of their advice proved useless. One or two people had methodically tried various therapies to find out what did or did not help their condition. The most believable among these came to one conclusion—rest helped. The CFS sufferer seemed to need far more rest and sleep than a well person. We already were aware that we needed more rest than formerly just to get through a day of reduced activity.

The forms of "music" the younger boys were choosing were appalling and Richard tried to provide healthier substitutes from his folk repertoire as he taught them about the dubious sources of what they were listening to. He was fighting a tsunami. Conflicts between us and the children were frequent. Dan and Eric were developing appetites for depressed and depraved noise. Furthermore, the boys had been drifting away from the family and doing a lot of visiting in other friends' homes. At first, that seemed a viable alternative because Sarah craved relative peace and quiet. But we were concerned about how to supervise the boys adequately when they usually were the ones sleeping over at other boys' houses, rather than sharing the reverse situation oftener. Not that Dan's or Eric's situations had been overlooked or slighted. Counselling, intermittent home-schooling, many trips to schools to talk to principals and teachers, encouraging the boys in such activities as basketball and cottage sports all were ongoing efforts. But they were not proving sufficient. And we were always extremely tired.

Meanwhile, Sarah had become terribly ill and could no longer attend school. A few weeks later, arrangements were made for a tutor to bring school assignments and to help her at home, but after a couple of months this sort of work also proved too exhausting for her. She could no longer think or reason as she had; her short-term memory was not functioning properly. By Christmas she was essentially bedridden. After numerous trips to our kind and patient physician it gradually dawned on me that "chronic" meant "medical science does not understand this sickness, there is no cure, and you are not likely to improve much, let alone recover." This was a chilling thought, and bad enough to contemplate in regard to Richard and me; but it was a tragic prognosis for Sarah.

Occasionally, I heard or saw radio and television programs and newspaper articles that explored chronic fatigue syndrome. Only one of these added to the information I had acquired and that was an article by Hillary Johnson in *Rolling Stone*,[1] which discussed the findings of a doctor in Rochester, N.Y., who associated the outbreak among his patients with a flu outbreak that had occurred in 1987–8. That I believed, because the spring of 1988 was when Richard and I and Sarah had first fallen ill, and we live 60 miles, "as the crow flies," from Rochester. We knew personally four other people who had developed CFS subsequent to the same flu epidemic and I had learned of several others in our area who were suffering similarly following the same epidemic. That did not explain why some people recovered from the virus while we and others did not.

In our private conversations Richard would wonder if Sarah could not "try harder." Perhaps he understood better her psychological reasons for unhappiness—certainly the loss of her friendships was dreadful for her. But Sarah had already demonstrated intelligence, flexibility, emotional strength, and endurance in the face of other challenges. Furthermore, I recognized in her illness parallels to my own serious and misdiagnosed illness during my teens. I knew that I had immediately regained health after a specialist, who discovered my tonsils were gangrenous, had removed them. After I learned about Dr. Tomatis's connecting the ear-nose-throat condition with physical and mental fatigue I realized that my illness in high school had entailed the symptoms of chronic fatigue syndrome, also. While Richard considered all the emotional issues assaulting Sarah—which by now included the termination of the writing friendship she had maintained with her boyfriend—and insisted she could just "pull up her socks" and get back to school, I knew Sarah could not handle more of my "you can do it" talks and that she had already pushed herself beyond her capacity. In essence, my knowledge of her personality together with my own experience of illness at her age told me that however stressed or depressed she might be, her sickness had an underlying physical cause.

Benchmarks

I happened to be the last patient during evening hours and Dr. A. walked with me to the clinic doors. Searching for a word of encouragement to support my efforts on behalf of our family he said, "I haven't seen this movie, but I have heard good things about it. Perhaps you would find it helpful to watch *Lorenzo's Oil*." That quiet word seemed important and I again watched *Lorenzo's Oil*[2] in which Susan Sarandon and Nick Nolte portray Michaela and Augusto Odone, parents who were fiercely determined to find healing for their son's incurable disease. I pondered Michaela and Augusto's tremendous determination and endurance and took from their drama the courage to press on through what might become worse yet.

Although Sarah's illness increasingly isolated her in her room, she was sick in the context of her family. In my triage assessment of needs I had been prioritizing Sarah that autumn and winter, but we were acutely aware, also, of the serious problems developing in the younger boys. I had been attending school meetings in regard to each of our younger sons, and was trying to accommodate my husband's special needs, whose illness entailed symptoms differing from my own. Before the CFS diagnosis I had thought of our family as fairly normal. Yes, we were poor, but I knew that there were many other families in Canada struggling financially. For educated people to become impoverished through the two recessions that had ravaged the Canadian economy was not at all unusual. Most of the people in our former circle of friends, in our family, and even among the more affluent in our area were not sufficiently knowledgeable about politics and the economy to realize that our impoverishment was not our fault:

> People tend to be poor, unemployed, or socially disadvantaged not through any fault of their own but rather because the social system produces poverty, unemployment, and structural inequality as conditions of its ongoing existence. Individuals should not be blamed, for instance, for being born into poor families or for living in communities in which factories close down or fisheries are jeopardized. Therefore, a proactive approach is required to ensure that all members of a population have access to a wide range of educational and social services and programs to guarantee them at least minimum standards of living and opportunities for success.[3]

The list could have included publishing houses folding and merging.

Our move to the country was not especially unusual, either. Some of the unemployed and under-employed had moved from an urban to a rural location as we had; but they were less accepting of the traditional culture they encountered than we had been. For example, I was surprised to learn from Sarah's tutor that at least 20 ex-urbanite families in our own area were entirely home-schooling their children. Occasionally, we edited books on some of the economic issues that had become personal for us, affirming how widespread and damaging to the nature of Canadian society the distribution of income had become. The dangers inherent in poverty were summarized in 1996 by the National Longitudinal Survey of Children and Youth:

> Even in intact families, being raised in lower socio-economic households is associated with negative behavioural outcomes. These include delayed school readiness, lower educational attainment, a greater number of serious childhood illnesses, childhood accident rates, premature death, high rates of depression, high rates of smoking and alcohol abuse as young adults, and more trouble with school authorities and the law, to name only a few.[4]

I had assumed that our faith and our unity as a couple would protect our children from such outcomes. I would discover just how mistaken I was.

The year in Virginia had contained a glorious glimpse of what education can be in a well-to-do culture that invests in its youth, and suggested to us what might have evolved as a matter of course for our children under such a system. For me, letting go of that imaginary alternative in order to find what else could happen in the context of our starkly different reality required strict self-discipline. On our return to Canada, we expected the situation to have improved in the parochial school and attempted to enrol the boys. I was informed by an employee of the school board that on no condition would our children be allowed back into the Roman Catholic school

I had read two newspaper accounts of women in less developed countries whose courage inspired me and gave me perspective on the difficulties I was facing. The first told of an African woman who had walked 20 miles with her child to get help for her illness only to discover she required abdominal surgery for which there was no anaesthetic. She chose to endure the surgery without anaesthetic, holding tightly to the hand of the woman journalist who later reported the event.[5] The second told of a woman in Peru who had run with her several children from the Shining Path invaders of their village, hoping to find safety in the nearby river. Her husband had been killed and she had been shot in the leg; her children in the water around her were crying to her for help, although some had drifted away beyond all help. She could reach only one or two.[6]

The loss of my mother, to whom I had been very close, still was a fresh wound in my heart and a painful loss to the children and to Richard as well.

My father sometimes needed more practical assistance and support from his family. My relationship with my sister had altered in ways I could not define. Lainey was unhappy to the point of instability in some ways (she talked very emotionally about wanting to become a "bag lady" like a woman of her acquaintance who more or less lived on the streets) yet she was closer to me than formerly. Richard and I mediated some of the bitter disagreements between her and Dad without being able to alter the fundamental problems between them. There were too many fires, and we could not attend adequately to any of them—I'm not sure we could have even in good health. We had become a family in crisis.

Luke had wanted to go to the U.S. before we made that move. He had spent a satisfying summer working with the premier restorer of traditional log homes in the eastern states. He fell in love with a charming classmate, then graduated from high school and went on to college, where he and his girlfriend were getting excellent marks. Luke and Kathleen had found part-time work and were getting on well with the members of her household where they were living. Certainly, there were challenges in his life, but he seemed to be meeting them extremely well. The news from the south was very reassuring both from the standpoint of Luke's achievements and as a confirmation that our parenting had prepared him for some of those challenges.

Despite the failure of the Virginia business venture to materialize, many affirmations of my abilities had occurred—from several business contacts and through personal relationships that bloomed. For this to have happened in a pressure-cooker environment of problems and uncertainties had renewed my confidence in my own abilities, despite unwellness. Furthermore, as I struggled with problems, I had reached out and found support from a small group of generous friends, who had provided ideas, encouragement, affirmation, and comfort. And C. had sent money. I might be part of a family in crisis, but I no longer felt isolated or alone.

Lainey gave me a large bound blank book, a typically thoughtful gesture, and I began journalling again. My writing became not only an important outlet for my worry and emotional pain, but a record of my observations in the continuing effort to find solutions to our health problems and financial difficulties, and to the problems developing in the children. When this "Opal Book" had been filled I went on to computer journalling, and then to e-mails and letters that provide a history of those difficult days that grew more challenging as the problems escalated into crises that would take more and more of our attention as we struggled to maintain our family under conditions of increasingly compromised health and an overall economic situation in the nation as a whole that never became advantageous for us.

Welfare, 1995

We had survived the summer at the cottage with some financial help from our parents but by fall we were frantic. Work from our main client continued, but at a trickle. The company that had been our principal client since we started editing in Aurora sold its college text division to another firm, with no guarantee that its freelance editors would find favour with the new owner. The few trade books that came in did not fill the gap. By spring, we had no work. We looked at what we were doing to survive. C. suggested we move to find better-paying work. I looked at newspaper ads. I recalled Richard's futile efforts while we were in Virginia to find teaching or other jobs in a much wider job market. What about our mysterious illness? Could we maintain regular jobs? I might, but I was certain Richard could not. Could Sarah make a move? No. We had no choice but to apply for social assistance. Richard knew better than I about the procedure and discovered more. Appling for welfare was frightening and depressing. The interviewer at the welfare office—unlike a different social worker who assisted us later—saw to it that the experience was humiliating, too.

As we walked back into the house, the phone was ringing. Minka was calling from Michigan. She gently lectured me on the pride that kept me from accepting with gratitude the sort of social network that people like Tommy Douglas and like-minded politicians had laboured so fervently to provide for people such as ourselves. Yet, how had we come to such a pass? A few weeks later, Joey Slinger's column in the *Toronto Star*[7] recounted similar crushing circumstances among his friends and acquaintances, suffering as we were through the second of two economic recessions in a decade, yet who had made all the "right" choices, done all the "sensible" and "responsible" things. The disintegration of the American Dream, though Canada is not the U.S., had become a recurring motif in the *Toronto Star*. Our fragile personal economic world had crashed with the recession in 1982 and while we had learned fairly well, I thought, to survive, I found it profoundly disturbing to read in Bruce Blackadar's column:

> For many voters . . . it is economic stress that has fuelled a decline in values. Parents must spend so much time fighting for economic survival that they cannot spend the time necessary to teach children values and to protect them from the chaos around them.[8]

We were finding the time to teach our children values, but we did not have the means to protect them.

Dan's Dilemma

Dan was in Grade 9 and not ready for high school. He wanted desperately to succeed and was increasingly at sea. He could not keep organized; he couldn't seem to keep track of assignments or stay focused working at them. Worse, the expectations of the system had changed over the last year or two with the introduction in Ontario schools of an entirely revised curriculum, a shift to "results"-based education with greater emphasis on standardized one-size-fits-all testing, which even teachers were balking at implementing. These changes were creating even greater pressures that Dan had not been trained to deal with. He was extremely bright, but there was something very different about the way he learned and it was out of sync with a system that had abandoned conventional teaching and expected the children to do their own research with minimal instruction, compose their own reports without assistance, and work in teams with almost no supervision. As an educator I was disgusted with the Conservative government for bringing in curriculum changes so precipitously. Even Sarah, who had stood first in her classes throughout grade school, had difficulty adapting to the new system in those weeks before she had fallen seriously ill. Interviews with her teachers revealed that they were unhappy implementers of a program of which they, also, disapproved and that Sarah was faring better than most of her peers in accommodating to the changes.

If Luke's decision to remain in the U.S. had been a severe loss for Eric, Sarah's absence from the high school once she became too ill to attend was a terrible loss for Dan. Her presence in the school, her interpretation of situations, and her familial companionship had been an important source of support at the outset of his high school experience. She had not only vanished from the school but as her condition worsened her ability to interact with her brothers diminished, so that she all but disappeared from his family life, too. As a beloved older sister, Sarah's diminished role in the family also meant the loss of an authority figure for Dan.

Dan was experimenting with drugs, as we suspected he might be, although he denied it and was secretive about his experiments. In meetings with the school I was assured solemnly in an assembly of school administrators and PTA representatives that there was no drug problem in the school. While I came away from the meeting only faintly reassured, I actually knew very little about street drugs, their effects, the symptoms of use, or why children who are already experiencing extreme stress turn in despair to something that will take away the pain that their well-intentioned parents and intransigent school system are creating. Children with dyslexic syndrome can fall into despair at a much younger age, but for Dan the downturn came in Grade 9. He was increasingly overwhelmed by the school's demands and our expectations and was pinning all his self-esteem on his friendships.

His room had become a quagmire to which his visiting friends added their own chaos. They had started smoking tobacco and I became certain marijuana was part of the mix although I didn't see any or smell it. My experience of marijuana was limited to one party two decades earlier where I was handed a joint and thought the person was asking me to stub out a cigarette butt; when I did, the roomful of people broke into laughter. Low notes of refusal and deception and defiance had been added to Dan's usually gentle voice. We had no idea that he had easy access to pot and lots of other drugs both in and out of the school and all along the main street of the town adjacent to the high school and the junior school next to it. Later, I discovered that the school was aware of this situation and regularly attempted to round up truant students from particular locations in the environs of the school.

Eric Pre-Teen

The autumn of 1994 had seen our under-employment descend into unemployment; Dan's growing rebelliousness; our own declining health; and Sarah's appearance to be dying. And people do die of CFS. In the midst of this, Eric was barely "holding his own" in a school of sharply inferior standards compared to the one he had transferred from. While Dan was sliding out of the school mainstream, Eric was experiencing stress within it. His teacher in Virginia had called my attention to Eric's exceptional co-ordination (his dad said he was "a natural athlete"), ability in art, and mathematics skills, all of which we knew but perhaps tended to take for granted in a family of artists and university-educated parents, grandparents, aunts, and uncles. Eric, who adored Luke, was traumatized by their separation and started to steal. Like Sarah, he was a very shy child, and as Dan's behaviour (especially towards me) got worse, Eric, also, started to withdraw from the family.

Eric had returned to a school system with much less to offer him. There was no sports program apart from gym classes. There was no music program, period. His next teacher did a good job of teaching Grade 4 math. His teacher the following year admitted during the winter term that he had chronic fatigue syndrome. This not only explained the teacher's frequent absences, but his inability to control the classroom or aggressively pursue a program when he was in the class. His plight was, of course, one with which I entirely sympathized. The substitute teachers were too many and too frequent to shore up the program adequately. Eric was being offered nothing in a school program to inspire his abilities. The class did Grade 4 math again that year.

His personable teacher the following year cheerfully admitted to me that he (and his wife) had dyslexic syndrome. He ran a free-wheeling class where the children were not asked to read consistently. Eric was encouraged in his guitar-playing within the classroom. The students did Grade 4 math again that year. The husband of one of my pottery students was so frustrated with the program that he (easily) gained permission from the teacher to go into the classroom and teach math for a week, for his own son's sake. His impact on the teacher's programming was minimal and the following year he and his wife moved to Ottawa and put their children into private schools. By such increments grows the gap between the haves and the have-nots in our country, as Forcese points out:

> Education was to have been the means of overcoming the inheritance of social class. However, as presently constituted, the educational system favours the already privileged, and screens out the already disadvantaged. Rather than defeating stratification, formal education is a cause of persisting and increasingly rigid stratification.[9]

For Eric's tenth birthday Dan had secretly given him a gram of hash. By the time he was 12 and in Grade 7, Eric hated school, especially math, was habituated to marijuana, and needed help. His teacher paired him at a desk with a lad whose parents, Eric reported, had been in jail for some reason. I was dismayed, but didn't see how I could object on those grounds without hurting the other child further. A new principal reached out in specific ways to be helpful when Eric was in Grade 8; Eric was on very fragile terms with school by then and was using pot regularly, but he still was responsive to special attention from us or from his teachers. I asked doctors, neighbours, friends in the city, educators, and other contacts what forms of help, such as residential or other programs for young people becoming addicted to marijuana, there might be. For rural children, there were none.

Daniel Starts High School

The discrepancy between Daniel's superior intellectual ability and problems in school caused high levels of frustration in him and in the other family members. The impression I gained at "parents' night" was that many of his teachers simply wrote him off as "one of those." It was evident to me that he had been streamed with students of little intellectual ability, less intellectual interest, or with learning disabilities that had done the same sort of damage to their self-esteem as Daniel's had to his. The school claimed he was in a program that was supposed to be supportive of his learning problem but that seemed to me instead to be a "dumbed down" curriculum indifferently supervised and little supported, and it had no positive effects on Dan that I or his teachers could see. In spite of several years' experience in educational research and assisting with curriculum design, I was unable as a parent working with school personnel (including administration, staff, guidance counsellors, and psychometricians and psychologists) to come up with truly helpful explanations for Dan's disability or with promising strategies for improving his increasingly poor school performance.

Dan's wobbly success during his first term at high school brought a succession of school meetings, each less helpful than the one before. In my presence the guidance teacher gave him a "tough-love" lecture that Daniel listened to with the patience of a beaten dog. I would soon see Richard display a similar attitude towards Sarah as she floundered—"you can succeed if you want to." The counsellor claimed to have been dyslexic himself, with the implication that one could pull oneself up by one's own bootstraps with the right attitude. But, as the black American comedian and social activist Dick Gregory said many years ago, "If you try to pull yourself by your bootstraps, you land on your ass" or, as Dan says, "the bootstraps break." The guidance counsellor lent me one of his psychology texts, which explained that dyslexic syndrome is the result of brain damage[10] and I tried to explain very gently something of this to Dan, now 14, thus no doubt hammering another nail into the coffin that held his self-esteem. After I had done that, why would he care at all to protect his own brain?

The one bright spot in Daniel's high school experience was basketball. He had worked with remarkable perseverance at shooting baskets for years, including the year we were in Virginia, and had attended a "basketball summer school," but though Dan made the school team, the high school coach did not let him onto the court as often as would have been fair or reasonable (typical of people overlooking the actual abilities of somewhat impaired persons). Dan's ungainly lope (typical of dyslexic syndrome persons) belied his genuine skill at shooting, rebounding, and otherwise participating in the game. He regularly benched Daniel, treated him unfairly (Richard was present at all the games and became very upset, but felt he couldn't do much for his son), and just destroyed the last of Daniel's self-confidence. So, what could have been Daniel's salvation became the last straw.

The subculture of drug vendors and users in the town around the high school found a willing experimenter in the despairing Daniel, no doubt recognizing onc of their own. At home, Daniel developed seriously disruptive behaviour, including obscene vulgarity, name-calling, defiance towards his father and me, and violence towards his younger brother. Before long, his impulsive rages, obsessions, and mood swings made us feel we were living in a war zone. He broke furniture, punched holes in walls, threw things, and occasionally hit someone when he was angry, including his sister on one occasion. Eric began to act out in similar ways. We did not have the money to pay for repairs to broken windows, smashed walls, and wrecked furniture that was wobbly to begin with. Dan and his friends spray-painted the barn interior with obscenities and turned the barn's granary into a sort of clubhouse where they smoked cigarettes and then marijuana. Since Richard still smoked cigarettes it was more difficult to campaign against Daniel's developing habit. Daniel's room was awash in garbage, dirt, laundry, rotten influences of a dozen kinds, and friends of similar outlook. I kept trying to work with Dan and his friends to effect some kind of changes. Keeping Eric away from this group became more and more difficult. Then impossible.

Levinson's Research

I read in the newspaper about Dr. Harold Levinson's *Smart but Feeling Dumb*[11] and *A Solution to the Riddle—Dyslexia*[12] and I ordered his books from New York, read them, and in accordance with Levinson's suggestions asked our family doctor to prescribe Ritalin for Daniel, which did not help him at all as far as we could observe. How much of the drug was bartered or sold on the street I do not know, but certainly a considerable amount of Ritalin finds its way into the local illicit drug culture. Our family doctor was not familiar with Dr. Levinson's work. I gave him my book, but he eventually returned it to me unread.

I had found incontrovertible evidence in Levinson's research (which included his daughters) that dyslexic syndrome could be treated, not just supported. I also came to see Daniel's physical symptoms (apart from those immediately involved in reading), both present and running back through the years, with much greater clarity. Other comparisons between Levinson's clientele and daughters and Daniel and his fellow-sufferers in our community were obviated by vast economic disparities. Nevertheless, I felt a surge of hope that something could be done for Daniel, whose aberrant behaviour had become a serious detriment to our fragile health. Richard and I talked about the possibility of placing him in foster care, a situation I contemplated with indescribable grief and anguish.

Unravelling

One morning Daniel refused to go to school because he had no pants to wear. He had by then adopted the subculture costume of ragged clothing partly because that fairly matched what we could afford to provide. Richard, for years, had owned exactly one pair of pants. I knew how much Daniel cared about his appearance; he had an especially keen awareness of the social importance and historical significance of various forms of costume and dress, I suspect as a manifestation of the dyslexic syndrome child's effort to establish norms in an unpredictable environment or to develop compensatory skills. But his clothing happened to fit the image of the drug culture, too, as I was well aware. On this day, I could see no way of holding together the tattered shreds of the only jeans he owned, however ingenious I might be with pins or needle and thread. Richard was hurrying me to get Dan to school. How could I take him to school indecently exposed in those rags? Everything that seemed to matter for Daniel's sake could not be supplied. My frustration and helplessness reached a critical mass. Something in my brain snapped. I stepped outside the back door so that no one inside would see me cry; but my crying suddenly turned into a wailing sound I had never made before and that I could not control. Richard and Sarah heard me and came running. Richard drew me into the kitchen and sat me on a bench by the window. Sarah prayed quietly, stroking my hand. Daniel heard the ruckus and said he would go to school anyway. Of course, he could not; his rags were indecent. He would have to wait for what his father and I would try to invent for the following day.

In a few minutes I stopped crying and regained my self-control. I explained to Richard that Daniel absolutely was correct—he could not go to school—and showed him the shreds of jeans. He had to agree. What had my hysterics done to Daniel? to Sarah? to Richard? to me?

The templates of civilized family living Richard and I had upheld were unravelling like Dan's jeans. We kept pointing our children towards certain kinds of hopes, values, and patterns of social interaction that seemed to us to be important. We had grown up in modest, not straitened, circumstances; we were not dyslexic; our failures in school or relationships, such as they were, were balanced by successes; the suburban society in which we grew up—a 1950s society on the bubble of Fordist economic growth with basic Judeo-Christian values still deeply embedded in the thoughts and actions of most people, churchgoing and churchless alike—was very different from the one our children were growing up in, full of irony and cynicism, and characterized by a lonely search for values, a blind acceptance of values without intellectual understanding, or a drifting without values. Although we had encountered serious problems in one church, we had not rejected its people; and we were churchgoers still under a different roof. In other words, we lacked the understanding Dan needed from his parents as to why he had begun to abuse alcohol and other substances. Poverty alone can provide the reason, but Daniel had others. Dyslexic syndrome children often give up, quit school, fall into deeper forms of despair, and self-destruct.

The crush of necessity, suffocating layers of problems, and the burgeoning needs of the children that could not be met were further eroding our imperfect strength. The skirmishes in the household against marijuana were becoming a prolonged battle, with many incidents where Richard and I confiscated what we found, flushed it or secreted it in garbage, threw it into the woods, or tossed it out of the car window on the way to town. Discussions about the evils of drugs more often turned into fruitless arguments. Threats of calling in the police sometimes had a temporary or short-term effect, but did nothing for our family relationships.

I remember the moment when Daniel and a friend stood by the back door while Richard gave our son money to pay for his entry to Spiderland Punkfest for the first time. We did not yet realize the event's host was a dealer, but figured "punk" meant wider tolerances than ours. His friends were going; including boys whose parents I was sure had decent values. Yet, I felt that something was terribly wrong. Richard should have been forbidding this adventure, not encouraging it. I questioned him about what he was doing, but he went right ahead. Was he thinking he was paying for good, clean fun? Was he looking forward to having Daniel out of sight for a few hours? But my husband was supposed to know more about teenage boys than I did. Wasn't he? I protested. I was overruled.

Johnson Expands on CFS

As I actively pursued every avenue I could in regard to chronic fatigue syndrome, I learned that an American investigative reporter had written a book not only on her personal experience with the ailment but about what she had learned of the efforts in the American medical establishment to understand the illness. I had seen Hillary Johnson interviewed on a TV noon news program promoting her book and talking about her illness, recalling the article by her in *Rolling Stone*.[13] She had titled her book in honour of the Canadian physician noted for teaching his medical students to listen to their patients. When I obtained and read her book *Osler's Web*[14] I learned that there had been honest and less-than-honest research into CFS. Her claims (Chapter 30) that the Centers for Disease Control and other agencies had diverted funds for research away from CFS and into other projects seemed well-substantiated. The considerable independent research of some doctors had been brushed aside by the CDC and other agencies that she claimed did not want to contend with another epidemic on the scale of AIDS. And, as Richard was quick to note, the implications for the insurance industry, and for the entire corporate welfare system, of the American medical establishment lending its imprimatur to CFS were such that this recognition would never occur. It became “yuppie flu” or a barely definable malaise “that women were prone to.”

Johnson had come to the tentative conclusion that the cause for the disease likely was a hard-to-isolate retrovirus. She described in the book two cases where health workers contracted CFS possibly from blood contamination and a number of CFS patients who traced their illness to blood transfusions. Furthermore, health workers exposed to numerous CFS patients also seemed prone to contracting CFS. I began to ask myself how dangerous we were to our community. I alerted our doctor to the fact that his assistant did not glove when taking our blood samples. How dangerous were Sarah and Richard and I to other family members? More fear; more grief.

Luke and Kathleen

The only good news in those days seemed to come from Luke and Kathleen. We were delighted that they had done well in finishing high school; but we were sad that we could not attend the graduation. Luke’s art teacher had stepped in to do the motherly things I could not, helping to find suitable clothes for graduation and the prom, and offering encouraging words to keep him on track into post-secondary education. I was deeply grateful for the kindness of this woman, even as I wept when photos of the young couple arrived by mail. Luke and Kathleen went on to a community college within commuting distance. His performance there meant Luke was encouraged to make his profession in as many fields as he had instructors—computers, history, geology, and so on. His impressive art portfolio could have gotten him into a couple of the leading schools in the U.S., as he discovered at the annual college signing-up in Washington, DC.

Then, Luke, who had taken to cycling, was proceeding through an intersection on a green light when a woman coming from the opposite direction made a left turn into him. As he lay on the ground with his neck injured she stood over him screaming at him for having been in her way. He told me later that he had felt nothing but pity for her. It would be months before his severe headaches abated. Some of the pain in his neck would remain. After he and Kathleen completed community college and were working to save money to go on with their educations, Luke found himself in an intolerable work situation. The counsellor he sought, a woman lawyer, assured him he was in a position to lay charges of sexual harassment. He and Kathleen decided instead to come north as soon as possible.

Motherhood

I felt that my own ability to discern and to teach discernment to my children had fallen far short of what they had needed in a cruel and evil world. I wrestled with this sense of failed motherhood, among the full range of problems we were facing, through my journalling. Lainey reached out in practical ways to cheer me. She spent money she could ill afford on pretty gifts—an embroidered curtain valance, pillow shams, and sheets to match an heirloom spread of my mother.

Richard was offered a little freelance work and gradually received more and then work came for me from other publishers or individuals; we were able to withdraw from social assistance after having used that means of help for about a year and a half. Luke and Kathleen came to see us. Discovering how well they had been handling the challenges in their lives helped me to feel a little better about my parenting and about how we were meeting the difficulties in ours.

Our doctor felt Sarah should see a psychiatrist in regard to her chronic fatigue symptoms, just to cover all potential exigencies. I was not opposed to the idea. Her appointment happened to fall a few days after Luke had quit his promising job and might be involved in a lawsuit, and when the whole family was deeply concerned for him.

The psychiatrist interviewed me, also. When I mentioned Luke's predicament to the psychiatrist she stated frankly that she lacked expertise about such matters. At least, she was honest. Her comment to me regarding Daniel's dyslexic syndrome revealed she knew next to nothing about Daniel's difficulties. And while she had encountered other patients with chronic fatigue syndrome, she knew so little about the disease as it affected patients differently that she had no useful input iregarding Sarah on that score, either. She did, however, drive a wedge into Sarah's mind by asking her, in my presence, if she viewed me as a "saint" (Sarah laughed and said she did), thus, conveying to our daughter a disapproval of our religion that helped to further undermine Sarah's respect for our values and at a time when Sarah was in very poor physical health and in a rather fragile mental condition.

Looking for Answers

Johnson in *Osler's Web* had cited Canada's leading investigator into CFS. In the library I found the results of a symposium[15] Dr. Byron Hyde had organized and I was able to learn from his publication what was known and not known medically about the disease. I subscribed to his Nightingale Foundation's newsletter: *The ME/Messenger* ("ME" for "myalgic encephalitis"). When eventually Sarah was seen by Dr. Hyde's associate in Ottawa, he did no more than confirm the diagnosis of our general practitioner and recommended the mood-elevators and nutritional supplements and aids to sleep any doctor might prescribe.

I had tried Zoloft (a selective serotonin reuptake inhibitor: SSRI), which had affected my mood for a time. However, it also had produced an alarming mental event in which one part of my brain produced an unstoppable, rapid stream of random words unordered by syntax or grammar while another part of my brain considered that it knew what intended meaning lay beneath the stream of words. I was stable enough somewhere in my brain to regard this event with sufficient equanimity not to become distraught, but I did not welcome further such events. From what I have since learned through Daniel's mental illness, I would identify my brief experience as a transient form of induced psychosis akin to a symptom of schizophrenia. I continued on the medication without a recurrence of this mental event, but neither could I discern much change in my health—it inflated my mood without healing anything—and I did not renew the prescription. Richard, on the other hand, received the same prescription, took one pill, and quickly was uneasy about its efficacy—he took no more

Did we have a retrovirus, an autoimmune disease, a virus, endocrine damage from a prior illness, stress-produced symptoms, or some other problem? By then, the term "yuppie flu" was being bandied about in the media with the implication that malingerers had invented a set of symptoms to give themselves an excuse for "opting out" of social responsibilities. I was in a position to know that nothing of the kind had happened to us or to any of the other people I was learning about who had CFS symptoms. The disease had struck people I knew personally who were happily employed, wealthy, under-employed, and unemployed. Johnson's description of the chicanery of people in high places in the U.S. did not encourage me to think an answer would be coming any time soon from the medical establishment. Nor has one in the intervening years.

The Summer of '96

For us, summer in the countryside is the season of greater comfort and relaxation. Children find happier ways to spend their time—camping out in the tent in the orchard, swimming and fishing in the creek, exploring the woods and fields. But the summer of 1996 brought a series of incidents that cranked up my stress level. First, Eric developed stomach pains and suffered for a few weeks before being diagnosed as having an intestinal parasite, likely swallowed when swimming in the nearby creek early in June. The medication carried a potential side effect of blindness. For the weeks of Eric's illness and treatment his needs were urgent and I watched him for dangerous symptoms until mid-August with an ache in my heart.

Our doctor thought Sarah and I might benefit from the effects of niacin. Minutes after Sarah ingested the first pill she turned scarlet from head to toe. Terrified, I sped to the clinic and squealed into the parking lot, with frightened Sarah glowing in the front seat, to accost our physician returning at the end of his lunch hour. Fortunately, I had brought the pill container with me and as he kindly prioritized Sarah he realized immediately that she had taken 10 times the amount of niacin he had prescribed. My error. He checked her over and assured us that the flushing and rapid heartbeat caused by the vasodilator (expander of the blood vessels) would subside. I was so relieved, I failed to mention that I, too, had taken the overdose and it kicked in as I was driving—more sedately—home. I spent part of the afternoon with ice packs on my prickling, flushed skin as I recovered more slowly than Sarah from too much niacin. The effect of the treatment on our CFS symptoms did not seem helpful. Another disappointing dead end.

In July, scabies was imported into the household from one of the boys' friends and the family was dousing with insecticides off and on for weeks. More laundry as *all* of the bedding had to be washed and dried *every* day; more careful attention to Eric to keep him from scratching himself too severely; even more trouble getting Dan, who tended to scratch himself raw for less reason, to co-operate. And I worried about how the insecticide might affect our poor health.

On a warm afternoon at the end of the summer I had left the kitchen to rest on the steps of the stoop at the back door when I had a sudden impulse to go out to the barn. The boys had been playing in the leftover hay and dishevelment of a barn falling into disrepair and misuse by the few aimless teens who often hung out there. I climbed the steep bank and stepped through the shadow of the great doorway. As I walked forward across the slippery loose hay and began to speak, Dan, from the platform top of the granary, upended a massive timber. I can't remember what we were saying, but Dan released the beam just as Eric began to run across the hay towards me. "Watch out!" I screamed as he moved into its trajectory—and he stepped back, the falling timber missing him by a few inches. I was weak with terror and relief, feeling I had co-operated in a miracle. The boys shrugged off my part in the drama while dismissing Dan's lack of care and denying Eric's narrow escape. I knew we had enacted a scene that captured something of the essential aspects of our personalities and relationships. I did not know it would symbolize the next decade of our lives.

3

Dan's Road

Dyslexic Syndrome

> *As I was driving Dan to the next town to high school, he started to describe the dream he had wakened from that morning. I was astonished at the similarity—both in composition and in the imagery—to one of mine that had puzzled me for several weeks and that I had discussed with my husband that morning. Both Dan and I had "seen" a road flanked by trees running left to right towards the sun with an abrupt drop-off at the end of the road. Dan and I talked about these elements with one another. Dan said, "On the left are two big trees. I think they might be you and Dad. There is a high road and a low road under the bushes, but of course I didn't take the low road. And the valley beyond the road is filled with mist, like in* The Gods Must Be Crazy."[1]

Prenatal

During my third pregnancy, I decided to build a "wall unit" for the living room. My plan consisted of two sets of four bookcases; the top units would be made from pine boards and the bottom ones from plywood. For some mechanical assistance, I borrowed my father's electric table saw and set it up in the garage of our townhouse. I brushed off both my mother's concerns and my husband's insistence that what I was doing with the table saw might harm the baby. Mother also had the idea that ultrasound examinations were undesirable, and I relegated that warning to the "old-fashioned" bin, as well. I might argue that I had known my husband a relatively short time and we still were learning one another's strengths and weaknesses, but I had known from childhood my mother's unusual discernment. However, I was today's woman and I assumed my own confidence was all that was needed to protect the child growing within me. The finished unit, measuring about 16 feet by 8 feet, provided the space needed for our small library and was a satisfying addition to the decor. I would not realize until 16 years had passed that I had accomplished the maiming of my unborn child.

Babyhood

Daniel was born two hours after the onset of labour, making his mom and dad very happy. We thought we were becoming quite adept at this birthing business and a more contented baby would be hard to find. That mid-December was bitterly cold. Richard had his hands full caring for four-year-old Luke and three-year-old Sarah at home, so I saw little of him during the four days Danny and I spent in the hospital. Furthermore, when Richard did come with Luke and Sarah, each bringing a rose, the car had broken down twice en route. Rinske was the only friend who came to visit, to find out how the baby was; and how very grateful I was to see her. Rinske was a nurse who had served in medical clinics for Native Canadians, worked in South America, and then settled into a northern suburb of the Toronto metropolis to marry and raise her family. I saw her rarely and heard from her only at Christmas—but she takes her place at the beginning of Dan's childhood like a fairy godmother with her marvellous red hair and in a green-blue-and-gold brocade coat shining like butterfly wings. She brought cheer, comfort, and love into my loneliness and the peace in her quiet demeanour suffused me with calm.

Calm was in short supply when we came home. Luke was an extremely high-energy child and Sarah, who still was very shy with strangers, had little opportunity to expand her natural quietness with her brother around. Instead, she became the tail to his comet, and I was forever beside myself trying to keep some kind of order and peace to their play, while caring for the new baby. In particular, I remember the frustration of trying to breast-feed him while the other two children were rocketing around the four-level townhouse. Attempting to control or discipline them, I would cover the baby's ears with my hands while I raised my voice. Not that I screamed, but I did not feel right about how I was dealing with a situation I did not know how else to control. The baby seemed not to mind—his distinguishing quality was his contentment. Unlike Luke and Sarah, who had required nighttime feedings, Danny slept through the night. As he grew old enough to become involved with his siblings there was always "something happening" and he revelled in it. He seemed the ultimately problem-free baby.

When Danny seemed slow to begin to walk and talk, I chalked that up to his older siblings' willingness to "do his talking for him" and to their making him the "passenger" of their toy trains, the "animal" in their zoo, the "baby" of their safaris. Danny learned to scramble up and down the carpeted stairs of the townhouse before he learned to walk—he had so much fun doing it that I wondered if that activity had delayed his learning to walk. The delay in talking concerned me more; I insisted often that the older children be quiet and let Danny express his own wishes—but frequently this effort was short-circuited by pointing, "give-away" looking, and giggles. He used single words, occasionally two-word phrases. One day when he was about 18 months old he said, "No, I don't want that toy give me the other one." I was flabbergasted. There was no sign of delayed development in his syntax, grammar, or ability to express his wishes. Evidently, he had been waiting until he was sure he had the language situation well in hand before he used it. From then on there was no difference I could notice in his speech development from that of the other children, except his slight lisp.

When we moved from town to country, the older children had real adjustments to make, but two-year-old Danny took everything in stride. One evening when Danny was about two-and-a-half, all the children were in bed and Richard and I were sitting in the kitchen when the peace was broken by a sudden scuffling, bumping, and crash against the door to the back stairs. We leapt to open the door, and there was Danny in a relaxed clump on the last step, slightly confused but full of smiles and giggles. He seemed unharmed and certainly was not upset. However, I was. Investigation upstairs revealed that the side of the crib had come apart, the bottom rail releasing its hold on the bars so that its small occupant had been able to slide off the mattress and out through the portcullis of spindles. But Danny was experienced on stairs—Why the tumble? We assumed he had slipped on the bare wood of the treads with the slipper feet of the coverall pyjamas. I roped the side of the bed in a masterful repair job and Danny was put back to bed, apparently no worse for his fall.

How stupid of me not to have checked the other side of the crib. The next night exactly the same thing happened again. The *other* lower rail was detached from its spindles—Danny must have found a way of standing or jumping on this part of the side, though we had not been aware of any unusual sounds through the ceiling grilles from the room above our heads. Again, he had made his escape through the spindles, and again he tumbled down the stairs. This time there was a blue lump on his head and next day I took him to the doctor, who confirmed my own impression that he had not been badly injured.

Several years later, Dan told us that on both these occasions he had not fallen but had deliberately jumped down the stairs because he liked the giddy feeling he had at the top of the flight, and the even giddier feeling of being in flight. I did not believe him then, but I should have.

Preschool

Danny's first notable unhappiness took place the September day both Luke and Sarah went to school. He sat in a shelf of the wall unit (by then, divided among different rooms in our farmhouse) listening to Danny Kaye's rendition of "The Ugly Duckling," and hugging a favourite toy—a plush duck given to him by his grandfather—with tears rolling down his cheeks. Had my efforts to reassure and distract him really consoled him? Even when his brother and sister came home and played with him after school, his sadness lingered.

There were so very many changes in my life when we came to the countryside that I could not guess what the children's feelings or reactions would be to the differences. Richard's academic folklore background, which I drew on frequently, and my own heritage of "survivors," such as the maternal grandparents who had raised nine children through the Great Depression, led me to believe our children's country upbringing was not likely to be more difficult in reference to their parents' middle-class suburban rearing than, say, the life of children whose parents are in show business and travel with them "on the road." So, when the school refused to let Sarah (whose birthday was at the end of January) into kindergarten even though she had been well-adapted for most of a year in junior kindergarten in our former locale, we resigned ourselves to her isolation with her younger brother and parents for another year. We had visitors frequently and came to know a neighbour with four sons close to the ages of our own children, but our family members were very much more thrown upon their own resources for company and entertainment than Richard and I had been as children, or they had been in Aurora. Sarah was thrilled to be back in school the next year. In spite of the arrival of a baby brother and playing with his siblings after school, Danny's last year before he started school was rather lonely.

Kindergarten

When I look back on the preschool registration I attended, I was answering questions during the interview that should have provided clues for me to Danny's anomalous development, but they were not explained to me as symptoms of dyslexic syndrome and I missed their significance. Such questions as:

> "Does your child skip?" No. (Do farm children skip on grass and rocks, or is that a city-kid thing to do?)
> "Does your child run?" Yes. (But a neighbour, who is younger,

Dyslexic Syndrome children have problems with balance; skipping is an activity that requires good balance.

runs faster; does that matter?)

"Does your child throw a ball straight?" Gee, I'm not sure; how straight is straight? I know I can't!

Dyslexic Syndrome children have problems with balance and eye-hand co-ordination; throwing in a straight line requires good balance and eye–hand co-ordination.

There were questions, too, about our impoverishment that I did not know how to answer.

"Are you on Welfare?" No.

"Should you be?" (The elderly lady was looking at me in a way I could not fathom.) I don't know. I don't think so.

"There's no shame in that, you know," she said.

I knew nothing about "Welfare" except that people with our educational background shouldn't ever have to need it. It had never occurred to me to learn anything about it. I didn't realize she may have been trying to tell me I should consider that option. She may have been trying to spare my feelings, but I didn't have the understanding to pick that up, either. Much later I learned we were more than eligible for that kind of help. I also did not see a connection between Danny's gross motor skills and our impecunious situation and found her question puzzling in its context.

Perhaps Danny's deficits would have shown up clearly in another environment, but he had been the youngest of the children in his household for quite a while and therefore always seemed to be at one age-related disadvantage or another, but quite a happy chap, nonetheless. When he did start school, little girls his own size were a novelty to him and he spent part of his first day in kindergarten going up to them in turn and hugging them.

Dyslexic Syndrome children develop strong peer dependencies and are hypersensitive to social acts with negative outcomes.

This did not bother the little girls, apparently, but the young teacher, who seemed very stiff and lacking in understanding of little children, was very annoyed. A few years later, in some school settings, that sort of spontaneous expression of delight from a four-year-old would be interpreted as a form of abuse.

A second way Danny's limited socializing affected him in that class was his confusion over "property." He had seen Luke and Sarah bringing home things from school (with permission, though he would not know that) and when he decided it would be nice to keep on playing with something in his class, he brought it home. His teacher informed me after school one day that she viewed this four-year-old behaviour as theft—immoral behaviour of a serious kind—and let Danny and me know that there was something very amiss in his upbringing. This was the same woman, I would learn much later, who shook Eric and led him by the ear sobbing into her classroom on his first day of school. (And later still, she would become a school principal.) Keeping my temper with this martinet, I shepherded Danny home and explained the situation, how some things stayed at school and some things usually stayed at home. He could take things to school, even give them away; but school things stayed at school. There were no more incidents.

In spite of his straight-laced teacher, Danny was thoroughly enjoying school, until it came to writing his name. Danny had not been interested, as his older siblings had, in my efforts to do some pre-school instruction in phonics and reading. I was so much busier when he became four than I had been with our first two children that I did not aggressively pursue the idea. He was obviously extremely bright, and had a phenomenal memory and a vivid imagination. He also had very sensitive feelings and warm affections—for which his godmother Rita teased him and called him "le petit mush." He would catch up at school, I thought. Though the teaching of reading was part of my professional expertise and I was aware of the problems some children had learning to read, like the rest of the educational establishment I did not know what caused dyslexic syndrome. When I discovered Danny was having serious trouble with his name, I found some clay from my pottery studio and we spent an hour or so making "name plates" of clay stating "Daniel," which he later fired and glazed and fired again. He no longer had trouble with his name. I wondered a bit just what that kindergarten teacher was capable of, but shrugged off as unimportant the necessity I felt for taking on her job from time to time.

Something I had missed in my publishing work for and research into dyslexic syndrome was the school of thought that considered the condition to be the result of brain damage. A publication from the files of my sister, who is a teacher, assumes that speculative stance:

> At some time during the course of the child's development—at birth, before, or even after birth—an accidental event has damaged some of the tissue of the central nervous system. It may have been a trauma occurring during a difficult delivery, or resulting from a severe fall or

> automobile accident. It may have been a high fever or an encephalitis or lack of oxygen. It may have been some condition of the mother such as a virus illness or blood incompatibility. The noxious event has produced changes in the central nervous system which interfere with its normal growth and activity. Often the child's physical growth proceeds quite normally; he does not appear to be handicapped; his vision and hearing are normal. We observe the effects of the central nervous system damage in the perceptual mistakes made by the child, in his conceptual confusions, in his behaviour difficulty, and in his learning disorders.[2]

The author goes on to say that such "minimal brain damage" need not spell failure in school, which was my professional emphasis, too, based on the success the publishing company where I worked (and others, such as Science Research Associates) had enjoyed with a multi-sensory approach to the teaching of reading. "Experience has shown that appropriate education given at the proper time can help a child to learn more normal ways of perceiving and thinking so that even if some deviations continue to be present, the gap between his faulty perceptions and normal ones can be reduced."[3] It was from such a point of view that educational writers provided their special insights and lauded their own skills for "getting through" to such children. As one author of that period commented, "the shelves are growing heavy with books about dyslexia."[4] I knew Danny was showing many of the symptoms of dyslexic syndrome, but his intelligence and creativity, his remarkable memory and engaging personality, deflected my apprehension that his condition could be serious. Other mothers who commented to me on similar problems in their children shrugged them off as an inevitable part of the inequalities in life.

Primary Grades

By the end of Dan's first semester in Grade 2, all the signs of dyslexic syndrome were plain in his workbooks: reversals of letters and numerals; confusion between the "tailed letters": "p," "q," "g," "b," and "d"; slowness and

Dyslexic Syndrome children become adept at picking up alternative routes to learning. Kinesthetic (interior awareness of gross motor activity) involvement is an important avenue of supportive learning. Dyslexic Syndrome children are processing much of their experience through their emotions, thus, with less capacity for self-control.

hesitancy putting letters together to form words; lack of motor control that led to pressing hard with the pencil (lots of broken pencil points) and to ungracefully formed letters; and purely phonetic spellings. His teacher, Verna Hughes, also recognized the pattern of his difficulties and insisted on a psychological assessment. The assessment found he was exceptionally bright, but with indications of a potential learning problem.

Dyslexic Syndrome children have problems learning to read and write. Their difficulties include auditory-perceptual co-ordination, motor control, eye–hand co-ordination, balance, and other problems. They may be extremely intelligent nevertheless.

Danny also was helpful, responsive, co-operative, well-mannered—but not able, for some obscure reason, to form letters and numerals correctly or to put letters together in the prescribed manner. Of course, the psychologists knew no more of the origins of dyslexic syndrome than did the rest of the educational system, and turned in a report that said in more technical language what we already had observed. I remember the day I met with teacher, psychometrician, and principal: the assessment meant Danny would get remedial assistance in phonics. Verna and I exchanged significant looks, knowing we were missing an important piece of the puzzle, but darned if we knew where to find it.

Verna ran a classroom with frequent large projects to which each child would make particular contributions. Along with the data input to their young memory banks, a great deal of self-esteem was engendered through the splendour of the finished project in which each child had participated. She had raised six sons of her own and was gifted at making each child feel responsible and special. The terrarium Danny made won high praises. Later, he planted at the edge of the garden the tiny spruce tree from it, which he and his father had found up in our woods. Miraculously, the little seedling survived. Winter after winter after winter.

Dan continued to thrive in school, despite his "learning disability." He did not demonstrate the sort of hyperactivity that some children demonstrate with such a learning problem.

Dyslexic syndrome children may have difficulty learning some forms of self-control.

However, hyperactivity in a dyslexic syndrome child adds to the discomfort of those wanting to control the child's behaviour.

Dyslexic Syndrome children are not necessarily hyperactive. And very active children are not necessarily suffering from dyslexic syndrome. Energy levels in Dyslexic Syndrome children may also be depressed. Hyperactivity in a child is often the indication of a poor fit between a high-energy child and a classroom geared to the sorts of quiet activity more comfortable for girls and more comfortable for the women who predominate in the teaching profession. Boys, in particular, are disadvantaged in this regard in most primary-level classrooms.

That Danny could absorb many kinds of knowledge, but not others, baffled me. No one we encountered in the school system—or beyond it—ever suggested Danny's difficulty could be treated, only that with "extra help" his learning might be supported. For the next few years I realized with increasing frustration and chagrin that the reading programs I had helped to develop, which assisted even dyslexic syndrome children in their learning, did nothing to address the underlying problem. I was face to face with that problem in our little boy and I did not understand it. No one else I encountered understood it, either. "Remediation" of various kinds would be tried. Eventually Dan learned how to read (but did not enjoy reading) and how to write (but not quickly and barely legibly). I was baffled by these inadequacies in a child so clearly of superior intelligence. I showered praise on him for his achievements, but he assessed himself strictly and found himself wanting. Like children in a cancer ward who know more about their illnesses than the adults have wanted to explain, Danny knew exactly where he stood relative to his siblings and his peers. My praises must have sounded hollow, and condescending or stupid.

Dyslexic Syndrome children learn better in environments where they are provided with alternative routes to learning. Dyslexic Syndrome children develop personality anomalies, for example, they learn to avoid activities they find confusing or otherwise emotionally or physically threatening. Avoidance behaviour can become habitual and lead to a breakdown in relationships with adults and to stigmatizing from peers.

We had noticed how hard it had become to attract Danny's attention. Certainly, he was not deaf, but unless he was facing me or unless I touched him to get his attention, it was very difficult to break into his focus on something else, particularly when he was involved in conversational play. It was tempting to think he was deliberately ignoring us to pursue his own interests; but I carefully checked his responses when I captured his attention by coming into his view (rather than simply speaking to his back) or by touching him, and was sure he had not been deliberately ignoring me.

Some Dyslexic Children are unable to pay attention to sounds in their environment, especially if they are trying to focus on an activity that involves listening or close attention. This tendency is called "Attention Deficit Disorder" (ADD). Drugging the child may provide the teacher and parents with a child unable to express his or her frustration, which may make the teacher's and parents' situation more comfortable.

Danny asked me just before his mid-December birthday in Grade 2 if he could have a birthday party. Our celebrations of birthdays had always been limited to family, but we decided we could accommodate his wishes and sent invitations to several classmates and prepared the traditional child's party. Several parents accepted the invitation and we were expecting a handful of youngsters. When the day arrived the weather was very unpleasant and blowing snow. The only person to attend was a little girl who came to our parish school from a village fifteen kilometres away. Those who lived in the village near to us must have decided our place five kilometres from the village was out of their orbit in that kind of weather. Danny and his young guest had fun as Luke and Sarah stepped in and rounded out the party. Games were played; the birthday celebrants went sledding on the big hill. Danny seemed fairly content with the unusual birthday and the guest went home with her good-natured father, whose huge pickup truck could have handled far worse weather than befell that evening. The rest of us were left to wonder why so many people who had said "yes" decided "nah." All of the parents attended the same church. Some local ethic was in play remote from my own social experience. It never occurred to me that Dan had tried to engineer an equalizing or normalizing social event for himself that would bridge any of his schoolmates' perceptions of himself as "different."

Sometimes, too, it was difficult for Danny to maintain his attention, particularly in regard to school homework. I could imagine why it might be more comfortable for him to brush aside what I was saying or to be daydreaming or to be impatient to run off and play. He did not have difficulty putting his attention to a project or other engaging situation for extended periods of time if he chose to, which beggared the value of such a term as "attention deficit disorder," which had not yet come into such common parlance then as it has since. On one occasion when he was in the middle grades I supervised his work on a school project about his rabbit Betty. Once I had shown him how to go about organizing such an assignment, he gave the project a lot of attention and did it well. It did cross my mind that possibly Danny was not getting sufficient teaching in his classroom about how to order and organize his work. A few weeks afterwards, I heard from another student that Danny had not gotten the mark he deserved because the teacher thought I must have done the work for him. She evidently had not engaged herself with Danny to find out what he might be capable of with a little guidance. Such events made it hard for me to assess how much of Danny's learning problems lay in the limitations of some of his teachers.

There were times, also, when I wondered if he had a *partial* hearing deficit, like my friend Ruth, who is deaf within a particular range of sound; and I would wonder about his prenatal exposure to ultrasound, the table saw, and to his infant peppering with my raised voice as I attended to him while trying to control the behaviour of his siblings. However, a hearing test indicated his hearing—like his intelligence—was superior.

Dyslexic Syndrome children with Attention Deficit Disorder have an audio-processing deficit that may *also* constitute a greater hearing problem. Hearing ability always should be checked in a child having problems learning.

Junior Grades

Until Danny reached Grade 6, and apart from his never really learning most of the multiplication table despite lots of efforts of assistance at home, he got on rather well. Slowly, other problems emerged. He had difficulty organizing his schoolwork—perhaps other things, too. I chalked up his habitual messiness to a lack of caring, or of being smaller than Sarah and Luke and feeling too often that he had to rush to catch up to his older siblings.

Dyslexic Syndrome children have trouble organizing because their listening has not become lateralized.

The audio-processing deficits of Dyslexic Syndrome children have exposed them to a cacophony of sound, confusing instructions, ideas that do not tally and form logical patterns, and hurtful responses from the important people in their environments. They take longer to sort things or may not be able to sort at all, causing many of their activities to take much longer than the same activity would for others. They are disadvantaged in a time-conscious environment.

Danny's friendships, limited through our own penurious circumstances to the students in his class, were with little boys who at that time did not excel academically and who apparently did not aspire to. It appeared to me that our son began to ensure that his success in school did not outstrip theirs by allowing his natural difficulties to flourish.

Dyslexic Syndrome children turn to their peers for affirmation when they do not find it from authority figures. They develop strong peer dependencies. Dyslexic Syndrome children find that their inner experiences and what the authority figures around them are saying differ so greatly that they lose confidence in these family members and teachers and may resist socialization, including a belief system.

I talked with Danny about not letting his friendships keep him from doing the best work he could do, and he made greater efforts for a while. At all times he got along very well with his fellow students and for the most part was well-liked—in spite of his dyslexic symptoms—by his teachers.

Dyslexic Syndrome children develop compensatory behaviours to win the respect of the adults they cannot please, becoming especially charming, ingratiating, dependent, and/or demanding. To maintain self-esteem with peers they may develop compensatory behaviours that are long-suffering, comical, unselfish, dramatic, subservient, too generous, or otherwise draw on their resources to ensure friendship, thus creating an unusual emotional burden that adds to their difficulties.

Outside of School

There were unbroken streams of lovable qualities in Danny that had nothing to do with school success. When the family came home late one evening to find our remaining laying hen in some distress, it was Danny who thought instantly to scoop her up and bring her into the light of the kitchen. He was tender-hearted with the dogs and cats, but his special pet was a real, live rabbit, April's idea of a Christmas present for her brother, which lived for several years in his bedroom. Eventually, I relocated Betty to the barn for having worked her way behind the furniture to consume remarkable quantities of the wallboard in the large bedroom, back hallway, and back bedroom. The barn seemed an unattractive home, so in the spring I encouraged a move to a thorn thicket over the hill where wild brown rabbits were known to homestead. Danny made frequent visits over the hill to feed his pet. A month after the relocation, the ground at the far edge of the thicket bore a patch of scattered white fur. A couple of days later, carrying a burden of guilt, I marched up and over the hill, hoping to find a renewed Betty, perhaps without tail. I discovered instead the square piece of wood painstakingly lettered in magic marker that 10-year-old Danny had solemnly placed in the patch of fur with the inscription:

Betty
was
Dan's
Rabbit

Daniel loved fairy tales, fantasy, and magic. He had composed a miniature medieval romance in a tiny booklet when he was no older than three. Onc of his classmates, who was just as motivated by fantasy, shared wholeheartedly in Danny's imaginative play. The two of them worked out storylines and details to a degree that charmed and involved even the hockey players among their classmates. For years the disagreement between Richard and me swayed back and forth: Would I continue to read J.R.R. Tolkein and C.S. Lewis to a kid who had stopped reading books for himself, but who had incorporated their worlds of fantasy into his reality? I did continue to read, but our concern deepened that Dan still was having difficulty reading for himself and was retiring too far into his self-created fantasies.

Dyslexic Syndrome children may use fantasy and other worlds of the imagination as places to escape from their emotional burden.

I understood how children for various reasons may become fixated or obsessive about these mental spaces where they can imagine they are powerful, loved, happy, successful, revered and respected, uninhibited, free from responsibility, and at peace. Many of us learned to use books that way when we were quite young. Role-playing games pose a particular danger for children and young people with fragile self-concepts. They come to favour the non-rational sphere of escape over solving problems in the real world.

Behavioural Problems

About the time he was being run through performance and psychological tests again, Dan's Grade 6 teacher commented that his sense of justice was acute. He took seriously the Catholic school's lessons of faith, integrity, and fair play. He threw himself wholeheartedly into his activities. The testing, which reiterated his need for language support, also confirmed his essential self-esteem and happiness. I was lulled into thinking there wasn't reason for much concern, except at home, where I was distressed that his older brother Luke rejected his tagging-around, hero-worshiping attention while accepting the companionship of the much younger Eric.

Dyslexic Syndrome children may be ostracized by family members or peers who do not understand that their differences are part of a genuine disability that can be understood while the disabled child is treated fairly and with respect.

Sarah became impatient with and critical of Dan's sometimes awkward ways of doing things and was annoyed by his dramatic ways of expressing himself. He was slower accomplishing some tasks, and less co-ordinated than the other children, and he began to find ways of avoiding situations that showed up his comparative lack of skill.

Dyslexic Syndrome children develop feelings of inferiority in reference to their non-dyslexic syndrome siblings and peers. Dyslexic Syndrome children are dealing with an encompassing disability that affects the way their brains function, their learning, their perception, their physical development, their family relationships, their peer relationships, their view of the larger society, and their concept of themselves in relation to everything in their environment. They may displace their emotional burden by abusing a sibling, a pet, or a neighbour, or may introvert the burden through forms of self-abuse.

All of our boys hid from us a pattern of violence between them and other children and among themselves. We intervened when we saw bullying, unaware we were touching only the tip of an iceberg. A bullying situation arose at school and we felt Dan had handled it very well. But his exercise of the pacifism we upheld was not supported by the school and he felt victimized. He won the admiration of his closest friend Len and some of his classmates, however.

Dan's brothers and sister were finding some of his habits annoying, especially his avoidance of the few household and yard chores that were expected of them, but also verbal habits such as frequent baby-talk and the persistent repetition of sounds that amused him and him alone.

Some Dyslexic Syndrome children may engage in repetitive language (perseveration) because they hear sounds repetitively. They may be trying to communicate that form of perception to others, identifying their uniqueness, or may be "playing with sound" between the left and right hemispheres of the brain. Rejecting or suppressing these verbalizations is not helpful.

Family meetings were held around these issues of responsibility to the family and to one another. I seemed to be the only family member aware of a sheer incapacity in Dan sometimes to respond in the terms being required of him. Again and again I was blamed for coming to Dan's defence. We established "court hearings" with Richard as the judge—and Dan relished the opportunities to defend himself. We would tease him in a friendly way about his tendencies towards the dramatic. Richard wasn't always sure it was a laughing matter, but my short career in theatre was founded on sympathy to such tendencies.

On the other hand, I had no sympathy (and insufficient insight) for Dan's increasingly physical abuse of his younger brother and of me. On one occasion, he was throwing things that included chairs and while I was trying to face him down, six-year-old Eric was yelling at me to leave, that he would control Dan and protect me. I strongly assured Eric of my ability to cope, but was aware, meanwhile, that some situations for Dan were going badly wrong and affecting his little brother very negatively.

Dyslexic Syndrome children experience high levels of frustration and displace their emotions from one context, e.g., the classroom, to another, e.g., towards someone at home.

How badly wrong Dan's relationships were proceeding was hard to know, because I had never had a brother, argued a great deal with my sister, knew my father had been too severe and that my husband was gentler. Since Richard had grown up with a brother, I tended to trust him, especially where the boys were concerned. However, he seemed to spend most of the times of conflict on the sidelines. And he was present in the house, where he had an office, virtually all of the time.

The worst occasion of conflict from my standpoint was one where I lost my self-control. The teenagers were watching a movie on TV. Dan—who at 11 was becoming adept at insisting on his own way—was keeping Eric at the TV although I had outlawed the film for him and his seven-year-old brother. Dan reacted violently and in the fray that evolved I snatched a belt and chased the giggling and laughing Dan all over the house, trying to smack him whenever I could get close until I cornered him in his bedroom, made contact several times, saw that he had become genuinely frightened, and stopped. And I, theoretically, am a pacifist. My self-esteem as a mother was taking a beating, too.

Dyslexic Syndrome children lose confidence in the authority of their parents. Because they have come to distrust them, Dyslexic Syndrome children may be less responsive to the socializing efforts of their parents, and may be resentful of their non-dyslexic siblings, who can be seen to enjoy a different relationship of trust with the parents. Thus, unusual conflict is created in the family.

One summer day, while cutting the grass with the lawn tractor, Luke sheared off most of one side of the little spruce tree Danny had planted. Dan was hurt and angry. I was relieved to see the degree to which the little tree had recovered the following spring. A couple of years later, Sarah, who was just learning to drive the garden tractor, did the same thing; the tree took longer to recover that time. I saw a symbolic connection between the damage to the tree and the failure of his siblings to respect it as his and to appreciate the qualities and needs of the brother who had planted it.

Dyslexic Syndrome children become accustomed to carrying additional burdens of emotional pain, but will reveal that through depression, sadness, anger, rebelliousness, "wildness," risk-taking, jealousy, withdrawal, and other forms of self-destructive and anti-social behaviour.

Richard and I had become perpetually exhausted—according to the doctor we had at that time, due to a recurring virus. Not that there weren't plenty of other reasons for fatigue. But something out of the ordinary was happening to our bodies. Richard was worried that he no longer had the energy or strength to play with the children, pitching softballs or shooting baskets. Many of the accustomed household tasks had become arduous for us. Luke, Sarah, and Eric regularly helped with bringing wood to the house, tidying their rooms and the living room, sometimes even mopping the kitchen floor; yet, our children had far fewer and far easier chores than what was expected of neighbouring farm children. They knew this and thought of us as "easy parents," but we could hardly turn our property into a working farm to give them more to do.

A scene I glimpsed through the pantry window is etched in memory. That autumn day when his father asked him to help with the hauling of wood from the driveshed to the house, Danny lay down under the big locust tree and refused to move. There was no humane way to force his compliance and in an impoverished family there are very few reasonable forms of leverage, such as promises of treats and outings and special time with the parents. Dan, aged 11, had declared his independence from a family already under stress. Perhaps, for Dan, this disobedience marked a culmination of sorrows; for me, it marked the beginning of a terrible distress.

Dyslexic Syndrome children may become so discouraged with their *inability* to perform to their parents' expectations that they dissociate themselves from the family's ethics.

Senior School

Dan's academic problems were exacerbated when he had to switch schools following the year in Virginia.

When we moved south we had imagined the shortened school year for the children would be compensated for by the educational values in travel and cultural exchange; but for Dan, with a further deficit of formal learning strategies, it meant beginning school in one of the top school systems in the U.S. at a serious disadvantage. Virginia's school system, especially on the verge of a university town, was lavishly supported and assiduously scrutinized, and opened possibilities and expectations exponentially to our children. For Dan, however, the experience was overwhelming. He became depressed, could not cope in the new learning environment, and suffered tremendous damage to his self-concept.

Dyslexic Syndrome children become very dependent on their compensatory benchmarks, peer relationships, and familiar environments; they often lack the confidence in their parents or in themselves to articulate their emotional burden, even when they do understand part of it. They lack flexibility and self-confidence when faced with social or learning challenges, such as a new environment.

This time, psychological testing done through the school revealed an unhappy child, intelligent to the level of genius, but with problems in processing information that were seriously impeding his success in school and producing feelings of low self-worth. It didn't help any that everyone else in the family was having his or her own serious difficulties. The relationship between Dan and his older brother deteriorated. At one family meeting Luke's entire contribution was a cruel diatribe against his younger brother, who sat listening quietly with tears streaming down his face the way he had years before when his brother and sister left him at home while they started school. I could not understand why I was the only member of the family outraged. I asked Richard to intervene. He did not. I ordered Luke to cease and desist. He ignored me. I spoke out to Dan, talking over Luke, to let him know more fully that the abuse pouring from his brother was utterly inexcusable, but neither of them was listening to me. I called an end to the meeting and attempted to console Dan privately. I kept much of my real feelings of outrage to myself as I assured him how untrue and wrong Luke's speech had been. But my words were of less importance to Dan than his brother's had been; the damage had been done and I could not undo it. My heated complaining to Richard after the event yielded no results at all, which left me aghast. To what extent had he allowed Luke to speak for him?

Over the next four months, Dan began to show signs of adjusting: his marks were improving; he had done well in a communications unit of study that included "radio" announcing on the school's public address system; he responded positively to the tutoring of a teacher who came to the house to help him with math; and he was having a terrific time developing his considerable basketball skills. That spring, when Richard and I decided to return to Canada, Dan was wrenched again in the changes that ensued.The small, poorly resourced village school, the comparative boredom of country

Dyslexic Syndrome children may become skilled at producing compensatory behaviour in changing environments.

life—even though Richard prioritized erecting a basketball net in the barnyard—the even more straitened financial circumstances, and parents whose health was compromised were gathering clouds in Dan's sky. His Grandma, the reliable source of unconditional love and treats just up the road, had died the month before his return. Perhaps his closest and certainly his most academically inclined friend, Len, had become quite ill and soon moved away. Family dynamics changed now that Luke—the problematical big brother from Dan's point of view—had stayed in Virginia.

I heard part of a radio interview with a male psychologist in the US Midwest who offered the opinion that parents of dyslexic children deserve medals; I was inclined to agree. For the first time I thought of myself as a member of an invisible host of frustrated, angry, sorrowing parents of both genders. Then, shortly after Sarah rejoined the family, she became very ill. Dan's family was taking a beating and he was drifting away from all of us.

Dyslexic Syndrome children may become vulnerable to drugs, alcohol, and other anti-social addictions. Impoverishment exacerbates their problems and increases the likelihood that they will become violent, engage in crime, drop out of school, become unable to support themselves, and form relationships based on their weaknesses and needs rather than on their strengths.

At summer's end, three pairs of Dan's friends, supposedly ready for high school, materialized on the sloped ground beyond the apple trees and engaged in "fencing," just like swashbuckling movie actors and at considerable risk to themselves with wood and metal "swords" that had been painstakingly and surreptitiously fashioned by Dan from lath, scrap metal, and duct tape and stored in the barn. Dan was stage-managing the fantastic battle. When I broke into the truly dangerous extravaganza, they desisted, but I heard later that they had just restaged it further up the hill and out of sight.

Dan had only six weeks to prepare for school-end testing and preparation for the jump to secondary school. The school decided to promote him without testing, as a student with special needs. In our last interview, the senior school principal offered his perspective on dyslexic syndrome children when he commented, "I just have the feeling that these kids are wired differently."[5]

4

Light Dawns

The Tomatis Method

Dan was "holding court" in the shambles of his room with friends and guitars, raucous music, and cigarettes. I was enjoying a rare moment of peace working on a watercolour of single roses, like wild roses, from photos in a soul-nourishing birthday book from my sister. As I painted a couple of rooms away from the racket, wishing the lot of them were more healthily out of doors, even with fake swords, an image came to me and I sketched it with ink. Daniel was standing inside a Gothic doorway, next to Jesus, who held a shepherd's crook, which Dan grasped, too. The entry where they stood was in pitch-black darkness. Outside the doorway, a "sea of faces," representing many nations and ages, awaited. I took the sketch to him; he was surprised and pleased. "Wow, we were just talking about religious things. Do you think this means I'm going to be a Christian leader of some kind?"

"Could be, Dan, but that's something you'll have to find out for yourself." But I worried about the pitch-black darkness under that arch.

A Sense of Support

I remembered that Len's mother had found some avenue of help for her youngest child, whose social behaviour frequently had been inappropriate. He was aggressive with other children at school and did not seem to respond in the ways his siblings had to his parents' teaching. The family had moved, but I located their new number and called her. The first thing she mentioned was the "listening" intervention she had spoken of before, which had taken place in Toronto and which I had observed to have made a revolutionary difference in their little boy's behaviour. The little boy, who was four when I first got to know the family, was by then about seven. His treatment in Toronto had involved listening to "special music like buzzing." His odd manners and serious behavioural problems had vanished over a period of perhaps six weeks. I did not understand what was involved in the treatment he had received and how possibly it could be related to his behaviour, but the results were impressive. He had rather suddenly become a normally happy and co-operative child. As Abby talked, I recalled having clipped and saved a newspaper article about an autistic child who had found help at The Listening Centre and whose mother had published an account of her daughter's recovery.[1] The cost of the treatment certainly was beyond our means. It would have been beyond Abby's, too, except for a small bequest that had come their way.

She told me, also, about a resource called "Beechgrove." She said this assistance was a branch of the county's counselling services, which I did not know existed. The services were free. Why had not the school personnel suggested this form of help to us in one of our many meetings?

Before long, we began weekly family counselling sessions with Lauder Smith and his associates that did a great deal to support our sense of worth as parents and that affirmed our values with the children. Even Sarah came as often as she felt well enough to manage the hour of conversation. Family members had a new opportunity to express their feelings in a more neutral setting. We began to see significant changes in Dan's self-control, self-understanding, and self-esteem. While Dan still required constant monitoring and support in regard to school attendance and was much more often driven to school by us than on the school bus (a further stress on our finances), an atmosphere of relative peace returned to our home and the health and happiness of every member of the family improved just a little.

The counselling reinforced for Daniel the idea that our family expectations, including zero tolerance for violence, were those of the wider community as well. The next time he expressed his anger through violence—he broke a window of the car—we called in the police, who issued a warning. I think it may have been then that for Dan the outer darkness became complete. Dan had learned his lesson—from then on he simply imploded.

Turning Point Counselling

Richard and I were in the process of turning our informal counselling into a formal business. We advertised and started seeing clients. At about the same time, Richard's expertise was noticed by another publishing firm and a trickle of editing work arrived. Within a few weeks he was as busy as he could manage and we were off the social assistance rolls.

Dan's behaviour was less erratic for a while, perhaps because he saw us, also, as counsellors, and possibly because a number of events in the wider community were impacting his circle of friends. A young woman driving a truck ran into a boy, who was on his bicycle without a helmet despite frequent warnings from his mother. The young driver knew first-aid techniques but was prevented by the bystanders from using them. The boy died. Eric knew him and his brother from riding the same school bus and their acquaintance was not friendly (a situation of which I knew nothing). I was having trouble understanding the nature of Eric's upset over the boy's death, since he said that he had hardly known him. Some of his friends went to the funeral, but Eric chose not to go. I discovered eventually, from the mother of one of his friends, that Eric had carried on a running quarrel with this boy on the school bus.

A few years before, some parishioners of the Roman Catholic church had made a pilgrimage to Medjugorje in Yugoslavia, a town where several children had reported seeing the Virgin Mary and receiving messages from her. The warnings of impending suffering the children conveyed were proven prophetic in the civil strife that subsequently broke out in Yugoslavia. One couple among the pilgrims returned and dedicated their farm to God, little knowing what might eventuate. A healing took place at a spring on the property. People from the Yugoslavian pilgrimage began gathering at the property to pray, where they were joined by other people from the Church. Someone had found a set of Stations of the Cross no longer in use at the local Catholic church and fastened them to trees along a path, then mounted a large cross on a prominent knoll. As more and more people made the walk to the Cross, healings and other sorts of miracles were reported. Once word of these events spread to Toronto, busloads of people—particularly among the Filipino community, which is culturally disposed to regular pilgrimages—were making their way to the farm and the stories of healings grew, along with the coffers of the local church and the prosperity of local businesses.

Simultaneously, a man who called himself "Spider" and was committed to punk music, marijuana, and "partying" began holding weekend bacchanalias on his sprawling, wooded property, which was within walking distance of ours. These private "parties," so-called to avoid such requirements as concert or festival permits, attracted young people and punk bands from as far away as Quebec, Texas, Michigan, and New York—and from all over Ontario. It was not unusual for hundreds of kids to show up for a weekend, fully costumed in leather and metal with green complexions, rainbow hair, and multiple piercing, their anger and social dislocation varying from tentative to brazen to ingrown. The "Spider mites," as Richard called them, populated the village on "Punkfest" weekends and became as familiar a part of the local landscape as the busloads of devout Catholics. Of course, Daniel and his friends were far more interested in "Spiderland" than in the pilgrimage site, although Dan actually walked the Stations of the Cross with me once. It was my only excursion to the property and I spent most of it trying to persuade a group of ladies that they would not improve their eyesight by focusing on the sun to find out whether or not it was "dancing."

I tried to form a coalition of parents who would shut down the operation at "Spiderland," or at least devise alternatives to it, but no one I talked to seemed to understand the seriousness of their children's indulgences and none was even slightly interested in organizing to combat it. I would discover that Spider was the convenient source of marijuana for some of the parents as well, so it is not surprising that they did not respond to my efforts. I wrote to the local newspaper about my concerns with a plea for others similarly worried about our young people to phone me. I had hoped that sports or community improvement programs might be improvised to engage the young people. The two women who responded were disabled, but promised to offer their prayers.

Someone torched the house and barn at "Spiderland." The owner collected his insurance and rebuilt. A girl hanged herself on the property. For a couple of years the parties moved to another venue.

A new priest was brought into the local Catholic church and the freewheeling developments at the place of pilgrimage dwindled. The church was re-roofed and the chancel redesigned and rebuilt.

The counselling services that had made a difference for Dan and supported our parenting were terminated at the location in the nearby village. We drove 50 kilometres to their new location, which was costly as well as awkward. Services would be suspended over the summer. By then Dan was becoming more resistant to the sessions and was of an age where he/we no longer qualified for the program. At about the same time he simply refused to attend.

Sarah remained very ill; Eric's problems were escalating under Dan's influence. Stealing—from us, from my father, from shops, and once from a neighbour—by both boys was becoming more frequent. We led our sons through confrontation, took them to apologize to the neighbour (who was forgiving and more than kind), made restitution to shopkeepers, took steps to limit the amount of money in the house, hid the very few things of value that could be pawned. During the summer the barn was spray-painted with new obscenities as the granary again became the preferred hangout. In the fall, as cold weather enclosed them within the house, both boys' rooms became filthy and chaotic. Their conflicts with one another were sometimes violent; when windows were broken or walls damaged, we had not the means to repair them. The house beyond their rooms was sustaining damage, too. We were living in a battle zone.

It became clear to us that few people in the community could afford counselling, even modestly priced. Especially among the young people who were coming to us, although their needs were critical, their financial resources were few or entirely lacking. We moved forward as we might with those in greatest need. However, the influx of drugs into the community was creating crises in many homes—and more frequent incidents in our home than we could handle. Although we turned down drug issues per se among clients (which we felt we had not the training and experience to tackle), drug problems, nevertheless, were on our doorstep. One night a friend of Dan's became psychotic and vomited all over our office; his parents were called, and he was sent to the hospital in an ambulance. Meanwhile, Dan was being peeled off the wallpaper, so to speak, and we were on the phone to the hospital emergency department because the boys had been experimenting with a tea made from the hallucinogenic seeds of the "angel trumpet" flower (not one we grew). Several times, drunken young men were accompanied from the property or driven home. At least one parent of these boys was hospitalized with a breakdown over what was happening to her son. After a year of trying to manage multiple involvements we stopped advertising counselling services and took down our sign at the gate, although I continued to remain available on an informal basis to several women who continued to share their problems and seek advice and support.

"When Listening Comes Alive"

When Len came to visit Dan at the end of the summer of 1996, he brought a book from his mother, who knew I would be interested in learning more about the listening intervention that had so transformed their youngest boy. She had bought the book from The Listening Centre in Toronto: it was Paul Madaule's *When Listening Comes Alive.*[2] The book taught me things about dyslexic syndrome I had not heard during my career in educational publishing—when I had produced reading programs designed to remediate the problem—or in my research during my career in educational research or even in Levinson's writing. Not only did it describe all of the symptoms so very familiar to me in Daniel that I had been unable to alter, it also offered a plausible, full *explanation* for them based in the neurology and development of the child. Most astonishing of all, the author explained his own dyslexic syndrome had been profoundly altered through a treatment that involved listening to filtered music, a method developed by a French physician, Alfred Tomatis. Dr. Tomatis's specialty was the French equivalent of the North American ear-nose-throat (ENT) specialty. He was renowned for his work with opera singers (for example, Maria Callas) who had developed voice problems, and he was awarded honours by the French government for his innovative research that disclosed the relationship between the ear and the voice: the Tomatis Effect.[3] Madaule's book suggested that the condition of dyslexic syndrome usually could be helped through the Tomatis Method of sound stimulation that corrected dysfunction of the middle ear, affecting in turn the inner ear (vestibule and cochlea), the brain, and the branches of the central nervous system.

I noticed, also, that Tomatis had zeroed in on the middle and inner ear in explaining dyslexic syndrome and that one of Dr. Harold Levinson's recommendations for dyslexia, that is, motion-sickness medication,[4] would affect the labyrinth of the inner ear. For two physicians so widely separated to have arrived at novel yet related ideas of the source of dyslexic syndrome lent credibility to their ideas.

The Middle and Inner Ear

Dr. Tomatis's observations and innovative treatments for conditions of the voice and ear grew into a new theory concerning the functions of the middle and inner ear (see Figure 4.1; the relationships among the parts of the ear are discussed in greater detail in Chapter 13). The mechanism of hearing and listening was thought to be comprised simply of the eardrum (membrana tympani), hammer (malleus), anvil (incus), and stirrup (stapes) bones that transmit sound to the hair-like structures (cilia) of the cochlea and thence into the auditory nerve that conveys the neural codes of sound to the language centre in the brain. Many of us learned that sequence in a high school science class. The best-known functions of the inner ear had been that the semicircular canals above the vestibular portion of the organ act like a gyroscope to create a person's sense of balance and that the cochlea processes sound vibrations. Dr. Tomatis knew that the tensor tympani and stapedius muscles in the middle ear also are affected by sound waves. They are the smallest muscles in the body. However, Tomatis told Nicholas Regush:

> These muscles are generally viewed as a system that can protect the ear from dangerously loud noise by diminishing vibrations of the eardrum and incoming sound. . . . But they're also capable of focusing the ear to tune into certain sounds.[5]

Tomatis discovered that these muscles are activated by the higher frequencies of sound, such as those made by violins. Tomatis learned that the middle and inner ear, far from performing a simple balancing and hearing act, mediates stimulation by sound to the brain in many ways. If these muscles are working properly they enable the feedback loop between the cochlea and the brain that adjusts the function of the cochlea.[6]

A third and extremely important aspect of the middle–inner ear's function is the way it establishes the individual's *preferred* route from the ear to the brain. The sound, transmitted through the stapedius muscle(s) attached to the stirrup and to the tympanic wall so that it impacts the vestibule, makes the passage (a) from right ear to left brain, (b) from left ear to right brain and then back across the hemispheric bridge to the left brain.[7] It would be difficult to over-emphasize the importance of Tomatis's observations and theory in this regard.

Tomatis carried out sound experiments to discover how the pre-born child hears sound and developed a theory about how it learns to listen. He theorized that the fetus moves its head near the mother's spinal column and later towards the pelvic bones in order to "tune in" better to the mother's voice conducted through the bones of her vertebrae and pelvis. In this way, the mother's voice calls her child out into the world of air and breath and new sensations of many kinds. Madaule writes, "This body-to-body connection made by the unborn child in his quest for the mother's voice is his first attempt to listen."[8] Thus, Tomatis thinks that even in utero the unborn child becomes selective in its responses to sounds received into the brain. In utero, the child most frequently moves in response to its mother's voice.[9] After birth, the infant will move towards its mother's voice selectively among the voices of others.[10]

Enlargement of the left hemisphere is evident four months before birth, indicating that reception of sound stimulation has already been established in "the language centre" there.[11] Although Tomatis was seeing confirmations of his theory in his clinical practice earlier, not until the mid-1960s "was it demonstrated that the eardrum, ossicles [tiny bones of the middle ear], and inner ear reach adult size and *become fully operational* midway through pregnancy.[12] However, it is not clear how, in utero, these ossicles function in relation to bone conduction through the skull and the bony covering of the inner ear. Neither is it clear whether the tensor tympani and stapedius muscles attached to the first and third of these bones, respectively, also are fully operational. Tomatis realized that conduction of sound through amniotic fluid and the fetal ear "air canal" that is filled with fluid must not be possible the way it is after birth when the fluid drains away from the orifices (ear canal and Eustachian tubes) and sound waves in air can vibrate the ear drum. He concluded that in utero the conduction of sound is primarily through the mother's skeleton and can be perceived through bone conduction by the fetus.[13] Possibly, also, fetal sensory cells in the skin, much like

Figure 4:1 Diagram of the Ear

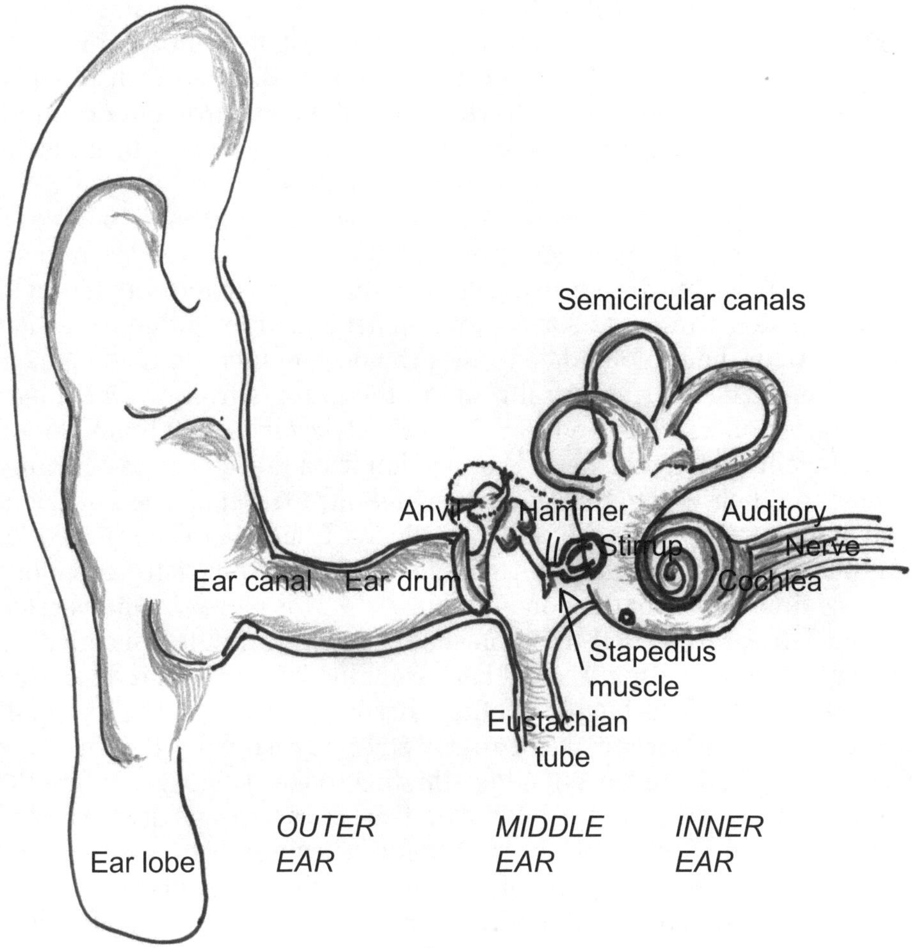

Imagine a flat rake resting on a water-filled cushion and the way the foot of the rake would move the cushion as you pull the handle back and forth. Thus, the stapedius muscle, fastened at the top of the stirrup, moves the tiny stirrup as it is stimulated by sound. The muscle pulls the stirrup back and forth, causing the foot of the stirrup to alter its impact on the vestibule.

the cells of the eardrum, respond to sound.[14] One cannot assume that the fetus organizes differently in its brain the reception of sounds received via those differing neurological pathways, for the fetus also demonstrates a developing laterality, more frequently sucking the right thumb and moving the right arm and leg more frequently than the left.[15] Madaule, with Tomatis, also suggests that the fetus is selective about the bombardment of sound, favouring the mother's voice, and demonstrates that by moving towards the "auditorium" between the ilea in preparation for birth.[16]

Turning to the child's listening after it has entered the world of air and unimpeded sound, Tomatis determined that for children who learn to attend (i.e., to pay attention, or to listen) with the *right* ear and with the right cochlear system attuned to sound, sound energy is transmitted to the left hemisphere of the brain and continues the development of the language centre there. The *shortest* aural route to the brain is from the right ear; therefore, sound heard by both ears is received *first* in the left brain. Sound waves travelling along the neural connection from the *left* ear pass around the heart on their way to the right brain and must travel through the right brain and on to the language centre in the left brain via the *corpus callosum*—the main neural bridge between hemispheres; thus, left-ear sounds "will lose time in the conduction of impulses."[17] This more circuitous route from the left ear to the left brain means that sound stimulation arrives in the language centre a split second later from the left ear than it does from the right ear.[18] The effect in the left brain of sound arriving not quite simultaneously from both ears makes a kind of staccato cacophony that the child must decipher in the process of understanding spoken language. At first the child reproduces the sound as double first syllables (ma-ma, da-da) and as babbling.[19] To simplify the noise, most children begin early to recognize and screen out the identical aspects of incoming sounds—they discriminate in favour of right-ear sound in their left hemisphere. Thus, they prefer their right ear and left brain for listening to sounds, while their right brain is used for the retreival of informations that allows them to interpret in a more nuanced way the sound input from their left-ear system.[20] I would find the influence of the right brain on this process extremely significant when I came to analyze Daniel's language during psychosis, at which point I would realize it had been significant from the beginning, too.

If the child develops a preference for attending with the left ear, the child is likely to develop more complex and confused responses to sounds; a "left-eared" child is likely to become left-handed and about 75 per cent of these children carry out the sorting process in their left brain and develop dominance there, with a sorting process that is a split second behind that of right ear–left brain children. The remaining lefties prefer the right hemisphere.[21]

If, for some reason, the child resists preferential listening with the right ear, several things may happen:

- the child more likely persists in paying attention with *both* ears;
- the child remains attentive to simultaneous input through both cerebral hemispheres and remains aware of sound information arriving in the language centre repetitively, which is confusing and may create the sound pattern the child reproduces as perseveration or the need to perseverate to "pin down" sound;
- the child cannot keep track of the *order* of sound reception, which confounds in his or her brain the normall concepts of "order" and "temporality"; this also affects the child's ability to organize, prioritize;
- the abilities of the right brain become more involved in listening and speaking (and in other ways);
- this conflicting receipt of information by the brain defines what the child considers "normal," but that does not conform to concepts of "normal" presented by the external world;
- the child develops the array of perceptual and behavioural symptoms that characterize dyslexic syndrome individuals.

At last, the learning difficulties and behavioural problems I had studied in dyslexic syndrome children were explicable and understandable. I had been aware of unsolved issues of laterality in dyslexic syndrome children. Here was the resolution for some of those issues. By the end of this story you will see that I have enlarged this understanding of dyslexic syndrome.

Other neural connections between the ear, other parts of the brain, and the rest of the body suggested an astounding correlation between the function of the ear and the health of the organs of the body and of health in general.[22] I felt I was entering a new country. And why had its exploration been neglected by North American medicine? Having glimpsed the difficulties in transatlantic medical communication during the beginning of the AIDS crisis in the film *And the Band Played On* (1993), based on the book by Randy Shilts, I imagine such difficulties have had a bearing on the spread of Tomatis's work, also. National differences in medical systems and knowledge predispose their medical professions to particular paradigms.[23] That Tomatis's writing, only recently translated into English, was not widely advertised or known in North America would have inhibited that process of communication during the time he was in Toronto speaking about his discoveries. Today, almost 200 treatment centres around the world use the Tomatis Method.[24]

The Dynamics of Gender

Why do boys more often suffer from ADD (ADHD) and dyslexic syndrome? To some extent, one can point to the interpersonal dynamics of the typical classroom that is organized by and for the women in charge of them.[25] However, the fact that boys have more nerves but fewer synaptical connections in their cerebral hemispheres than girls have may give them less neural flexibility for the sorting process in auditory reception.[26] Other theorists hypothesize that hormonal differences ensure that a boy's brain takes longer to mature to the stage where he can perform reading tasks as easily as girls do.[27]

The ability of the person to pay attention with one ear or the other reduces mental cacophony and makes learning easier. Therefore, children who find it difficult or impossible to attend—hence the term "attention deficit disorder" for their erratic listening—have difficulty learning efficiently. They have difficulty distinguishing particular aural stimuli from a general jumble of sound in the brain. People trying to get their attention may have to take comparatively extreme steps to do so: raising the voice, a touch on the arm, etc. Then, too, a noisy environment (such as an open-concept classroom) will create higher levels of confusion in the bi-attentive brain. If you factor in additional issues of social upheaval in the life of the child, such as recent arrival in the country, living in a different linguistic setting at home, impoverishment, and/or being punished physically or emotionally for not succeeding, you soon will have a child who feels that he lives in enemy territory. Even among teachers and family members with the best intentions, he is oppressed and is trapped in a state of oppression—and this leads to any number of self-defeating behaviours or affects: loss of self-worth; a fawning self-identity with his "oppressors" to the exclusion of recognizing or realizing the self more fully; blaming of self for the problems inherent in the larger system and beyond his control; striking out against his "oppressors" in acts that sabotage the relationship.

For a child having difficulty attending, many of the steps—the very concept of "steps" requires an awareness of discrete events ordered in time—involved in learning to read become muddled. Such children read poorly and when they attain little success and less praise, they try to avoid reading activities. Children with dyslexic syndrome frequently have difficulties learning to speak the language, have problems with balance, posture, and co-ordination, and with activities combining these basic abilities, because the middle–inner ear is not functioning fully. Dyslexic syndrome children may not be able to navigate a crowded room without bumping into other children or adults or desks and chairs. They may be misunderstood as having created such encounters deliberately and may be treated harshly and unjustly as a consequence. People who have tried to remediate balance-related symptoms in these children with special exercises and training routines usually have only modest results, because the underlying cause of the problem is not being addressed.

Girls can more often sort out their own dyslexic syndrome difficulties, possibly because of their greater numbers of synaptic connections in their cerebral hemispheres, an aspect of their brain that may explain their advantages in language development overall.[28] Thus, we have the phenomenon of the girl who is "slightly dyslexic" or who "grows out of" dyslexia. I had observed through a number of the people I had counselled that for those girls who cannot adapt, the social consequences of their "learning problems" are just as devastating as for dyslexic-syndrome boys.

You will notice that I use the term "dyslexic syndrome" rather than the more usual terms "dyslexic" or "dyslexia." As one psychiatrist told me haughtily, "'Dyslexic' means 'can't read' and your son can read, so he is not dyslexic." The problem with this stance is that "dyslexic" has always been a misnomer for the cluster of behaviours and symptoms surrounding malfunction of the vestibular–cochlear system. The use of "dyslexia" that has been popularized refers to children who display a range and variety of problems associated with learning to read. A "syndrome," of course, refers to an observable cluster of symptoms apparently related to some underlying cause that may or may not be fully understood. Persons unacquainted with Dr. Tomatis's research and findings may not understand how a child who nevertheless has learned to read can be called "dyslexic"; similarly, chronic fatigue syndrome comes with an array of symptoms that one or another sufferer is *not* experiencing, including, oddly enough, fatigue (see Appendix A). *"Dyslexic syndrome loosely defines a cluster of symptoms with an underlying cause."*

Music and the Brain

The perception, interpretation, and production of music require both hemispheres of the brain and a synthesizing of the various elements of sound that make up music: beat, rhythm, melody, harmony, form, lyrics, and so on. Children who engage in music activities develop more complex neurology and flexibility between hemispheres. The neurological bridge between the hemispheres has been found to be much thicker than average in adult musicians.[29] The healing potential in music was to become central to Daniel's story, as you will discover at many turning points in his journey.

The Perception and Manipulation of Objects

Ambi-listening children may become ambidextrous, attending with both ears and developing manual skills on both sides of their bodies. Leonardo da Vinci produced remarkable examples of this form of development; he was able to write with both hands at the same time. His cerebral hemispheres may have developed two language centres and may have functioned disparately for separate tasks carried out simultaneously, a sort of bi-unilaterality. Also, his writing was mirror image as he performed ambidextrously.

One of the things about Daniel's perception that had surprised me was his ability as a child to read as successfully when the book was upside down as when it was right-side up. The same skill was displayed by his grandmother Jean, Richard's mother. In the process of relating visual stimuli (objects seen) to sound stimuli (words heard) children have to concentrate on the position of objects in space. The easily accepted notion that children learn first while they are learning to speak their language is that of "object constancy in space." A pencil, for example, is a pencil, whether held up, down, or sideways, whether it is lying on a table, upended in a holder, or suspended at any conceivable angle in mid-air. However, in the world of letters and writing and reading, which they encounter a few years later, the rule the child has absorbed through learning to speak and to listen is broken. The "ball-and-stick" shape of a letter "p" cannot be thought of like a pencil. It is to be considered a "p" *only in that configuration*; if you move it relative to the line of writing or of type, it becomes a "b" or a "d" or a "q" or even a "g." Hence, dyslexic syndrome children must pause and struggle with context to try to figure out how "the ball-and-stick shape" is functioning in a particular word and, then, to see how it fits into the words in the sentence.

Since their stapedius muscles are not functioning properly, neither are their labyrinth, vestibule, and cochlea; and, as a result, they have great trouble perceiving "up," "down," and "sideways." In some cases, the person's brain interprets in reverse order what is seen, as well, as does my friend Nancy's. After hearing and seeing in dyslexic mode for 60 years, she now wears glasses containing a prism to re-sort the image of letters that will be transferred to her brain.[30] Dyslexic syndrome children's experience of standing upright, for example, may include a kind of giddiness they have learned to live with or even to enjoy. The giddy feeling you or I might feel standing on a dock while watching the waves or when standing at the top of a flight of stairs looking down that would register as a caution not to fall into the water or not to fall forward down the stairs is not processed the same way in a dyslexic syndrome child, as toddler Danny had demonstrated by twice jumping down a flight of stairs. Those "acts of daring" were not merely symbolic of what would eventuate, but intrinsic to his dyslexic syndrome perceptions and judgement. For these and related reasons such children are disadvantaged in the classroom, endangered in their environment, and the ignorance of their parents and instructors and other school personnel quickly creates situations that isolate them emotionally and socially from their successful peers.

Those dyslexics who have found a means of compensating for their learning disability may learn to read, but usually not with the efficiency of their "normal" peers. (Although a few learn to overcompensate in reading.) The brain has many ways of compensating for difficult external circumstances and confusing internal ones. Many children learn how to sort out a certain amount of the "noise" in their brains and learn how to function according to their own developed strategies. Some teachers using multi-sensory approaches to the teaching of reading—such as colour-coding the letters of the alphabet and making letters or phonemes from pleasantly tactile materials—assert that virtually every child (and adult) can be taught to read. However, by the time dyslexic syndrome children have attained physical maturation they may have developed such serious emotional scars that their learning becomes possible only through access to highly sensitive and personable teachers.[31] Some dyslexic syndrome people have become spectacularly successful in finding compensatory strategies, among them Thomas Edison, Albert Einstein,[32] and Charles Dickens.[33]

Dyslexic syndrome disability is illustrated in the experiences of the woman who had been my secretary. She has suffered in many ways from dyslexic syndrome. She struggled to maintain her academic status in school where she knew she was smart but "felt dumb"[34] (in Dr. Harold Levinson's phrase) compared with her university-graduate brothers and sisters who regarded her as stupid and inept because of her "learning problem." I and my office associates could see that she was performing some secretarial tasks with great effort despite her superior intelligence; my friend and co-worker Margaret, a teacher, first recognized her dyslexic syndrome.

In her mid-forties Nancy was tested to determine how she perceived spoken sentences. It was found that she heard a word or two with her right ear, then a word with the left; then she switched back to listening with the right, then the left. Depending on the length of the sentence and on the complexity of the comment, she might make six or seven switches during the hearing of a single sentence and dozens of switches during a conversation. No wonder she so often asked people to repeat things and had trouble expressing her own thoughts. Many years following her hearing and listening test I noticed a sudden and dramatic change in the coherence, clarity, and grammar of her e-mails. Shortly thereafter, in the course of a visit to me, she mentioned that she had developed a tumour in her left ear that was causing deafness. Evidently, the loss of function in her left ear had ended the perceptual "racing from hemisphere to hemisphere" in her brain necessitated by her ambi-listening, which had produced the constellation of symptoms called "dyslexic syndrome." She had also become a much calmer, more content person who "no longer felt angry all the time."[35]

Fatigue, the Immune Response, and Listening

As described below, the vestibular–cochlear system mediates the transmission of sound energy to the brain. If the stapedius muscle attached at the oval window ceases to function properly, the person's brain is deprived of the movement of the vestibule and cochlea that also conveys sound energy to the brain.

> For him [Tomatis], the cochlea and the vestibular system have the same role. They both help us to perceive movements. The vestibular system is in charge of the slower movements—those that we see, feel, and call movements. The cochlea specializes in faster oscillatory movements—those that we don't see and can hardly feel, but those we can hear. . . . Reaching a certain speed—sixteen to twenty vibrations, or cycles (Hertz) per second—a movement starts to be perceived as sound. . . .
>
> The cochlear–vestibular system of the inner ear must be viewed as a single apparatus doing one job—analyzing movements—not as two different entities with separate functions. . . . I will call the vestibular end . . . the "ear of the body" and the cochlear one, the "auditory ear." This double antenna largely contributes to our awareness and communication with our own body—our body image—as well as our awareness and communication with the world around us through the dimensions of time and space. . . . it is this [time] quality of sound production that brings the "time dimension" into the realm of the auditory.[36]

Those movements of the vestibular–cochlear system also are essential to physical and mental health; they ensure a properly functioning immune system and freedom from depression. Although we do not yet know all of the ways in which this movement is translated into the feedbacks that produce energy, the clinical evidence abounds.[37] Madaule's *When Listening Comes Alive* contains an intriguing account of exhausted and depressed monks who regained their health when Dr. Tomatis recommended they resume the traditional chanting that they had abandoned post-Vatican II.[38] Dr. Tomatis not only discovered how a person's overall health is affected by the listening process, but describes in detail the neurological routes by which sound moves through the body to affect all of its organs.[39]

During Daniel's interview process, the Director mentioned that a listening program also would help with my chronic fatigue syndrome. As Paul Madaule would like to specify, he "used the word 'help' because he was aware the *cure* was not the purpose (or the goal) of the program but he still believed that without cure there was still a possibility of improvement. This [approach] is fundamental to the way The Listening Centre deals with many of the problems they help, including dyslexia."[40] I appreciated his cautious optimism, but I had not identified with the French monks he had written about. On the contrary, I considered myself rather expert on what could *not* be done for CFS. During the nine years of my illness I had read every available publication on CFS, including magazine articles, books by laypersons who were experimenting on themselves with alternative therapies, the investigative journalism of Hillary Johnson in *Osler's Web*, and the contributions of allopathic medicine to the symposia convened by Canada's then leading expert in CFS, Dr. Byron Hyde of Ottawa, and the newsletters (for several years) of his Nightingale Foundation. I assumed that the Listening Centre director, like most people I met, did not really understand the disability that I and other members of my family were living with or know what I had learned, namely, that "chronic" did not mean "off and on"; it meant "medical science cannot cure this and you are not likely to recover." In other words, I had no expectations whatsoever that Paul Madaule's suggestion that my condition might be improved could possibly be correct. But I was very curious to know how part of the intervention Daniel was being exposed to sounded.

Dyslexic Psychology

Madaule's *When Listening Comes Alive* and "The Dyslexified World" cite considerable detail about the psychological effects of dyslexic syndrome—the alienation and distrust of the adult world that develop in a child who constantly is being asked to do things he or she cannot perform and who is being told that reality is something other than what his or her mental and physical experience of reality can confirm. Such children live in a state of confusion and may become dominated by sadness, fear, anger, or dismay and may fall into despair while very young. These children are vulnerable in many ways.

The cause of malfunction of the inner ear may or may not be treatable. A congenital absence of part of the inner ear or its destruction will render the syndrome untreatable. However, most often the tiny muscle attached to the vestibule may have shut down for such reasons as trauma from too loud noise, from an ear infection, from a generalized trauma such as a car accident (like Yolanda, the fellow CFS sufferer my doctor had referred me to) or warfare, or even for psychological reasons because of which the child has felt a need to shut out the outer world.[41] In these cases, there is hope that the middle ear's tiny muscles can be stimulated and exercised into a return of normal function. When someone has experienced normal development and the mechanism is interfered with—as happened to our neighbour whose eardrum was pierced by a metal splinter in an industrial accident, and the subsequent infection was inadequately treated by his physician— the effects on the brain can be devastating. This neighbour had been an exceptionally well-liked, caring, ethical, and hard-working individual. Following the surgical removal of his entire ear mechanism, including the cochlea, he developed escalating mental and behavioural aberrations, and eventually he took his own life.

The Puzzle Pieces Come Together

As I read about Dr. Tomatis's research, I felt my suspicion had been confirmed that my use of a table saw while Daniel was in utero had caused some kind of trauma to his middle and inner ear, a condition that likely was exacerbated when I raised my voice to the other children while holding him during his infancy. My sense of something having gone wrong very early in Dan's life had not been mistaken. Everything I had observed about Danny—his slowness to speak the language, his preference for crawling over standing, his frequent inability to attend, his emotionality, his avoidance of preschool reading and learning, his relative lack of co-ordination, and then his school experiences that reflected discouragement with some teachers but not others (depending on their teaching strategies and classroom control), plus all the effects of those difficulties in his family and other social relationships—at last made perfect sense. And everything I had read, observed, and heard about dyslexic syndrome in my professional career fit seamlessly into the Tomatis Effect theory. My hope surged that Dan could be transformed into a normally productive and self-confident individual who could realize more of his intellectual potential. At the same time, I acknowledged that Richard's (and my mother's) sense that I should not have been using power tools while I was pregnant was true and accurate and that I would have been wiser to have listened to them.

Preparing for Treatment

Dan was willing to go to Toronto to see if The Listening Centre could help him, but on investigation I found we could not afford the treatment. After a few months, Richard's mother surprised us with a Christmas gift of $5,000. We immediately prioritized Dan's treatment, sought an interview at The Listening Centre, and confirmed the program set up for Dan for February. Richard also thought he should use some of the money to go to Louisiana to visit his recently widowed mother.

During our interviews, Paul Madaule gave me a copy of a presentation he had made at a psychology conference[42] and I found his further insights meshed precisely with my observations of Dan. I was very excited about the prospect of Daniel's being healed of his dyslexic syndrome and was steering firmly towards that intervention. The few weeks of reading, learning, and observing prior to his treatment at The Listening Centre had shown me that I had been correct about aspects of his "differentness" that the family did not understand and revealed how deeply wounded a child he had been. I was not inclined to favour my husband's advice that the intervention be delayed at least until he had returned from his trip to see his mother.

Richard and I were not knowledgeable enough about the effects of street drugs—nor were we sufficiently aware of the extent of Dan's secretive drug abuse—to appreciate how fragile his mental stability had become. We soon would be dealing with the effects of these currents. I wrote to Paul Madaule in anticipation of Dan's appointment, explaining to him that Daniel was struggling somewhat with the concept of entering into the program of "listening retraining." Having read his paper on "the dyslexified world" and noting the importance of Daniel's making this decision on his own, we were careful to let him come to his own decision about entering the program.

The day after Dan's first interview, he came to me and said that he wanted to enter the program, but that he needed "to see a psychiatrist" because he was sure he was insane: "neurotic or psychotic or something." I suggested that he might like to read Mr. Madaule's paper on dyslexia, or that I could read it to him if the task appeared too difficult. He agreed and asked me to do the reading, insisting, though he had not eaten all day, that we begin without taking time for food. As Mr. Madaule's paper went on to describe the inner world of the dyslexic, Dan's interruptions became grim: "That's me. That's what it's like. That's why I took drugs. That's what I've tried to tell you. That's what you don't understand. That's what you've never understood." I had to keep pausing for his reminiscences and explanations. Halfway through the paper we took a break for dinner, with Daniel already showing an inner peace we had not seen for many years. He was eager to go back to the reading after we had eaten.

As we continued, Daniel began disclosures about his drug experiences. He was afraid he had destroyed his sanity with drugs, but now, he said, he was realizing that his awareness of his own mental processes was simply dyslexia and "normally abnormal." Daniel had said something profoundly important, only one-half of which I could appreciate at that stage of my own learning.

The information he poured forth about his use of drugs hardly seemed credible. I could not believe he would still be as mentally alert and as functional as he was if all that he was telling me was literally true. Daniel said Mr. Madaule's observations were giving him back his sanity, and, indeed, they seemed to be turning aside his darkest fears, restoring his sense of worth, opening new vocabulary and expressive language to him. He acted out with his face and body how his paranoia feels. Again and again he affirmed what Mr. Madaule had written about the neurotic and what he termed the "schizoid" psychology of the dyslexic, up to and including suicide. Dan said:

> People say that I'm slightly dyslexic and talk about it as if it's something little and unimportant, just the reversal of letters. It's total, and it's huge. I know people who are worse than I am, but even for me, it's terrible.

He spoke knowledgeably and with compassion about a peer and how badly people have treated that person. He talked about how close his unstable emotions bring him to suicide.

I didn't finish reading the concluding couple of pages to Dan; he had too much he needed to tell us. During the rest of the evening he made little decisions about what he would be doing—picking up my dulcimer to strum it, walking his girlfriend home, going to the study to talk with his father—with a confident, calm, natural decisiveness that was absolutely remarkable for its contrast to the tension that had become constant both in him and between himself and others. As he talked to us late into the evening, his body relaxed in a new way and his facial expressions subtly altered—becoming quieter and more expressive. We could see that light had entered his darkness and he was drinking it in, swimming in it.

I wrote to The Listening Centre about Daniel's reactions, adding: "Of course, this is only the beginning of a long road. But there is hope that it will be a long road, not one cut short by his own hand in the agony of his 'differentness.' He wants a little more time to contemplate change. He is full of fears that he will change and become so unrecognizable to himself that all his values will shift, that he will lose the enormous security of his friendships; that he will become someone he won't like." Just how important Dan's peer relationships, especially those bonded by drug use, had become to him as a substitute for a more broadly based self-esteem we had yet to discover.

Richard added to his motivation for seeing his mother a profound "thank you" for a treatment we expected to heal Daniel. When Luke and Kathleen learned he would be driving south, they invited Sarah and Eric to come as far as Virginia and stay with them. Sarah felt she could make the trip if she limited her activity. She was especially anxious to see again the friend she had stayed with while completing Grade 10. Eric was excited about seeing Luke.

Richard's trip was scheduled to begin just after Dan's treatment at The Listening Centre, which added to other reasons we felt Daniel should stay at home. Dan had already missed school for the interview process in Toronto. There was not much room in Kathleen's mother's house, so Dan was asked to wait until another visit could be arranged. Dan preferred to stay with his friends in any case, and for his own reasons—we would come to know much further down the road—that had as much to do with his involvement with drugs as with friendship.

Richard had wished that Dan's schedule could be postponed until his return. I was reluctant to delay what might help Dan. Before he left, Richard mentioned he had spoken to a golfing buddy, Peter, who would be willing to help out if I found I had too much to deal with by myself while he was away. Everyone in the family was looking forward to the end of February.

5

Light and Darkness

Healing and Psychosis

> *My restored energy was tested rather dramatically. At about ten o'clock the following evening, the sister of our hostess in Toronto, whose family was non-English-speaking, was haemorrhaging seriously. Furthermore, as I learned from her sister, she had been bleeding abnormally for a month, following a D&C. With a kind of boldness I seldom display, I took charge of the situation and insisted—against her family's fatalistic despair—that she be taken to the hospital. She was fainting in the hallway and had to be half-carried to the elevator and parking garage. At the hospital the family was having trouble communicating in English, so I stepped into the role of language facilitator and mediator for the family with the staff. When the doctor was about to send her home "to be observed" I suggested to him that none of us had the medical expertise to "observe" whether or not she was bleeding to death. What would he do if his wife were in this condition? He paused a moment, and decided to operate. My renewed ability to act with energy and confidence may have saved Maria's life. Although tired the next day, I was able to put in a full schedule. For the first time since I had contracted CFS I slept restfully—instead of waking with a sense of depleted resources I faced the day with a sense of renewed energy.*

My Listening Intervention

While Daniel was engaged in his listening program at The Listening Centre, I was listening, too, to a parallel but less intense (without bone conduction) program designed for parents who were interested in what their children were experiencing. I found the music pleasant, especially the half-hour of Gregorian chant, and, as the Mozart segment became progressively more filtered and unmusical, I found the listening just a bit odd—a curiosity—and to the musical ear a trifle irritating. During the listening I read much of what was at that time available in The Listening Centre's shelf of fascinating books on music, sound, kinetic-music therapies, and learning environments.

During about three-and-a-half sessions—a total of about seven hours of listening—I did not notice any particular effect in myself. Very suddenly, halfway through the fourth session, I felt as though someone had flipped a switch somewhere in my head. My former sense of "normal energy" returned instantly. I was amazed and delighted and a bit euphoric. And puzzled. It was very difficult for me to relate what had happened to me as being like what had happened to those French monks Dr. Tomatis had analyzed and Mr. Madaule had described—probably because their symptoms had not been given in much detail and because I had not been chanting. I did not know if the change would last, or to what degree my whole array of CFS symptoms would be altered. (Eventually, I discovered that I should have continued listening, singing, and making music to maintain my healing.)

I presumed a latent synapse somewhere in my head had fired. Neurons fire in an "all or nothing" way; until the stimulation to a neuron achieves its "action potential" the neuron remains quiescent. Although Tomatis did not explain the recovery of the sick monks in quite this way, I believe the stapedius muscle must attain a sufficient level of conditioning (about seven hours for me in this instance) before it can induce the action potentials that will convey to the nerves of the body the sound energy it is receiving. "The action potentials . . . do not grow smaller however far they travel along the nerve. They are pulses of equal intensity, with varying frequency—which is the basis of sensation and of control of muscle."[1] Thus, once the neurons start to function they carry the energy from sound not just to a particular location in the brain but thence throughout the body.

Furthermore, the nerve fibres "conduct electrical signals from the senses at about the speed of sound."[2] The entire nervous system is designed to convey impulses from sound! We can see that, on account of the way the nerves function, the whole body is poised like a battery of tuning forks to receive the entire range of sounds (starting just above the sound of air molecules hitting the eardrum) that will activate all of the components of the ear, including the stapedius muscle. A healthy stapedius muscle induces the action potential of particular nerve bundles conveying the impulses from sound, via the brain, to the entire body.[3]

Simultaneously, Daniel was undergoing what seemed to me no less than a miracle. During the first four days *he lost all of the symptoms of dyslexic syndrome that had been displayed in him in one form or another from birth*. I was so focused on the changes in him that I paid relatively little attention to what was taking place in me. However, there were clear signs that I had regained all of my former energy and mental acuity.

Following my healing—although the stress in my life increased exponentially—my energy remained high, my immune system functioned normally, and my mental processes regained clarity. I probably did not gain full restoration of my muscles and related physique eroded over almost a decade, as I was post-menopausal and had some arthritis and other signs of aging; but my muscles were much improved and mild exercise no longer caused permanent muscle damage. In addition to the unusual needs of Daniel and other family members, I returned with enthusiasm to editing, enjoying the challenges of texts in logic, environmental issues, international economics, and so on, without mental fatigue, fogginess, or confusion, and with renewed confidence.

Daniel's Pre-Intervention State of Mind

Daniel had been experimenting for many months with street drugs and spent much of his time at the apartment of a friend who lived near the high school, mostly without telling us what was happening. He had been extremely secretive. That the high school was in another town made it difficult for us to know how he spent his days. The fact that the high school denied to us the presence of drugs in and around the school established a false sense of security as to how Dan was spending his time outside of the classroom. We could tell that he was getting a certain amount of marijuana and we were terribly anxious about that. My experience years before with friends who used it made me very cautious about its touted benefits; I was opposed to its use in general and had never tried it myself. Additionally, I relied on my husband's reiterated remark that he had tried it in college and found it affected him little. This was a benchmark in his mind of its potential for harm, and neither of us knew that the potency of pot had increased significantly since the sixties. I knew too little at the time about other street drugs and associated severe mental problems to realize that Dan had lied to us, was using drugs other than pot, and had experienced drug-induced psychotic episodes. The following account is essentially the notes I kept in a journal during the weeks leading up to Dan's listening retraining, and while Dan and I were in Toronto during that intervention.

Daniel was first interviewed on the 12th of February 1997 at The Listening Centre in Toronto and was judged an acceptable candidate for listening retraining. He said he felt uncomfortable, awkward, and "neurotic." I think he was pleased and a little relieved to be accepted into the program. The following day, Richard and I read "The Dyslexified World" by Paul Madaule, which explained Daniel's state of mind more fully than anything else we had read. We were aware that Dan should make up his own mind about whether or not, and when, he would go to The Listening Centre. When Daniel came to me the following afternoon and said he needed a psychiatrist—"I want to go to The Listening Centre, but I think I'm neurotic or psychotic, or something"—I suggested we/he read Paul Madaule's paper first (see Chapter 4).

Richard's impatience with Dan regarding his going to school the next day created a pressure Dan could not stand. He tore off his dad's sweater he'd been given and was crying and shouting in frustration before he was out the door. If that day began unhappily for Dan, the next one gave cause for alarm. Dan called to be picked up from the town where he went to high school. He had been at Kurt's place and he was afraid (later, he said not intensely afraid) that he might jump out of the window. He described having what he called "flashbacks" to how he felt as a child when he was hurt by his playmates. He felt sick. He was twisting the fabric of his coat with his fingers, struggling desperately not to cry. By the time we had driven the 20 km home he was calm and feeling better. He had supper and played his guitar in his room. Later, he asked if Kurt and Tonya could come to his house for a change. I welcomed an opportunity to meet this couple with whom Dan was spending (I had no idea how much) time.

Two days later, Dan brought me the following account he had written of what happened to him the night Kurt and Tonya were visiting. Daniel's room and the guest room form a sort of separate apartment at the west end of the upstairs, so all of this took place without our knowledge. The first thing that struck me was how lucid and well-expressed an account he was giving of an experience I did not understand. The second thing that occurred to me was that he apparently had recovered completely, judging from his appearance and coherence as we discussed this episode.

> I went to bed around 12. It was like I slipped into a coma or something. I was spitting up foam and choking on my tongue. My legs were shaking. My eyes were rolled back. I just lay there. I couldn't do anything.
>
> When I would think about things I would normally care about, I wouldn't care. My mind would block it out. I wasn't able to think anything.
>
> Suddenly I shot up [got up] gasping and panting and then I walked around like I was in a flu dream. I sat there shaking. As I walked around I felt I'd gone from neurotic to psychotic. I smoked 3 or 4 cigarettes in a row. I must have been lying there for hours like that.
>
> I lost my sense of time.
>
> I went and got a drink of water. I went to the middle room asking for Kurt and he woke up Tonya. (I had felt like I was on some drug. I had been hallucinating, thinking Kurt was beside me: weird stuff, I can't describe it.) I was shaking and looking around, my eyes darting

around. I was scared and didn't care any more. I'd gone over the brink.

Kurt and Tonya and I talked for a while about what it could be, but I knew it was about my dyslexia. Tonya guessed it was related to the flashbacks, or getting the flu, or choking in my sleep.

Then I got another drink downstairs and smoked another couple of cigarettes.

Dan did not mention that he had ingested hallucinogenic mushrooms the day before. With Dan's permission, I shared his writing with Richard right away. He identified many of the characteristics of the episode as part of the "being ridden by the old hag" dream motif, which had been elaborated in an article by a colleague, expert in folk medicine, who was attached to the University of Pennsylvania medical school. I felt this observation linked Daniel's experience to one common worldwide in folk experience during the dream state. It was a concept I would come back to again and again as we walked with Daniel through his mental illness. When I asked Dan to let me record his experience he was somewhat reluctant. "Why would I want to remember this?" he said. I look back on this gem of self-expression and introspection as holding the clues to all the mysteries that would launch my odyssey into Daniel's interior world, seeking for explanations.

Daniel's Listening Retraining Experience

Day 1 February 27

Ten days later, Daniel and I went to Toronto to begin his Electronic Ear™ exposure at The Listening Centre. Daniel's listening ability was tested via the Tomatis Listening Test, which is a modified audiogram, not the sort of hearing test administered by an audiologist. Paul Madaule explained that the listening test graph illustrates the way that the person hears the sounds from the environment (air conduction) versus the way he or she hears his own voice (i.e., via bone conduction). Daniel's responses showed high sensitivity (low thresholds) on the bone conduction measure compared to his responses on the air conduction measure. So, in Daniel's case, the goal of the listening program was "to reduce the sensitivity through the bone conduction with the purpose being to allow him to achieve better sensory regulation."[4] When Paul Madaule told me that the object of the listening retraining is to normalize these forms of reception, which would bring the red line into conjunction with the black, or even below it, I did not really understand what he meant thoroughly enough to appreciate what they signified for Daniel's ear function. I had no idea how his ear related to his brain function, either. And I also did not know enough about these processes to formulate questions so that The Listening Centre staff could enlighten my ignorance.

A clear summary of the listening retraining process is that of Nicholas Regush, the well-known reporter for the Montreal *Gazette* and critic for ABC news who specialized in medical science analysis. Mr. Regush writes of his interview with Dr. Tomatis:

> Tomatis spoke about tiny muscles connected with two small bones, the hammer and stirrup that sit in the middle ear. "These muscles are generally viewed as a system that can protect the ear from dangerously loud noise by diminishing vibrations of the eardrum and incoming sound . . . . But they're also capable of focusing the ear to tune into certain sounds." The English ear, for instance, could be trained to adjust to the range of pitches of other languages. As it happens, the nerves controlling the hammer and stirrup muscles are also involved in speech, which helps explain why attuning to sounds would allow one to reproduce them vocally with more ease. Delving deeper into the ear's physiology, Tomatis was interested in the phenomenon of the dominant ear. Humans are either right-ear or left-ear dominant, which indicates through which ear they primarily take in sounds. Tomatis came to believe that listening predominantly with the right ear is a more efficient means to tune in and respond to the world. The right ear, he reasoned, has more direct and numerous connections than the left ear to the brain's left hemisphere, the primary control centre for speech and hearing. Listening predominantly with the left ear, according to Tomatis, delays the processing of sound and reduces the quality of sound reception. Because most nerves in the brain lead to thc oppositc sidc of the body, signals must first be routed from the left ear to the right side of the brain, and only then are they forwarded to the left. This view led Tomatis to postulate further that a delay in sound processing could well affect speech, leading to inarticulate or monotonous expression and even stuttering. He also studied the structures of the inner ear, specifically the cochlea, conventionally associated with sound perception, and the vestibule, which handles the perception of three-dimensional space such as body position, movement, and balance. Tomatis didn't view cochlea and vestibule as functioning separately, as was widely thought; rather, he saw them working together, their anatomic connections strong throughout the nervous system. Together the cochlea and vestibule analyze movements, jointly contribute to the awareness of ourselves in space, and link up with all forms of sensory information. In a sense, such inter-

> connectivity enables the entire body to be involved in the processing of sound.
>
> . . .
>
> At this point, it was hardly a leap for Tomatis to conclude that poor listening skills could have an impact on the entire body, meaning that human imperatives such as speech, coordination, and body image could be disturbed. By the end of our meeting, I began to realize that perhaps this is partly what was occurring with the monks Tomatis was able to help. It wasn't merely a matter of sending energy to their brains by means of sound stimulation. The therapy Tomatis had designed was calculated to rouse the body. In this sense, he pioneered not only a therapeutic approach but also an understanding of what it may take to live life more fully. . . .
>
> . . .
>
> The core of the Tomatis Method, the Electronic Ear™ is designed to manipulate sound frequencies. Tomatis believes that high-frequency sounds rich in harmonics, such as those coming from a violin, energize the brain, while low-frequency sounds, such as drum beats, sap energy. As Madaule explains it, a membrane in the inner ear is covered with cells that translate vibration into sound. There are more cells transporting high-frequency sounds than low-frequency to the brain. High-frequency sounds carry more vibrations and therefore pack more energizing power. The Electronic Ear™ essentially is a sound amplification system that can increase or decrease the frequencies, or ranges, of sound in both ears. An automatic switch, or "gate," in the machine can turn up one form of sound (such as the high ranges of a Mozart sonata) and then the other (the lower ranges), thereby giving the hammer and stirrup muscles of the middle ear a good workout. The machine can also reduce volume to the left ear, forcing the right ear to work harder to listen.[5]

Although most of this information was made available to me, I could not absorb or organize it because the components of the brain and its function in feedback to the ear and to other aspects of behavioural function were missing. Doubtless, my chronic fatigue syndrome initially impaired my ability to absorb this new information. My very limited high school knowledge about the mechanism of the ear provided my only visualization. My memory is strongly visually oriented and the only imagery I had to help me was the highly stylized picture accompanying Regush's article that began and ended with the ear; again, not much to help me picture the route of sound from the ear to the brain and its effects there. Then, too, I was nervous about the "newness" of Tomatis's discoveries vis-à-vis the "stable" body of knowledge with which I felt comfortable. The mere idea of comprehending brain function was intimidating to me (much less was known and disseminated about the brain 12 years ago); I imagined that understanding the workings of the brain was on a par with nuclear physics, which likely was why my father (conversant with nuclear physics) had a shelf of books on the subject. Daniel's incapacity for learning was my overriding concern—I confess that I was not seeking knowledge, so much as the expertise of people who might be able to effect positive change in Daniel.

Daniel would be listening to increasingly filtered violin music of Mozart through headphones that delivered sound preferentially to his right ear through the air in the ear canal "and through a bone conductor device to stimulate both through the bone and air conduction at the same time. . . . The purpose of the progressively filtered Mozart music is to train the listening function at every level of the sound spectrum in order to restore a harmonious reception of sound; especially in reference to Dyslexia, the sounds of language located in the mid-range of the auditory spectrum. Gregorian Chant is also used to provide rest following the very stimulating effect of the filtered music of Mozart."[6]

I found the additional device in the headphone confusing. My profoundly deaf great aunt had heard by bone conduction with the help of an "ear phone." She was the second person in Canada to wear the device, which would later be called a "hearing aid" and be greatly refined; she became an instructor in its use to other deaf people whose hearing was partially restored via this invention of Alexander Graham Bell. I was familiar with several of her hearing aids and how they worked; I could not understand how Dan's hearing could be superior but that he would also need bone-conducted sound stimulation. Later, I would find these two modes of bone-conducted sound were for somewhat different purposes. My aunt wore a sound amplifier behind her right ear; Daniel would hear a tone sounded from a tiny device in the headphones positioned against that area behind the right ear to "stimulate" via bone conduction, thereby emphasizing the sound he received through the headphones by air conduction and making him focus on that avenue of input to his left hemisphere rather than on sound input from his left ear.

For those readers already feeling sensory overload in regard to the ear and the brain, be assured that the same information will be presented in many contexts, and with some diagrams.

My voice was recorded "and modified by filtering. In Tomatis's theory this allows memorization [recall and reinforcement of memory] of sound perceived by the child in utero. At The Listening Centre (or in the listening program) it is used to reinforce the initial step of the development of language which starts as early as the acquisition of the mother tongue or—as early as the child's ear and the brain are able to perceive the sounds of language (at about 4 months of pre-natal life)."[7]

Later phases of the listening retraining program that Dan did not complete would have involved feedback work with the voice and reinforcement listening. The listening retraining program included tapes of my voice filtered to sound as Dan would have experienced it in utero. This was the one aspect of the program that at first I had reservations about (see below and Chapter 13). I had thought at first, also, that his drawings were simply a reference to that aspect of the program; but Dan revealed that the Nirvana CD *In Utero* (e.g., #2, #3) was probably the greater influence. Once I fully understood Dan's (and likely Cobain's) brain function I could see that Dan's art resonated both in connection with his "idol" and with his own condition from before birth.

Day 1 February 26

During the listening sessions, clients are free to use art materials that The Listening Centre makes available, to read, to relax or to sleep on comfortable mats and cushions provided, or to sip a drink. Eating or chewing is not acceptable, for the activity affects the area around the ear and creates distracting noise. Dan occasionally slept, but spent much of his time drawing and painting. The accompanying images are some of the art Dan did while listening, keyed to the chronology of his program, with his explanations of subject, process, or method. At the time I could see that they revealed the changes taking place in his motor control, his sense of balance and awareness of gravity, his ability to order concrete things in space, his visual perception, his ability to distinguish different states of consciousness and changing concepts of the self, including improved self-esteem. I also saw that the singer Kurt Cobain, who died (officially, by suicide) in 1994, had influenced his work. What I could not have guessed was that these paintings, drawings, and writing were records of themes and of conditions in his cognitive processes that would recur for many years, leading me to a new understanding of how the brain develops and functions, not only in dyslexic syndrome individuals but in most people.

The first night Dan said that he was "more tired than I have ever been in my whole life. I can hardly lift the blankets." Of course, during the 10 days of his treatment in Toronto he was completely cut off from drugs. He probably was showing withdrawal symptoms at the outset that I did not recognize, as well as "flashbacks" from the drugs he had been using. However, he was "in withdrawal," also, from the way he had perceived reality from before birth, as we would discover.

My sister had made arrangements with her friend Anna for Dan and me to "camp" at her apartment during the 10 days of Listening Centre appointments. We met her late in the afternoon and I brought a pizza in for dinner.

Day 2 February 27

After Daniel's second day of listening retraining he had a slight headache, which he thought came from the second, bone-conduction signal behind his right ear.

Instead of going right back to the apartment that afternoon we went to see *The Empire Strikes Back*[8] in a theatre and were astonished by the impact of the soundtrack. As George Lucas comments in the commercial before the film starts, "The audience is listening!" Both of us felt we could do so with new abilities.

Day 3 February 28

In the car, immediately after the listening retraining session, Daniel said, "I feel like something's happening. I know I'm changing." He felt a little sick and panicky, which I later realized are typical symptoms of drug withdrawal, but also of near-psychosis. Later he said, "I felt panicky after I woke up" [he was sleeping during most of his retraining session]. "Whenever the music cut off I wanted it back on. [In the interval] I wouldn't like it. I would get panicky. It felt like mental confusion. I was coming to the reality that something *will change* and it was strange."

We went to Chinatown to sightsee in the afternoon and had a late lunch there. We spent the evening with our hostess. Dan was much attuned to Anna's conversation and was charmed when she sang in her native language for us. She enjoyed his guitar playing and I was surprised that she could coax him into playing for us.

Daniel sketched a castle (#7), a typical subject, but done differently from any others—much less like a fortress and more like a residence. He placed it in a realistic landscaped setting, which also struck me as unusual, and with mountains that were attainable or even pastoral. Dan was familiar with the idea of an image of a house representing the self-concept. Anna, who had been an artist by trade in her native land and continued to sell her work in Toronto, picked up Dan's drawing and added some shading and highlights. He did not seem offended by her intrusion into his work and we talked about notions of communal art production, such as the collaboration of Daniel's godfather Ken St. Onge with the Canadian artist William

Kurelek. By 11:30 I was exhausted, as usual, but Dan would have stayed up talking if I had allowed it.

Day 4 March 1

For the first time since we had arrived in Toronto, I had some trouble persuading Dan to get up. He was tired during the listening session, but continued to draw. A teacher working at The Listening Centre showed me some of Dan's art (#6), which again I did not recognize as a recapitulation of art on the Cobain CD. Despite its origins, the content was significant: "I want to die" had been crossed out and "I want to live" had been left as a statement. I wept when she shared this with me. I had not realized the depth of Dan's despair nor how close he had come to embracing the idea of suicide. After the listening session each day when we went back to the car, Dan would run ahead of me down the parking garage ramps to the car. He seemed to be expressing his elation and energy following the listening sessions. And independence.

Day 5 March 2

After the listening session in the morning, we spent a long, slow afternoon at my sister's apartment, with Dan watching music TV while I read.

That evening Anna cooked stew for us and talked at length about her childhood—both of her good and of her painful experiences. Dan was attentive; Anna presented a rare cross-cultural experience for him. Anna talked

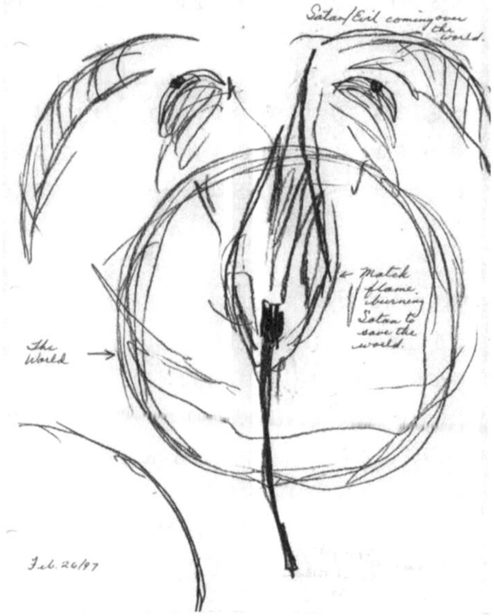

#1 February 26
"The circle is the globe of the earth; the eyes surmounted by horns represent Satan/Evil coming over [overcoming?]."

about her experiences in non-violent resistance of the dictator in her country, including having seen her girlfriend bayoneted in the eyes by the soldiers and having taken sweets to the bereaved parents of the resistance coalition. I wondered how Dan would respond to her accounts of such horrific suffering, but he seemed better able than I to hear what she had endured and from what she had escaped.

Day 6 March 3

Perhaps Dan had had more trouble processing the previous evenings' conversation than I had guessed because he woke up crabby. On the other hand, staff at The Listening Centre had told us to expect Dan to feel mood swings. When I went out for groceries in the morning, Dan waved to me from the 14th-floor balcony of Anna's apartment. I had noticed the locked door and the furniture in front of it, thinking they would be adequate deterrence for Dan to stay off the balcony. I was mistaken. Remembering the incident at Kurt's apartment when Dan had phoned me to pick him up, I was terrified. As I looked up, waving and smiling, I decided—partly on the basis of something I'd read the previous day at The Listening Centre about placing confidence in one's child—that I would only make the situation worse for Dan if I were to show anxiety. So I drove away—in tears. When I got back from a hurried trip to the grocery store, Dan admitted sheepishly that he had gone onto the balcony to test himself and that he'd had to sit down in the corner of the balcony; but he was OK. After this incident, he would sometimes go out on the balcony to smoke and I gained some confidence in his self-control in the situation, but was silently anxious while he was out there. If we were picking our way through a mine field, neither of us was aware of the dangers.

During Dan's listening session in the afternoon he painted boldly in acrylic, a new medium for him. He made (#8) a brilliant orange and red and yellow flame of a blue candle and (#9) an eye in darkness. When he showed them to me the following evening, I asked him, "What does the eye mean?"

Dan replied, "I dunno. It's a vision I had yesterday." Later, he tried to describe the vision: "First, I saw the eyeball. Then, there was a white Being [or two Beings] surrounded by golden light rising up inside me but I had to turn my eyeballs inside to see it; I couldn't draw the Beings because they were on a different plane or slope. So I just drew the eyeball." A while afterwards he said, "I feel so cleansed today for some reason. I just feel good." And shortly, he repeated this.

My journal was lying open on the big glass coffee table that evening and Dan started to write on the page Sophie had returned to me (#6). Although I wasn't too happy about his taking my book, I didn't interfere.

#2 February 26
Daniel in Utero

#3 February 26
Tree [like Cobain tree] sprouting fetuses.

#4 February 26
Match flame as in previous art work contending with dragon. "The world. The match flame is burning Satan to save the world." [The sketchiness, with many lines on an outline, is typical of Dan's drawings; he could not place a line confidently.]

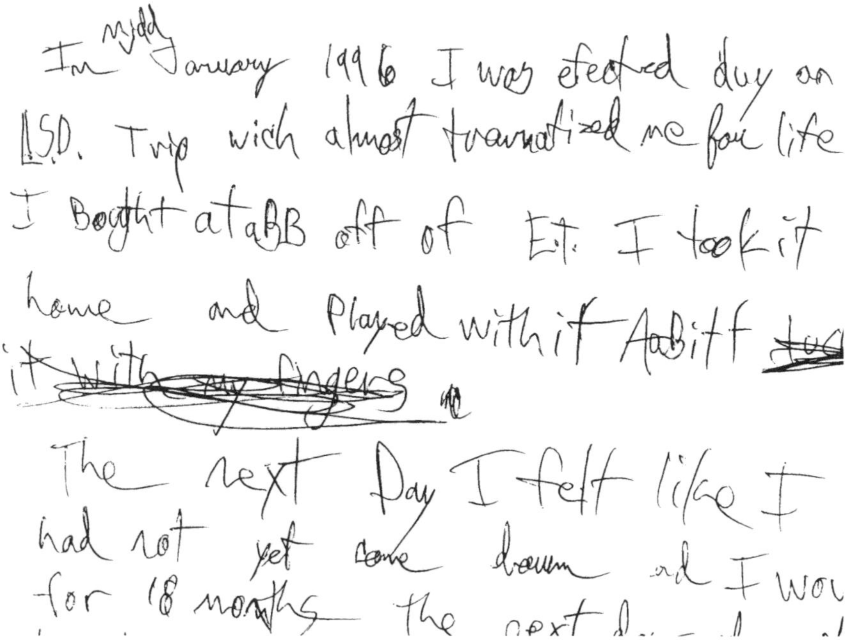
In myddl January 1996 I was efected day on LSD. Trip wich almost traumatized me for life I Boght at aBB off of E.T. I took it home and Played with it AaBitt
The next Day I felt like I had not yet come down and I wou for 18 months the next

#5
Daniel's writing in a style typical of his pre-listening experience.

Suddenly he exclaimed, "This is fucked up!" When I bent over the table to see, I was amazed to find that he had moved from his typical scrawled printing (e.g., #5) to neat, legible cursive writing. He proceeded to write a note to his Grandmother (#10), for which he was eager to correct the spelling. His rude exclamation, for which he immediately wrote out an apology, had meant he was shocked to notice that his writing was *not* a mess, but that he could write clearly. Dan's writing generally had been difficult to read unless he made a great effort to be legible. The drawing and writing on the page seemed to me the documentation of another rebirth, like that final comment on the page, to which Sophie had called my atteniton—circled and crossed out, "I hate myself and I want to die", followed by "I love my self and I want to live." It was then he told me that the circled and crossed-out statement is the name of a song by Nirvana, the Cobain group.

He showed me the drawing he had done of a beautiful girl (#12) and said, "Do you think it's the girl I'm going to marry?"

"I don't know, Dan; could be." Edna, the listening therapist, thought it was also an expression of the happy Dan and I told him so the next day.

As I have mentioned, during some of the listening sessions Dan had been

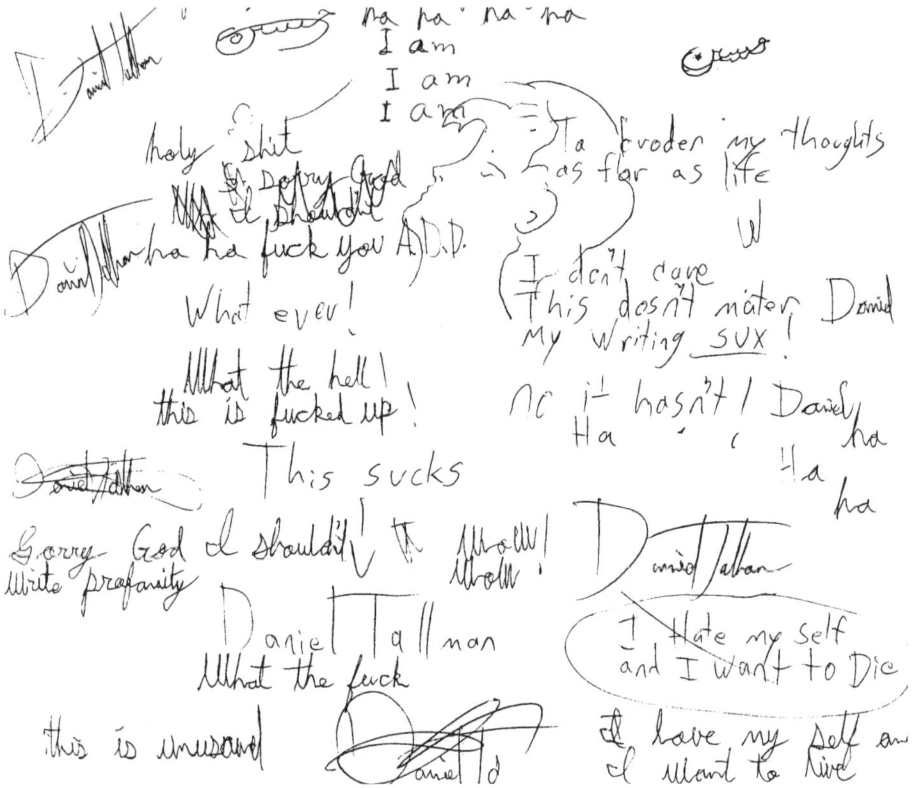

#6 March 1
Daniel suddenly can write legibly in neat, cursive style; his pre- and post- intervention styles appear consecutively as he tries both.

hearing my voice, filtered. I had some questions about this phase of the listening retraining theory because I was not sure Dan needed to be "taken back to the womb." He had a relationship with me that I considered quite normal and that shouldn't be interfered with; however, I did not suppose arguing psychology with Mr. Madaule would help anyone at that point and Dr. Tomatis was convinced of the incompleteness of the psychological birthing process for some inner-ear damaged children.[9] Furthermore, Dan was accustomed to hearing me read to him. What was unusual was that he had started reading silently from Tolkein's *The Return of the King* during the listening sessions. Dan rarely read alone, silently, to himself. I had read most of the *Lord of the Rings* trilogy to him. Here was another act of independence and a new ability to focus on reading material.

That night, after we had bedded down in our sleeping bags on Anna's living-room and dining-room floors, he spoke through the darkness to me, "I feel like I'm turning inside out—like all the bad stuff on the inside is turning outward."

"That's good, Dan," I answered, without having any idea what he meant.

Day 7 March 4

Today, Dan asked to have his hair tied back. We couldn't seem to get a ponytail to work, so he parted it and put it behind his ears. After a year of "hiding behind his hair," it was wonderful to see his face again. And to see him smiling. That evening Dan drew pictures of the denizens of hell.

Day 8 March 5

Daniel mentioned during the evening that he had written a poem/prayer of gratitude for the changes taking place in him. He wanted it kept private along with the rest of his notebook.

#7 March 1
Dan's Castle

#8 March 3
Candle Flame

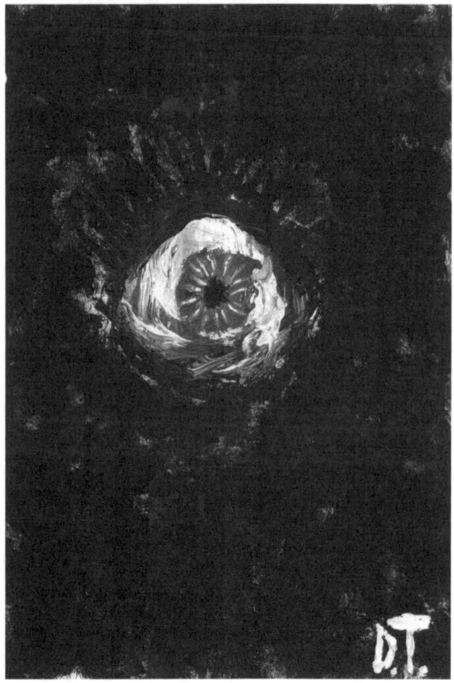

#9 March 3
Looking into Myself

At about the time we were thinking of getting into our sleeping bags, Anna wanted to see her sister, so I went with her to another apartment in the building. I learned that Maria was bleeding quite profusely. Her family seemed to think that whether she would live or die was in God's hands. I suggested she should be taken to the hospital and the family agreed and helped her to dress as I rushed back to Anna's to make sure Dan was settled down to sleep. I explained to him that Anna's sister needed to go to the hospital and that we might be very late returning. A certain amount of risk was entailed in leaving him alone and I made my decision on the basis of the very positive changes in Dan's outlook and the gravity of Maria's condition. Dan was concerned, but very level about all this. Anna and I spent half the night at the hospital. Dan hardly wakened when we returned.

Day 9 March 6

Dan inadvertently called attention to something about himself that I hadn't noticed: "Now that I'm standing straighter . . . ." He was right; his posture had changed. For the past year he had said he wished he could stand straight, a complaint to which I would reply, "Well, just stand tall," not realizing that he could not. Now he could. He added that he used to have to stand with one leg turned and extended to keep his balance.

That evening Dan spontaneously came and hugged me. That he had become so polite and gracious most of the time made me aware of how an underlying unhappiness must have motivated much of his behaviour. He remarked, "I don't have to not like myself any more" and "My short-term memory's better; I can go into my mind and find things better." While he was strumming his guitar he observed, "I am carrying myself better and I can do things better with my hands."

Day 10 March 7

After Daniel's tenth session of listening retraining, we went to a mall. As he explored the place, he commented, "I feel more intelligent. I feel smarter." He enjoyed choosing cards for Luke and Kathleen and actually wrote and mailed them to Virginia before we left Toronto. I was very surprised to see him follow through on this activity.

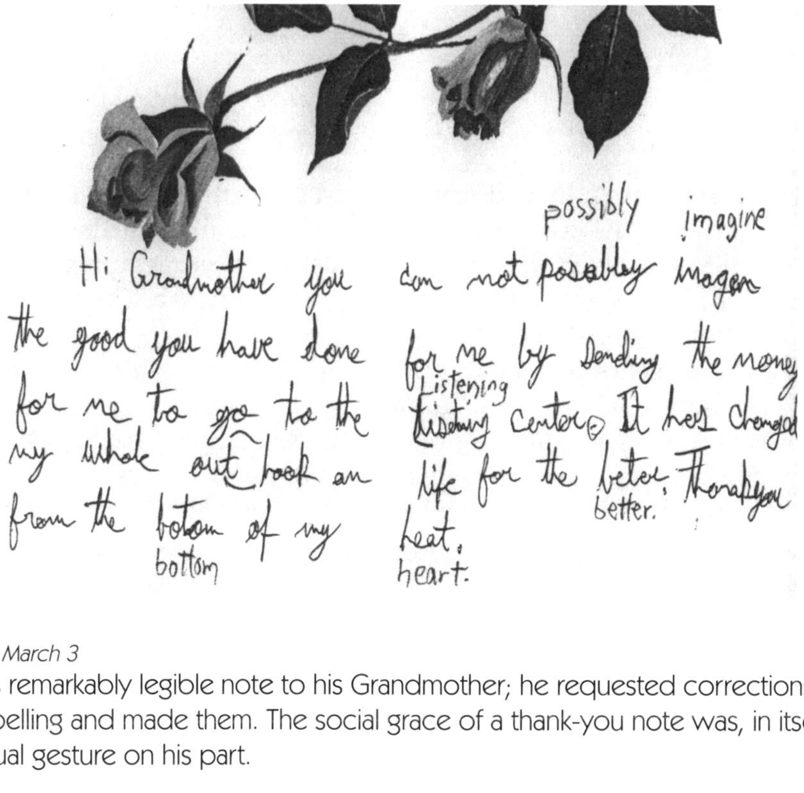

#10 March 3

Dan's remarkably legible note to his Grandmother; he requested corrections of his spelling and made them. The social grace of a thank-you note was, in itself, an unusual gesture on his part.

 # 11 March 1
A man smoking /A sword
I was thinking of God's sword/
The angel represents
the Listening Centre/Jesus, the
Road (my life), and the baby
is me.

#12 March 3
A Young Woman

During the evening, he and Anna shared and laughed about their tendency to "put themselves down," i.e., to be in denial about their positive characteristics. We talked about their abilities. Dan was beginning to be able to converse with Anna despite her limitations with English. The talk strayed to spirituality and ghosts, and when I made a reference to my mother (whose death was three years previously) Dan said quietly, "I can see her beside you." This troubled me a great deal. Anna said something about also being able to see Jesus Christ standing beside me. Anna might be following his train of thought with a little teasing; I did not know her well enough nor was her English sufficiently nuanced for me to be certain what she intended. But I decided the talk had gotten out of hand, and I was really concerned about Dan.

Day 10 March 8
Dan and I spent the following afternoon with my friends Charis and David, enjoying their warm hospitality and bending their ears about what we were

#13 March 6

"Jesus holding a baby, me. It's a bad picture of Jesus, though. No one can draw what he looks like. I don't think he's supposed to be drawn. He can be drawn symbolically, like your tree, or like someone who doesn't look like him." [Daniel has discovered verticality and gravity.]

learning and experiencing at The Listening Centre, while they showed divine patience. Daniel was completely involved in adult conversation for the first time in my memory, partly as a result of the graciousness of our friends, but also as an expression of his new capacity for attentiveness.

Day 11 March 9

Again, we visited friends, a family Dan already knew well: Chris and Claire and their children. Dan remarked, "I can listen to adults now. I am learning

#14 March 11
"My dyslexia cured through my art. A little troll like [my friend] Adam [can draw]. I can draw just as well as he can, now."

so much." Again, I was astonished at his ability not only to listen to the adult conversation but to contribute to it. Formerly, he would have chosen to spend all of his time with the children, although they were much younger than he. I could see that Daniel was reassessing how he wanted to relate both to those younger and to those older than he. I have read that dyslexic syndrome children tend to relate as strongly across age groups as to peers; I think that observation may refer to a compensatory tendency some dyslexic syndrome sufferers have developed with all age levels, but perhaps at relatively supperficial levels of attentiveness and of communication.

Day 12 March 10
Dan awakened unhappy and grumpy. That morning Dan slept during the listening retraining session and mentioned later that he had had "heavy-duty dreams." Shortly after the session he stated, "I'm at peace with myself and with God. I feel so refreshed." Then, "As I see it, it's not a question of 'It will be OK.' It is OK." As we were driving east along Bloor Street after his listening session, Dan suddenly exclaimed, "Now I understand why the earth has to spin, how people on the other side of the earth can stand up straight!" Most certainly, his interior gyroscope was functioning in a new way.

Anna and I went grocery shopping and left Dan alone for awhile. When we returned he said he had been crying about Luke and the hurt that his brother had caused him, but that Jesus had appeared to him and comforted him. Since Anna had said she had seen Jesus in the apartment, too, I began to wonder if these people have a gift, or a problem, or if this phenomenon is some dyslexic mode of perception (visual projection?), since Anna, who is a devout Muslim, not a devotee of Jesus, says she also is dyslexic.

In the evening, Dan surprised me by quickly sketching an artificial rose before him on the table (#16). This ability was in sharp contrast to his previous mode of drawing, which I had noted five years before in my journal and of which I had saved an example. I was able to share this with Mr. Madaule, who had said to me that this item was the one that had led him to consider Daniel as an appropriate candidate for listening retraining. The statement Daniel had made regarding his art production when he was 12 years old was a response to my question as to why he always drew pictures from his imagination rather than drawing something real that he could see, such as the plant on the table. He had replied, "I can't draw anything by seeing it. I have to put something in my own head—I have to put it [the plant] in my own head—then I don't have any trouble doing [drawing] it."

#15 March 8
Markham Street, opposite The Listening Centre.
Daniel has extended his new sense of verticality into three dimensions with two-point perspective.

Day 13 March 11

On our last day at The Listening Centre we met individually and together with Paul Madaule. He ascribed Daniel's dramatic improvement, which he said was unusual in their experience, to the acuity of Daniel's hearing. (Evidently, Dan takes after my mother in this ability. Her hearing, tested in her eighties, was reportedly as acute as that of a young person.) He showed me a graph of Daniel's preliminary and post-treatment modified audiograms

#16 March 10
Rose drawn from artificial rose at Anna's, without "putting it in my head first".
[The line work is bold and assured.]

from The Tomatis Listening Tests, which showed the improvements in Daniel's right-ear listening thresholds for certain frequencies in Hertz. As explained above, the listening test graph illustrates the way that the person hears the sounds from the environment (air conduction) versus the way he/she hears his or her own voice (i.e., via bone conduction). "Daniel's responses initially had showed high sensitivity (low thresholds) on the bone conduction measure compared to his responses on the air conduction measure. So, in Daniel's case, the goal of the listening program was to reduce the sensitivity through the bone conduction with the purpose to allow better sensory regulation."[10] In other words, his ability to receive sound by air had improved relative to his ability to process sound by bone conduction. Apparently, the ability of the tiny muscles of the middle ear to work together in the transmitting of sound through that stage of the listening process had improved to a remarkable degree. Paul also cautioned that Daniel would vacillate between his former and new modes of perception and that this changing mode of perception would be accompanied by mood swings. Daniel was to return in a few weeks for audio-vocal sessions.

As events transpired, Daniel did not return for these slated phases of the program, which would have involved audio-vocal feedback sessions to reinforce his improved means of listening. Morana Petrofski, now the Co-Director of The Listening Centre, corrects me by adding, "as well as work on the auditory dominance of the right ear. . . I am not sure if you are aware that we did not work on the dominance of the right ear in Daniel's program during his first session, this is done only during the expressive, audio-vocal

#17 March 11
My [Dan's] Ear

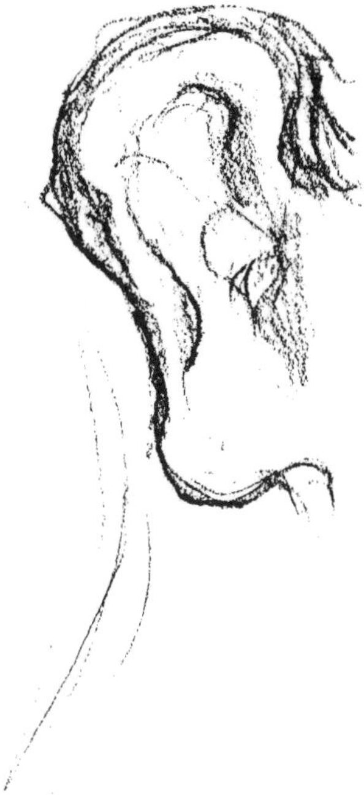

work."[11] Her comment suggests that neither did The Listening Centre fully appreciate what had been accomplished in Daniel's listening experience.

I will mention briefly ideas that I came to understand only many years later and that I shall develop further along in this story. Although Dan's laterality had not been strongly or purposefully established, I think a profound shift in hemispheric dominance, which affects not only laterality, had taken place in Daniel's brain; but that it was a shift he could not yet sustain. This shift would have been progressively reinforced by the subsequent audio-vocal sessions that had been planned, if he could have attended them. At that point in 1997, however, I had no concept of how Daniel's experience of sound stimulation would affect his brain or how the brain would respond to ambient sound reinforcement. His astonishing display of normalized behaviour did not raise any issues or compel any questions from me in that regard. Obviously, his brain was functioning better than I had ever seen it perform! If I had understood laterality, hemispheric dominance, the relationship between the ear and the feedback routes of neural stimulation between the "opened ear" and the brain, I would have appreciated better not only the method by which Daniel had become normal, but also the fragility of that newly established neurological network and how it must be supported and reinforced. And protected from drugs. I am not sure that even Tomatis appreciated all that he had set in motion by his discoveries, although he already was treating autism, also called "infant schizophrenia,"[12] during the 1970s, work that has been furthered by The Listening Centre and other practitioners of the Tomatis Method since then. I was going to need my newly restored energy for the learning process that lay ahead.

Home Again

Once home, Dan set up overnight meetings with his friends. Richard was hurt that Dan didn't want to spend more time with him. Richard was concerned, too—especially since he was about to leave on a trip to Louisiana—and with half the family—that he wasn't being given an opportunity to make his own observations of the changes in Dan so that he could judge whether or not he should be leaving, and because he wanted to see to what extent my reports and Dan's of his amazing cure were valid in his observation. I was busy showing the notes I had made and Dan's drawings to the family.

Dan spent much of the next four or five days with his friends. On the 13th he went to spend the night with Kurt and Tonya. He had expressed annoyance at "going in and out" between his dyslexic state of perception and his new mode. At about this time he expressed confusion about the "old Dan" and couldn't seem to absorb that we loved the dyslexic Dan just as much as the non-dyslexic Dan. He seemed to feel that his dyslexic self was somehow evil and should be rejected and that we should know that.

On the 14th, he wanted to have friends to his place and we responded with a firm, "No, it's time for your own family to see you." He complied very reluctantly.

How naive I was! I would come to understand that Dan's friendships were inextricably entwined with the drugs available through their enterprise. The drug-use bond, like any other physically related bondage, had spread its roots far beyond the dyslexic syndrome where it had first taken hold. Daniel, however, must have had some idea of what he was facing, judging from his last few drawings at The Listening Centre—the Ring of Power, Tolkein's mountains, himself still "in utero" as a fetus in the birthing position, and in his final drawing of a terrible world of confusion with an unseen angler offering him a hook, and he the fish about to bite. Dan could already fathom, in his depiction of that fish and hook, the temptation that destroys.

#18 March 11

The Ring of Power, a Fetus in the birthing position, and Tolkein's Mountains.

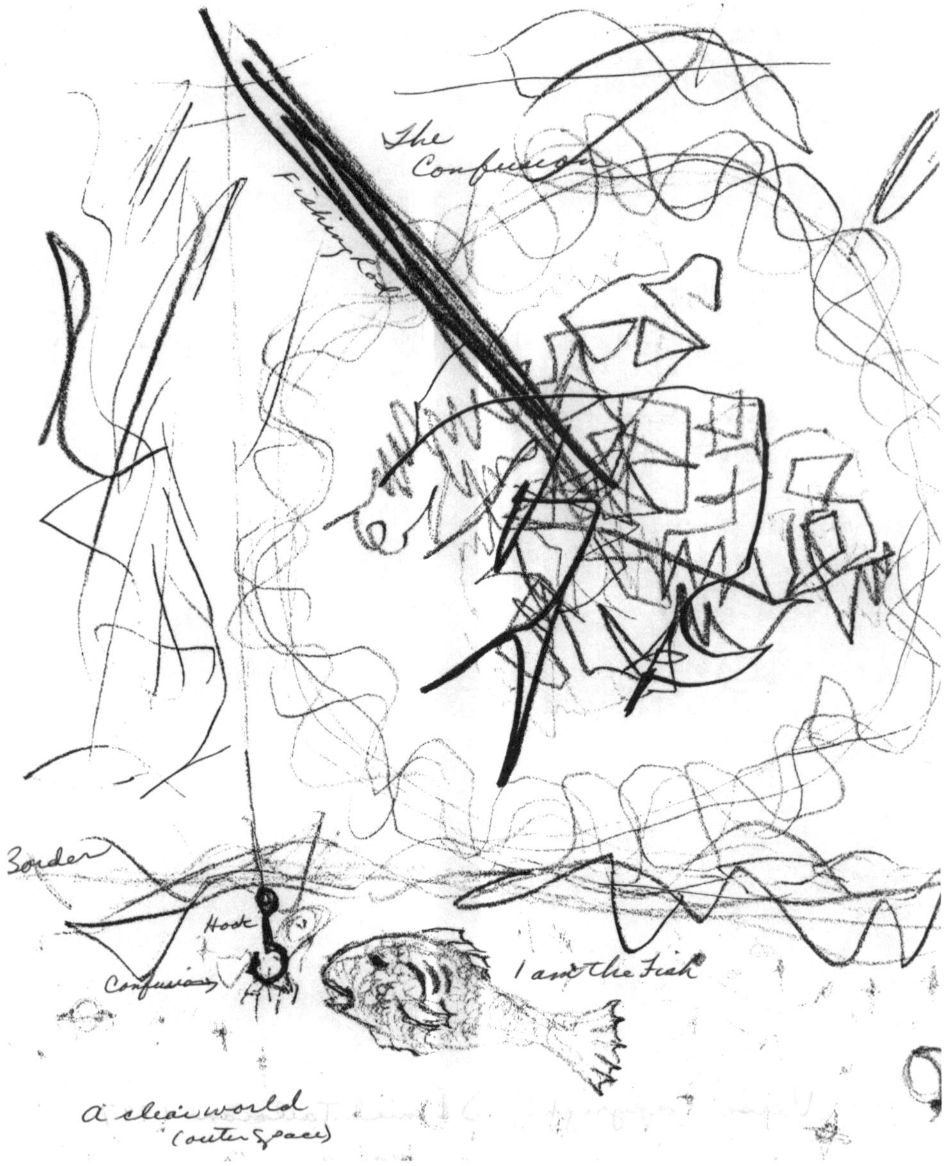

#20 March 11

The Confusion
"The fish is in a not understood world. It's a clear world. I guess I'm the fish. But the fish is living in a world of his own. To the Confusion it's not clear; there's a barrier up. To the confusion the fish is confusion; it wants to catch it but it can't."

In retrospect, I understand how Dan's repressed guilt about his drug-taking lifestyle was deeply integrated into his concept of the "old Dan," and that his connection to the drugs and to that lifestyle had become full-blown addictions. He could not sort out where to ascribe guilt and he could not distinguish what constituted "normal brain function beyond his control" from ways of thinking about himself over which he might exercise self-control. Neither was he helped by living in a society (perhaps especially the subculture) that had largely decided self-control was an old-fashioned and outmoded concept, except for very self-centred purposes. I had yet to appreciate the degree to which he had become subject to his nervous system in ways that might or might not respond readily to his efforts to modify it.

In School

Dan returned to school, i.e., his one class, on St. Patrick's Day. We had agreed that I would talk to his teacher to explain The Listening Centre intervention and effect. Dan was able to express to the teacher, though briefly, the changes that he saw in himself. He had neither textbook nor notebook in class, but he brought home the assignments and seemed quite willing to tackle them—a remarkable change. Since he couldn't do schoolwork in class, he wrote the following:

> I have been gone from this class for three weeks. In those three weeks many changes have taken place. These changes have made it

#21 March 11
The Beast, Satan, has been killed. All the evil that I've drawn is being killed."

> possible to do things I never have imagined. I find that I have become able to concentrate with ease. Little things no longer distract me. When I listen to someone, my eyes no longer wander and I can focus on what people want to tell me. I can actually listen now. I can express my views without frustration. As a result of all this I feel more at ease. I feel whole. I feel like me in a way I was never able to before.
>
> I feel more at ease, but I really don't like the way I continue to "go in and out" of being the way I was before and the way I am now. It alters my focusing ability. One minute things are fine and the next I feel the way I did weeks ago. (This is why some of my writing is hard to read as I write this.)
>
> I notice the effects of my "in and out reactions" to the listening retraining when I am right one minute. It will appear to me that I am feeling focused and I will know what I am wanting to [convey] on paper and the next I will find myself lost and confused [not knowing where I am or what I am doing]. This same is true in reading and also in normal everyday conversation.
>
> It's almost more frustrating than before because I feel positive and at ease within, I feel as though I am leaving a good impression and then it backfires on me and I feel like I'm not coming over [across] the way I would have liked.
>
> I am probably confusing people now. They're getting parts of what I want them to hear, but then I'm lost but they don't know that. They probably think I'm insane. I get confused, angry, frustrated, and then I'm back [after the dyslexic moment] to feeling relaxed, at ease, full of joy.
>
> I feel as though I am confusing people just through my appearance, my eyes will change focus as well. One minute they will feel random, my vision will become confused and blurred, and the next it will be clear and attentive. One minute I feel straight and balanced and the next I feel awkward and unsure on my feet. I have discovered the reason I slumped the way I did was to control my balance.

He also had done a sketch of my painting, "The Tree and the Wave," he said, to comfort him.

For my part, I was utterly amazed at Dan's ability to express so clearly and eloquently what he was experiencing. It sounded as though he now had the platform he needed to recoup so many losses.

Daniel awakened us at 3:45 the next morning with a series of revelations that were occurring to him about how he related to his friends and about his very emotional day. The nighttime observations were linked in his mind to his LSD experiences and as he "went in and out" of the dyslexified world he more and more identified his whole dyslexified outlook with the damage done by the LSD trip. He wept and shook. We talked to him and soothed him, until he slept a little. Then he was up again and Richard, after vacillating about whether or not he should set out for Virginia and Louisiana, decided at least to get ready to go. Dan was in no state to go to school and finally slept in our room; later, he didn't remember being asleep.

I was out in the afternoon for a couple of hours. Most of the rest of the time Dan had been talking about the process of accommodating to the changes in his brain. He talked about his new insights into his friends, their forms of interaction with him. Notable among them were Shawn for having introduced him to LSD; Jake, for his neutral or supportive role; Kurt and Tonya, for being more understanding and supportive people than Jerry, Shawn, and others. Dan kept coming back to his sorrow over having been beaten on some occasion by Luke (which I had never heard about) and his guilt over having introduced his younger brother Eric to pot. The afternoon culminated in Dan's saying he needed to go out and his returning about 15 minutes later clad in only his underwear—he said he'd had that off, too—and with his feet bright red and scratched all over from the ice. Apparently, "God had told him to put his clothes back on." I was hearing a lot of "God said this and God said that" and a lot about flashbacks to the LSD experience. This performance was not like Dan or anyone else I knew.

Richard and I did not know where to peg Dan's excitement and euphoria. I came from a cultural background that included some very expressive French people. Richard's family was much less emotive, although his mother was exceptionally outspoken. Had The Listening Centre intervention affected Daniel more extremely than it usually did its clients? Or were we just seeing the "mood swings" the staff had warned us about? Richard's expertise in folklore gave him a broad understanding of cultures that place high value on those who step over the edge into insanity as being enlightened and of religious traditions (Christian, Amerindian, Hindu, etc.) that permit or encourage psychosis induced through drugs, fasting, and mortifications to achieve greater awareness and enlightenment. He was trying to figure out where Dan's thinking and experimenting had led him.

From my standpoint, Dan was edging over into emotional expression beyond anything of which I had personal experience. We became less and less comfortable with what was happening in Dan, but for differing reasons and from different standpoints. I tried to compare what Dan was reflecting with some of the incidents that had taken place during the previous summer in reference to angel trumpet tea. Dan seemed in better control now than he had been then. He seemed much more alert and much more analytical and rational. Except for his disrobing outside in winter weather, most people would not have immediately recognized any differences in Dan's behaviour from that of an ordinary but highly expressive person. Certainly, Dan was sufficiently himself most of the time for Richard to feel that he could leave for Louisiana without abandoning me to a serious problem, and he left on the 19th of March for points south with Sarah and Eric.

That night Daniel "prayed away a demon." He alternated between confusion, sorrow, grief, and euphoria. Richard called from their motel room; I brought him up to date and expressed my concern. We agreed that we must talk to Paul Madaule the next day. I left an urgent message on The Listening Centre answering machine. Dan ended up sleeping in our bed, a situation that forced me to stay up and unable to sleep; but it seemed to comfort him and I was afraid in any case to leave him alone or unobserved. The next afternoon, Daniel wrote:

> In mid-January 1996 [the previous year] I was affected by an LSD trip which almost traumatized me for life. I bought a tab off of E.T. I took it home and played with it a bit, touching it with my fingers. The next day I felt like I had not yet come down and I wouldn't for 18 months. The next day I explained this to Shawn, Jerry, and Jake. Jerry said, "So you caught a buzz." Shawn laughed. Jake said nothing and now that I look back on that I see the real effect.
>
> Jerry saw me as a Joke, of course. The three of them, they all knew what had happened to me. Jerry didn't feel for me. Shawn didn't feel for me. Jake felt like shit. I knew this, because he was quiet. He said nothing, he has kept this a secret from me, not knowing why.

The next morning Dan seemed able and eager to go to school. He made it through three-quarters of his history class and had to leave because he was afraid he was going to cry. Fortunately, I had arrived early and was waiting for him when he left class. He spent the rest of the day moving back and forth between the new, confident Dan and a Dan who was reliving and reprocessing everything significantly painful that had ever happened to him: his hurt from Luke, Jake, Mattie, Shawn. At some point, this no longer seemed to me to be "processing" towards a resolution, but had become circular and obsessive.

I talked to Paul Madaule in the afternoon and Daniel, when he got on the phone with Paul, sounded calmer and more adult than he had with me all day. Paul said that he himself had never done drugs and couldn't speak to Dan's reactions; that he had told me that Dan's emotions would fluctuate, but that I "hadn't wanted to believe him." (This seemed unreasonable to me in light of the extremes in Daniel's behaviour, which, of course, Madaule could not directly observe.) He said that Dan had not revealed his drug abuse to him and that if Dan didn't come out of this we would have to find someone else in the community for him to talk to—presumably a doctor or psychiatrist. I asked him if anything was to be gained by having Dan go back to The Listening Centre for further treatment sooner than planned and he said, "No." He recommended strongly that Dan go to school and that his life be routine. I thought that would be a tall order for me.

The chief sources of Dan's expressed pain had been his brother Luke and his friends; then his complaints switched to his drug experiences, especially LSD, PCP, and pot. He vomited twice when he tried to smoke cigarettes. He said that he knew people who had done other drugs and smoked pot to cover up the damage the harder drugs have done. I thought there could be considerable truth in his revelation. I thought possibly he was trying to tell me why he had continued to smoke pot and to defend its use so strongly from time to time over the past year. His friends, in the same order, were his points of struggle to reassess and understand, always in the context of how they had treated him—both in fact, and in terms of how he had perceived them. He seemed to identify more with Luke's classmate Doug in Doug's Christianity, whereas before, Dan had been critical of "Doug's preachiness." His emotions levelled out a little. He said he had "learned to let some sorts of memories roll up without panicking." He was sometimes breathless; sometimes put his hand over his heart to express pain.

An image came to me of a shattered wall in a castle Dan had inhabited. Through the hole in the wall was a rocky shore of shallow water—the kind of shore that makes for the most difficult sort of wading—and the sun was rising over the open sea. It expressed the idea that Dan's Listening Centre experience had shattered his dyslexic (and/or drug-related) prison, for good rather than for ill, and that, although the walking might be rocky for a time, he would attain the place for a smooth launch. As the day progressed, Daniel's emotions appeared to be overpowering him. His mood swings became wider and some of what he was "remembering" simply was incorrect. He frequently asked me to help him to know what part of his thinking was correct and what was "right." He was driven manically to use his new facility with language to analyze and reanalyze his relationships. His mind would not give him rest. Not all that he at first thought were "insights" seemed so to me, but I judged some of this rambling to be truly integrative learning.

It was hard to get him to focus on tasks. He announced that he wanted to "clean everything out of his room." So together we went at it. He decided which of the Nirvana posters to save, which pictures and posters of bands he wanted to keep in a file. Many pictures he enthusiastically trashed. The only poster of his erstwhile idol Kurt Cobain that he kept on the wall was the one that looked most like Dan himself. The others he kept in a file in his dresser. He had trouble keeping on task and thought for a long while over some of the pictures.

Later in the day, I was concerned that Dan was tired; he was eating very little and I knew he had slept poorly the night before. But he talked to Jake and wanted to see him, so we drove to Jake's to fetch him and at the same time picked up the mattress he had offered Dan for his room. Dan seemed oddly dreamy and emotional with Jake, who left around 10. Dan settled around 11 but was full of barely rational ramblings about me and my smoking (I had quit 20 years before) and about my relationship with my father.

I talked to Richard on the phone and we agreed I should get Dan to Dr. A. and/or to the hospital in the morning. At 3 a.m. Dan wakened me because, he said, he wanted "to pray for me," which he did with great and charming gusto. In his conversation he started jumping from topic to topic; then, he seemed to slip from personality to personality, perhaps four personas, from moment to moment. I was very frightened and tried to get him to settle back to bed. I don't remember precisely how things next unfolded. This is what I can recall.

I went to the kitchen to make tea and Dan came downstairs half-naked. In one of his personas he called himself "a druggie" and moved aggressively towards me. Without thinking, I picked up a bowl of water from the counter and tossed it over him. He said, "You never did like that sort of thing, did you?"

I told him firmly to go and get his clothes on. He came towards me in a friendly way, put his hands on my shoulders and tried to convince me of the importance of my having smoked 20 years ago and how that had ruined my relationship with my father. I sent him to get dressed and he reluctantly went upstairs.

I ran to the study, locked both its doors, and phoned Peter to come immediately. I waited for at least 15 minutes, hoping Dan was in bed and Peter almost to our house. Peter phoned back and said he was having car trouble. I begged him to come as soon as he could. Just as I hung up the phone, Dan smashed one of the upper panels of the study door with his foot to open the locked door handle on the inside. He was naked. He became completely different personalities at about 20-second intervals. One persona surfaced as he came at me, forcing me back against the other door of the room and, excessively pleasant, tried to convince me about the smoking issue. His hands were around my neck. Another persona appeared and he cringed in a corner of the room and apologized, looking sad and confused. At some point he told me mothers were supposed to have sex with their sons. I moved to phone Peter again, but Dan wrested the phone from me. I think he believed I was calling our friend Paul, or perhaps the police.

I kept talking to the essential Dan, the gentle and respectful and morally responsible Dan. I suspect that if I had been dealing with a child who had been raised differently or who did not have Dan's essentially gentle and loving personality that the outcome for both of us in that crisis might have been tragic, but at some level (or during some intervals) the "essential" Dan was responsive. By the time Peter arrived I had only just gotten Dan upstairs to his room. Peter took over, got him partially dressed, and persuaded him to sleep. An hour later, I spelled Peter off, keeping watch. Two hours later, Dan woke up just as irrational, but not as aggressive, as he'd been during the night. We had trouble convincing him to get in the car to go and see Dr. A., and had trouble keeping him in the car. That we succeeded indicates that the rational Dan was still momentarily present to respond to our coaxing.

Dr. A. gave Peter instructions for calming Dan while in the car: do not look at him directly in the eyes; refer constantly to Dr. A. as the authority who wanted him to go to the hospital. Dr. A. stressed that I should keep a low profile because, being his mother, I would elicit more emotion from Dan and that would tend to be destabilizing. Peter managed Dan, applying the doctor's instruction, while I repeatedly reached behind me to relock the back door on my side after Dan had pulled up the knob preparing to exit. We arrived at the hospital after a harrowing hour's drive, and Dan was admitted around 8 a.m. He did not respond to my coaxing him to get into bed, perhaps because I was too tired to sense the "rational Dan" moments. He was restrained and sedated with Haldol (haloperidol, an anti-psychotic) and Ativan (anti-anxiety) medications and his blood sample was taken perhaps an hour later.

I had forgotten the image of the darkness under the Gothic arch. How could I have known that Daniel was about to step through that door?

6

The Lion's Den

Daniel among Psychiatrists

From late March to mid-April the Hale-Bopp Comet moved mistily in the evening sky as we returned from visiting Daniel in the hospital. Enormous efforts of observation on the part of a relatively few astronomers have brought such phenomena into public awareness. To what ends can that sort of awareness be directed? I might write poetry about the comet or find some kind of inspiration in it; but late in March, 38 members of the Heaven's Gate cult in California committed suicide in a uniform that included Nike footgear, purportedly to enter a spaceship on the far side of the Hale-Bopp Comet. We may label this event "mass psychosis" without answering any significant questions. Suicide was their way out of some kind of painful reality. For me, the comet was a "wonder in the heavens" that reminded me that "miracles" discovered in the heavens might be paralleled by "miracles" discovered on the earth. As I pondered what might be done for our son, I sided with the astronomers: I believe that the burden of observation rests on all of us— to see, to notice, to suspend judgement, to compare, to discern, to learn.

What else could I do? I trusted the medical authorities into whose care I placed Daniel, despite a mixed experience with the medical profession that ranged from death-dealing to life-saving. When I was seven, the prescription of antibiotics in the early days of "wonder drugs" killed my infant sister. In my teens and twenties, medical decisions had almost claimed the life of my mother several times—through unnoticed allergies, by ignoring the signs of impending heart attack, and through unhelpful medications. The prescription of an inadequately tested drug plunged my father close to death and into long-term disability from liver damage. A tonsillectomy had saved my life. Medical knowledge ensured that my Rh negative blood did not kill my children. A doctor sexually assaulted me when I was a teen (and other women; he eventually was removed from his job at the clinic). The ENT doctor at the same clinic, who had saved my life, saw me a few years later when I was carried into his office at the beginning of my severe reaction to a typhoid shot he had administered. His "prescription" this time was to have a couple of shots of whisky. I was naive and trusting enough to take his advice, which affected my breathing, distorted my sense of time, and caused central nervous system damage from which I was four months recovering. Other doctors had proven trustworthy, insightful, helpful, and fully professional as well as courteous and compassionate. In short, I was not naive about the possibility of medical ignorance and errors, but I was not experienced with those who ministered to psychotic illness.

My "encounters" with psychiatrists had been second-hand, through friends who consulted them or whose relatives saw them or close friends who nursed in attendance with them in hospitals, both on psychiatric wards in general hospitals and in Ontario mental hospitals. One doctor I knew told me he had known about 50 psychiatrists in his career and, among them, would have recommended only two, one of whom had died. The woman psychiatrist my roommate had consulted in reference to a mutual troubled friend provided helpful guidelines for us in that situation. On the other hand, that unfortunate young woman's aunt and uncle both were psychiatrists and their approach to their niece (my friend was about 20 years old and alleged that her uncle had spanked her) had left me with many unanswered questions. My roommate's mother had suffered severe depression and had been treated several times with electric shock therapy, rendering her unreachable psychologically, largely because her memory had been destroyed in an unpredictable pattern, like the holes in Swiss cheese. While in university I had read more widely in the field of mental illness than most laypersons, perhaps, but until Daniel became ill, I thought psychosis was synonymous with schizophrenia; I did not realize that other types of mental illness that include psychotic symptoms are not subsumed in medical literature under the general term "schizophrenia." Through my aunt's accounts of my uncle's mental illness I was aware of the shift towards medication and away from electric shock therapy, although he was subjected to both. I had known (but had forgotten) that lithium had freed particular patients from psychosis and liberated them from asylums. In short, I assumed that Daniel would be given good, knowledgeable, and up-to-date care.

The Psych Ward

Once Daniel had been admitted to the hospital, I was taken aback to find how much control the hospital had over my access to information about him. He had turned 16, so was deemed "capable of deciding for himself" whether or not his parents should have access to information about him from medical authorities. This law would not be abrogated by Dan's mental condition unless a court of law should declare him mentally incompetent long-term, and if that were to happen those authorities would gain even more far-reaching control.

I was told that morning that I would be allowed to visit Daniel, depending on his condition, according to fairly generous visiting hours in the afternoons and evenings. I took our friend Pete back home, then returned to the hospital during visiting hours. I was very surprised to have been able to do all the driving, waiting, and unusual caregiving with Dan on little sleep and under extremely unusual conditions; clearly, the changes in my CFS initiated by the listening program at The Listening Centre were enduring, despite stress. While I was waiting to see Dan, hoping also to talk to a psychiatrist about his condition, I had ample opportunity to read all of the pamphlets about mental illness and Ontario's psychiatric resources in a literature display offered to people in the waiting room. I read something new to me: that schizophrenia often improves over the lifetime of the schizophrenic individual. The literature explained that doctors did not know what caused schizophrenia. This led me to think that, nonetheless, they knew more than I did about its causes because research was ongoing and some thought a failure of the immune system was involved in the disease. Obviously, Dan's drug abuse was a factor. But no one would suggest to me in the days, weeks, and years to come that his drug abuse symptoms and schizophrenic symptoms could be one and the same thing. I was left thinking that the street drugs Dan had taken had triggered a mysterious disease named schizophrenia, in much the manner that my friend Michael's cigarette-smoking was said to have triggered Goodpasture's Syndrome, the auto-immune disease that had taken his life.

My contact with Daniel had shifted from a more-or-less continuous motion picture to film clips. Within the clips, the backdrop of the psychiatric ward and how it functioned had set in motion another learning experience. At first, the kindliness of certain staff members was very reassuring. The nurse who led me to Daniel's room explained to me that he was in four-point restraints: around his wrists and ankles were leather bands set with metal clips attached to stout chains fastened to the bed frame. She was trying to cushion any shock I might feel at Dan's restricted condition and I thanked her, although I already knew how he would be restrained. In fact, Daniel was by then heavily sedated, although he recognized me and was foggily responsive to my questions. He appeared to me still to be vacillating between his two forms of perception: as I supposed, between dyslexic and non-dyslexic. In fact, he told me that he was still "flipping in and out," apparently a confirmation of what I was observing. A pleasant nurse chatted with me. In a few minutes about eight staffers came and stood in a semicircle around Dan's bed and listened courteously to me as I described The Listening Centre intervention, Daniel's mood swings, and then his shift into unmanageably erratic behaviour.

My own sincerity, caring, and utter naiveté about the potential effects of street drugs and prescription drugs would have been fully obvious to people who were admitting teen addicts on a daily basis. Neither then, nor at any point during Dan's hospitalization in that facility or any other, would any of the personnel trouble to inform me about the street drugs Dan had taken or about the medications that were even then coursing through Daniel's veins. Eventually, I would discover that many of the patients on the ward were young people, like Daniel, who had pushed themselves over the limits of their rationality by using drugs they obtained on the street. At no point, as far as I could determine, was he counselled regarding his drug abuse or pointed to drug counselling or to such help as Alcoholics Anonymous. Nothing of that nature was suggested to us, his parents, either. I think now that the staff presumed that if our child had gotten into drugs, it was because we were part of the problem.

Two days later, Daniel was deemed "competent" to grant me permission to receive reports on his medical procedures. I was told what medications he was receiving but without any information about their purpose or side effects. I borrowed my father's *Physicians' Desk Reference*,[1] and as I took my shreds of information to this resource I began to learn how Daniel's psychosis would be controlled and at what potential cost to his body.

During the first few days of his hospitalization, I brought Daniel a small tape player and tapes of Gregorian chant for him to listen to; cards with messages of love and concern from the family and friends; and my watercolour sketch of the image of him sitting beside the blown-out wall of his castle. Dan asked me to fasten the painting to the wall beside his bed. He was responsive and rational for short intervals and unresponsive and irrational for short intervals. After about four days, this "flipping" process that had been quite obvious to me at first became harder to discern; Daniel was more uniformly confused and irrational. His responses in conversation were consistently undependable or nonsensical. I assumed his intrinsic condition had deteriorated and that the progress of his disease had worsened. A nurse who sat near him for long intervals was keeping notes of some kind; I was not allowed to know what she was recording and the psychiatrist said to be in charge of his case was not available to answer my questions.

A few days later, I read that the psychotropic or anti-psychotic drugs also cause psychosis. How did that work? I could not find explanations, but today I know why. The part of Dan's brain that had not maintained its rationality at the outset of his psychotic break would become "subdued" by the medications to become less psychotic; the part of his brain that did *not* require medication would respond to the chemicals by becoming confused and psychotic. At length both parts of his brain would adapt to the medication to approximate more normal function. However, at the time, I simply thought he was becoming sicker with the mysterious, underlying sickness. A nurse mentioned to me that Daniel was constantly in restraints because after I left, he would try to follow me. The nurse left me with the vague impression, from the tone of her voice and the way she snapped at me, that Daniel had an inappropriate relationship with me at all times. It did not occur to me to probe for her complete thinking on the subject and I was, in any case, in the process of leaving the ward at the end of visiting hours when this comment was tossed at me. Later, I would have reason to think I had interpreted her comment accurately.

I tried on each visit to contact the psychiatrist in charge of Dan, but the elderly doctor never was available. I left my number and asked her to call, but she did not. I stayed late and queried the nurses, to no avail. I saw at least two other psychiatrists making rounds on the ward and I discovered on a subsequent occasion when Dan again was in this ward that all of them saw all of the patients. Thus, I learned that my assumption—that a specific doctor "in charge" of a patient developed a personal relationship with that patient to effect improvement—was incorrect. About a week later, I was informed just as I was arriving that Daniel was being transferred to the Ontario hospital, which would be about a two-hour drive from our home. If I had not arrived at that moment, would I have been informed? I was frightened. Did this mean his illness was intractable? Would his incarceration be permanent? The reason given to me was that there he would not have to be kept in four-point restraints but would be allowed to walk around the ward. A doctor was working with charts in the nursing station and by asking a passing orderly I discovered that she was Daniel's psychiatrist. As I overheard her talking to a nurse I found she was signing the papers for his release to the Ontario hospital. I interrupted, "Excuse me, Dr. B., but can you tell me how Daniel is?"

Her reply was peremptory, "Daniel is acutely psychotic." She turned and left the nursing station by a door into an interior room.

The Provincial Hospital

I gathered Dan's personal items from the bedside table and ran to the parking lot just as the ambulance was pulling away. I followed it to the Ontario hospital and, after a wait, was allowed to see the young psychiatrist, serious to the point of solemnity, who was admitting Daniel. For the first time, I was questioned carefully about Daniel's drug abuse and I informed Dr. C.

of the little I knew about that and something of the context of Dan's life. Dr. C. explained to me that the sooner Daniel was adequately medicated, the sooner his psychosis could be controlled. He told me that allowing Dan to continue in the psychotic state he had been in was the worst thing that could happen to his brain because it would increase a tendency to further psychotic breaks. He encouraged me to ask questions, and I did, but my background for what was happening to Dan was inadequate. I had too little information about drugs and about the drugs Dan had taken to know how to provide really useful data and to know what to inquire about. Dr. C. told me a little about the possible side effects of a drug taken by one of his patients who had "done mushrooms" and four years later was still having "flashbacks." I took from that comment the notion that someone less damaged than Daniel still might require regular visits to a psychiatrist for years. A new world of drug abuse and potential treatment was opening to me, but I had few landmarks. I was allowed to see Daniel.

No longer restrained, Daniel seemed much more himself. We were together in a "quiet room" furnished with a mat like the ones you see in a high school gym and with a window barred with very heavy mesh. The door, with its plastic observation bubble that ballooned into the room, was open. I had brought Dan a couple of oranges and a stuffed toy; I told him the teal rhino represented the dyslexic Dan, who was loved, too. He was surprised and bemused by the shift in venue. I could more easily judge his "non-dyslexic" intervals and we joked a little, tossed the oranges back and forth, and made conversation about the stuffed rhino. He certainly wasn't recovered, but he was more coherent than he had been a week before. He fumbled the oranges but could catch them, too; was calm and slowly responsive to my comments. His answers indicated that his mental world was fantastic or dream-like. He moved in slow-motion. He smiled. When I explained to Daniel that because he was now much further away from home I could not visit him as often, he seemed to understand and did not seem particularly distressed. It was hard to judge how much he really understood, but he was calm and seemed reasonably comfortable. I had not been into a locked psychiatric ward before. Like most people interested in drama I had seen such films as *The Snake Pit*,[2] *David and Lisa*,[3] *One Flew Over the Cuckoo's Nest*,[4] and *Awakenings*,[5] and I had read several books that described psychiatric hospitals, although I had not yet read Erving Goffman's *Asylums*.[6] I wondered if Daniel's environment would prove to be anything like what I had seen portrayed on film or read about in books.

The admitting psychiatrist I had spoken to, Dr. C., was not in charge of the ward.

On my second visit I met Dr. D., one of at least two (Dr. E.) ward psychiatrists who saw Dan, and I tried to learn more about how Daniel was and what was wrong with him. I assumed that Daniel was being exposed to some sort of counselling or "talk therapy" and I said something like, "You probably can't tell this very well from how he is now, but Daniel is actually very bright."

"Oh, Daniel is very bright alright!" Dr. D. replied. I was so taken aback by the tone of innuendo in her voice that I was afraid to ask what else she had in mind by her remark. She seemed to be implying that Daniel's intelligence was a foundation for mischief or that he could not be very smart to have gotten himself into this plight. She told me the medications Daniel was being given, which may have differed somewhat from what I had been told at the general hospital; I cannot recall.

When I got home, I received a call from Paul Madaule, who expressed sympathy and concern for Daniel and told me he had been trying to reach us. Of course, I had been away at the hospitals. His advice was to follow to the letter what the psychiatrists would recommend. I wondered if he might feel that his business was in some jeopardy and that I had given it a kind of publicity some would consider pejorative. But it was too late now to change that circumstance. I was unable to give him much information about Daniel's condition. I was grateful for his kindliness. From my recent reading of Dr. Tomatis's writing, I imagine his warm regard for some psychiatric theories could possibly have informed his protégé's advice.

Richard arranged with Luke and Kathleen to extend the stay of Sarah and Eric in Virginia, while he shortened his visit with his mother and drove back from Louisiana. He went directly to the hospital to see Daniel before coming home. A very long granite-floored corridor, with dingy yellow walls, extended from the locked entrance/exit past the glass-enclosed and locked nursing station, past patients' rooms, a smoking room, showers, bathrooms, and "quiet rooms," to a large recreation lounge that the patients were far too drugged to use for its designed purpose. Dan was standing by the door of the rec room, wearing a red, hooded shirt, and seemed the image of loneliness. When he saw his father come through the locked door, he stretched out his arms to him and shuffled the length of the long, long corridor into his father's embrace.

We were no better informed about Daniel's condition when, a couple of weeks later, we were told that he was being discharged. How did we "think Daniel is?" inquired Dr. D. We saw that he was improved and said so. We were told that he was "very fragile," but had no idea what that might mean and no further explanation was given. We were cautioned that changes in patterns of eating, drinking, sleeping, and socializing would be warnings that all was not well. Nothing further about Dan was discussed with us—not his condition, not his use of drugs, not his responsiveness to medication, not the protocols or philosophies of the doctors regarding mental illness or how he was responding to treatment. What we could see was apparently all we were entitled to know. We were not instructed to check back with the general hospital to complete the discharge process, although later we were told that we were supposed to have done that. Daniel was sent home without a prescription for medication. Again, perhaps someone expected the general hospital to provide one. Continuing medication was not discussed with us. Knowing what I now know, I have to ask myself, what possibly could have been the agenda of such a doctor for such a patient as Daniel?

Hiatus

Daniel came home to the hurly-burly of a busy household. His siblings, especially Eric, and his friends thought of him as being in recovery the way someone who has had an appendectomy is in recovery—you need to take it easy for a few days and then gradually resume your normal activities. I could see that Dan's home environment was more stressful in some ways than the hospital setting had been and I did what I could to limit Dan's activity and Eric's and his friends' access to Dan, with only moderate success.

Within three days of Dan's return home we saw small signs that all was not well with him. For one thing, he was hardly eating. To me he seemed heartbroken, the way he had when Luke and Sarah had returned to school and he had sat in a niche of the bookcases holding his favourite toy duck with tears trickling down his cheeks. Now, Dan was doing all that we asked of him: studying as he had never been able to prior to his Listening Centre experience, performing without protest any of the household chores requested—an unprecedented happening—and trying to set up meetings with friends that for ordinary reasons were not materializing but that disappointed him too much. Friday he began to lose his grip on reality and that night he kept waking us in reference, first, to a bad dream such as he had had as a child, and then with racing thoughts. Saturday we took him to the medical centre in the village and Dr. F., not our family doctor, gave him some lorazepam (Ativan). The tranquillizing medication didn't seem to help and Sunday we took Dan to the hospital, imagining a better medication would be available and that some help from the psychiatrist was in order. No one had explained to us that such medications usually do not act quickly; it takes from several days to several weeks for the medication to build up in the brain and achieve the desired effect.

Daniel probably had "coasted" at home for three days because he still had a huge amount of medication in his system. As the blood concentration decreased he cascaded rapidly towards psychosis. I also understand that when the psychiatrist said he was "fragile" that meant he should not be required to do anything but rest and wander, eat and sleep. In fact, the mere alteration in his environment would have been a huge adjustment for him to make. If I had known half of what I know now about addictions, Daniel would have been prevented from returning to the same school, the same group of friends, or to any of his favourite hangouts. Over the next few weeks, I learned that the doctors' position was that at 16 Daniel was not likely to be compliant with those sorts of lifestyle changes; a program for recovery for such young people didn't exist, either. Such helpful programs were proliferating at the time in all major U.S. centres in addition to the expensive private rehab programs; but knowledge of all such facilities was beyond my sphere; and access was well beyond our means.

Back in Custody

The general hospital's psych ward staff instructed us to return Dan to the Ontario hospital, which we did. Dan was quiet and co-operative when we left him at the hospital. We hoped for better explanations and answers this time as to what was happening to him. We expected to see him the following day. After many tries I was able to speak to Dr. D. on the phone. She would not answer any of my questions. "Daniel is sixteen and has not given permission for you to see him or for me to talk to you about him. Yes, I will meet with you; my secretary will phone with an appointment." Click. The next day I phoned and was told, again, we could not see him yet, which struck me as very strange in the context of the visiting allowed in the general hospital and during Dan's previous sojourn at the Ontario hospital. However, we were given an appointment for the following day, Wednesday, late afternoon.

We found Dr. D. in her office. She had us sign a consent form for Daniel to be medicated, which was a trifle puzzling. As she led us down to the locked ward, she mentioned that Daniel was an extremely sick young man and that he had been violent; that he had refused to eat; that fluids had been forced into him. We were shocked. What could have happened?

Daniel was locked in one of the quiet rooms. He was lying on the mat dressed only in a skimpy hospital gown open down the back. As he caught sight of me and the psychiatrist looking at him through the bubble I could not be sure he recognized me. I wondered how the bubble distorted what he could see. When he had been admitted the first time I had been in that room with him, joking about the stuffed toy I had brought, tossing oranges back and forth for fun, and he was demonstrating many normal sorts of responses. Now, he raised one arm weakly towards me—in defiance? in frustration? as a plea for help? What had the doctor done or not done to bring Daniel to this diminished state? Questions flooded my mind. Apparently he had not been medicated. Did her proposal for treatment that we had just signed represent an unorthodox protocol for someone with Dan's mental problem? For what had we given our permission? Dan could not possibly have used illicit or prescription drugs prior to his entry, so he had no need for a "washing out" period. Furthermore, drug-screening can be done on urine and blood samples. What was her point in forcing him deeper into this hellish state of psychosis by withholding anti-psychotic medication?

Dan tried to raise his arm again, but had not the strength to hold it up and it fell to his side. His arms and legs were blue. He was emaciated and opened his mouth to form words but could not. He was trying to say something with his eyes. He appeared to me to be dying.

The psychiatrist said something then about going to another office. I gazed at Daniel; Richard looked, too, mouthing the words, "I love you." Neither of us could imagine what had happened to our son.

In the office, we were introduced to a social worker. Neither she nor the psychiatrist made social comments or informative comments except that Daniel was a very sick boy. The doctor recited the drugs Dan might be given and their side effects. As she rhymed them off I wrote their names in my notebook: ativan, lorazepam, perphenizine, risperidone, haloperidol. Each unfamiliar name registered like a body blow. Why so many? Why could there not be a simple antidote to the drugs Dan had already admitted to his body? Richard began to cry quietly, his shoulders shaking. Neither staff member so much as twitched a muscle or said a word. I went to my husband and put my arm around his shoulders and said to him, "At least we know we've done the right thing by bringing Daniel to where he can get help." Richard got control of his emotion and the doctor pointed out that the social worker would be available to answer our questions and gave us her phone number. For the moment we had none; we were numb with grief and horror. It seemed remarkable to me how stony-faced both women continued to be. We would be allowed to see Daniel when he started to improve and the visiting hours were mentioned. I think Richard asked if Dan would be allowed to walk outside and was told the equivalent of "we'll see."

Richard and the social worker and I were preceding the doctor out of the room when I turned and impulsively thanked her, "Thank you, Dr. D. Thank you for all you are doing for Daniel," I said warmly. The doctor looked at me and in that moment I could see in her eyes that she was evaluating my words—that she had "done a double-take." She was having some trouble responding in a forthright manner. Why? If she was considering whether or not my sincere thanks were appropriate, that would mean that she knew I had reason *not* to thank her. Or, was she considering telling me something else? We are reasonably intelligent people capable of learning what she might impart to us, but nothing further was forthcoming. I had seen that sort of calculating look before and on a regular basis. Did this woman, like someone I had tried to work with for a year, have a second agenda? And if so, what possibly could that be? Had Dan become part of a research study? As I moved again to follow the others from the room, ice settled around my heart. This woman had the power of life and death over our son, was not being frank or open with us, and I no longer trusted her.

Only when we were driving home in tears of grief and fear did I recall the words of Dr. C., the admitting psychiatrist the first time Daniel had entered the hospital: "Allowing him to continue in the psychotic state is the worst thing that could happen to his brain because it will increase a tendency to further psychotic breaks." Dr. D. had not needed to wait three days to medicate Dan—she had that authority automatically once he was admitted. The general hospital—for that matter, the general practitioner—had the authority to dispense antipsychotic medication. Furthermore, Dan was 16 and could be considered to have admitted himself. Why had she withheld the medication from him when she knew he had been stabilized on it a few days earlier? Why had she not provided a prescription for his continuing on medication when he left the hospital? Why had we not been given instructions regarding how Daniel might respond if he were taken off medication precipitously? (I would learn that such medications should be titrated slowly and withdrawn slowly.) If she thought that Dan might have returned to street drugs (a) she could have asked us; (b) she could have tested him for that. What possibly could be her motivation for deliberately allowing Daniel to become much more profoundly psychotic than he ever had been? Especially in consideration of what Dr. C. had told me about prompt medication?

I pondered the permission form we had signed, wondering about the meaning of "informed consent." Why would she tell us that she needed our permission to medicate Daniel unless she was following an exceptional protocol? Dr. C. had talked about medicating Dan immediately and he hadn't needed my permission to do so. The Consent for Treatment form we had signed, of which I had been given a copy, listed categories of drugs, not the specific ones the doctor had recited to us: antipsychotic med.; anticholinergic med.; tranquillizers; laxatives; ECG; necessary blood/urine tests. The sympathy I had read into Dr. D.'s words that Daniel was a very sick boy echoed now like a deception. We had taken a sick boy to her and he had become a whole lot sicker. And what were we to think of the discrepancies between what Dr. C. had said and what Dr. D. had done? Had Dr. C. been trying to warn me about something and I hadn't the wit to receive that caution from him?

A mental hospital, like a prison or the armed forces, is a total institution.[7] Once a person has been admitted on a certain level or "form," that person has very little in the way of civil rights. The doctors have the power of life and death and mutilation over their patients, as well as the potential for assisting healing. My husband, before and after the events described here, had edited and read books[8] on the topic of Ontario's mental hospitals and had a better idea than I of what perhaps had befallen Dan.

Once Daniel was back on medication, he stabilized in a few days. This means he was minimally functional in terms of being able to eat, drink, dress, and go to the bathroom. But the confusion in part of his brain was deeper and his ability to hope or to trust was deeply wounded. It took him much longer to regain the same level of stability he had demonstrated after his first hospital stay. He was profoundly frightened, a condition that could be "dismissed" as "typical schizoid paranoia," but he said he felt that he had been victimized like a prisoner of war. His father and I wondered if he might understand better than we did some aspects of his treatment on the ward. We could not access his assessment of the effects of the medication he was being given; he said it was much stronger than any drugs he had done except LSD. Much later, we would learn through our own observations how unnecessary were the massive dosages of medication that had been administered to Daniel; our observations were confirmed by psychiatric studies being carried out at other hospitals in Canada at that time.[9] We had led him into this. We could not get him out. He had gotten himself into this. He could not get himself out. All that any of us could do now was walk through it together. Dan asked us to bring his Bible to him, even if he couldn't read it; it was a fairly large one Lainey had given him. He kept it in one of the large leg pockets of his camouflage pants; as he said, he was holding on to Jesus.

Thanks to the gift of money from Daniel's grandmother, not all of which had gone to treatment at The Listening Centre and Richard's trip, we were able to establish a routine of one or both of us making the two-hour drive to the hospital every day to stay for as long as possible (10 a.m. to 8 p.m.) to support and encourage Dan through the behavioural stages that would allow him to be discharged, then driving home. Richard took on the major portion of this effort because we had noticed the validity of our family physician's observation that Dan would find my presence more emotionally stressful. His hospitalization was grim for Dan and exhausting and frightening for us. Minka's phone calls and emails and Charis's were very important to me. Richard had no relief from his CFS symptoms and his contact with the germ pool of the "public" on the ward kept him feeling ill. Daniel's progression towards normalcy was slower this time; for some reason or reasons, he had greater losses to recoup, or greater deficiencies caused by the medications he was being given. Sarah and Eric visited their brother two or three times. Once, Eric brought his skateboard and his guitar, although Dan couldn't balance very well on the skateboard on the walkways on the hospital grounds.

We came to see that the nursing staff was there for restraint, which primarily meant doling out "chemical restraints," that is, medications. They also performed a minimal monitoring function, filled out reports, and locked and unlocked doors. The physical care of patients, and, less often, restraining of violent patients until they were medicated, apparently was the responsibility of orderlies. Dr. D. seemed capable of cruelty. For example, in a brief meeting in the ward hallway with Dan and us when a blood test was mentioned, Dan looked distressed. "Don't worry, Dan," I said, "they don't take much blood."

"Oh, no," the doctor interrupted me emphatically and without humour, "they take a whole lot." A comment like that by a doctor to her frightened mental patient admits of a very narrow set of interpretations. What notion in Daniel's "conditioning" was she reinforcing? Was her comment a sort of threat? Was it a test of how her patient would respond to emotional pressure? Was it a test of how compliant her patient would be—to his death? And what relationship of "trust" or of control did she aim to establish by this means with her patient?

Dan's dark brown eyes were pools of fear, although his father and I were standing beside him. His gaze flickered uncertainly between us as though he doubted our ability to protect him. When we mentioned Dan's presumed future progress, the psychiatrist said with slow and pointed emphasis, "As long as he manages to stay out of the little room." I was confused by her reference to the quiet room, as if she were a stern parent and Dan was a particularly disobedient child. Dan in a normal state of mind would be an exceptionally polite, co-operative, and sociable person in most social settings. On medication, Dan would tend to be even more compliant. It is difficult for me to imagine what would have had to happen to Daniel for him to become unco-operative and resistant to staff requests. Once Daniel was out of doors with us, he burst into tears. He was terrified at the idea of "being put in the hole" again. It seems that, in addition to his first three days of horror in the quiet room, he had been locked in there again. As he explained it, when he had held his arm and fist, as he had been instructed, to stop the blood flow after a blood sample had been taken from him, someone decided he was making a threatening gesture and wrestled him into the quiet room (leaving a bad bump on his head) where he stayed for another day. In other words, the doctor and the orderlies had turned what was initially for Dan—when he was with me—a quiet and non-threatening room of peace and of refuge into a punishing torture chamber. To top it off—all of our children and their father detest egg salad sandwiches—the designated food for anyone in the quiet room, for every meal, was egg salad sandwiches. I saw them there for other patients similarly confined. I remembered the smear of egg on his forehead, the large blue lump, and his confusion about what was supposed to be happening on the ward when I would ask him if he had seen the psychiatrist or if anyone was talking to him about his malady. "No," Dan said, "they only care about behaviour."

Behaviourism

If Dan's observation was correct, and he reiterated his observation once he was fully rational, he may have been "treated" for his psychosis or for his addictions according to the principles of behaviourism. "The central tenet of behaviourism is that thoughts, feelings, and intentions, mental processes all, do not determine what we do." (And a short step away is the idea that they do not matter and have no value.) "Our behaviour is the product of our conditioning. We are biological machines and do not consciously act; rather we react to stimuli."[10] This school of thought is profoundly antagonistic to psychoanalytic, psychological, and religious modes of analysis of human behaviour. It also assumes the brain is a single entity that responds predictably, a misconception about how the brain actually functions that I have discovered and explain in the later chapters of this book.

The psychology department at my university became aligned with the theories of the noted behaviourist B.F. Skinner and I worked with people trained in its school, a viewpoint I had rejected for ethical reasons. The effectiveness of behaviourist practices in altering behaviour does not, to my way of thinking, justify the manipulative methods that may be used. Thus, philosophy comes into the equation after all as a matter of ethics. Ethics come down to one's system of belief. Behaviourists believe the ends justify the means—that the "desired" behavioural outcome justifies whatever negative thoughts and feelings may be created to achieve the result that has been determined upon. *Even if the goal cannot be achieved.*

Most of us hear about behaviourism in connection with the healing of some "neurotic" tendency. The theory is that a "phobia" stems from a conditioned emotional response, which is assumed to be the result of some traumatizing event that has "become linked with a fear-producing unconditioned stimulus."[11] Somewhere beneath this terminology is the assumption that society is a benign environment and that phobic persons hold no solid, rational reason for such a fear. Furthermore, behaviourists develop strategies for "removing" such fears through a variety of techniques including (a) desensitization, (b) flooding, and (c) modelling. If I were afraid of enclosed spaces and were willing to enter these therapies I would be subjected to (a) incremental exposure to smaller and smaller enclosures, (b) exposure to an extremely small enclosure, or (c) a mentoring kind of relationship with a person who is not fearful of small enclosures. The reader will notice the social contract of trust implied in my relationship with the behaviourist; and presumably I am free to choose whether or not I want my phobia tampered with. I enter into treatment willingly and presumably can withdraw if and when I choose, although fear (phobia) breeds vulnerability. Daniel, of course, was imprisoned and with no choice whatsoever in his treatment protocols, including whether or not he declined to be treated by behavioural measures. Nor had he any advocates.

What happens when the society in general chooses to remove freedom of choice from certain individuals? Criminals and the mentally ill and those with contagious diseases are routinely deprived of their rights to choose, and of other civil rights. (Actually, the erosion of the civil rights of the rest of us is a matter of grave concern, too.) Insofar as the society is willing to relegate persons with "unmanageable" problems (such as persons suffering from psychosis) that impinge on people in the society to behaviourist psychiatrists and within the guidelines provided by the society through law (or through high status conferred upon people who can freely flaunt the law), society is saying that the ends justify the means. The foremost and time-honoured principle in medical ethics is that the ends *do not* justify the means, as enshrined in the Hippocratic Oath: "First, do no harm." Although physicians have inadvertently been doing harm in every generation, the growth of rational medical progress can take place only within a system that sets as one of its goals the avoidance of harm to its patients. When the society that regulates medical practice gives doctors the message that it no longer cares about harm to patients as long as freedom from "harm" can be guaranteed for the populace at large, the most vulnerable person in the equation—the mental patient—is placed at risk. He or she is about on a par with the lab rat or the victims of the prison camp hospitals of Nazi Germany.

Our society has decided through our political decision-makers that mental hospitals are not worthy of funding that would provide certain kinds of care. Chemical restraints, that is, drugs that will reduce symptoms, have become a substitute for the compassionate listening and teaching—and learning on the part of the doctors—that "psychiatry" once implied. Drugs have reduced the interest of the caregivers in the human factors in mental illness, apart from controlling them. The inquiring psychiatric mind, even in regard to the nuanced use of chemical restraints, has thus been obviated. The inquiring psychiatric mind has been further muted by the availability of funding for research, which is being provided overwhelmingly by the pharmaceutical companies that sell the chemical controls. This situation has not yet blanketed the Western world; the writings of Gerald May, to provide an easily accessible example, combine the knowledge of neurology and behavioural reinforcement with an appreciation for other dimensions to the human brain, which he calls "spirit"—the "psyche" about which psychiatry once purported to have knowledge. However, one need not be a theist or to believe in the "soul" to have respect for the personhood of an individual with a brain problem.

Staffing and Funding

Further on in Dan's hospital stay one of the male nurses became confidential with me after we had sat with Dan during a meal. "I like Daniel," he said. So, apparently others did not? Who? Patients? Doctors? Other staff? And why not, since Daniel tended to be sociable when he was well and seemed to be performing as pleasantly as possible from 10 am to 8 pm every day even in his drugged, zombie-like state? And how did that antagonism felt by some others affect their treatment of him? "You know," the nurse added, "someone Daniel's age should never be in a place like this with all these serious older cases." He pointed out a building on the far side of the property. "That was the children's and young people's pavilion. They just closed it down with the government cutbacks a few weeks ago." How might those government cutbacks have had a bearing on Dan's treatment? Well, "many practical applications of these [behaviourist] methods have been developed particularly in the treatment of deteriorated psychotics."[12] Although Dan, in the midst of his first or perhaps second episode of psychosis could not by any stretch of the imagination be deemed a "deteriorated" psychotic, he might have been caught in an institutional squeeze where "one size fits all." We could not know and we were never going to be able to find out.

I will pass over the other denizens of the ward—the nurses who never leave the nursing station to pay attention to the patients except for dire necessity and the patients who come in screaming in psychic pain and who re-emerge from the quiet room drugged out of their need or ability to protest anything. Who brought them into this system? Someone as trusting as I had been? Someone of malevolent purpose?

During Dan's time in the Ontario hospital he said that he likely had saved the life of a fellow patient—a young man who was lying asleep when another inmate came at his jugular vein with a sharp pencil. Dan wrestled the pencil away from the attacker. Almost anything can happen in such a place, where the patients' prescription drugging makes them unlikely to initiate a court case and the psychiatrist in charge is considered to have an unassailable expertise and ethic.

Since the doctors have no understanding of what causes the disease, their only attempt to treat it is by ameliorating symptoms with drugs. The nurses remain in the isolation cubicle. The patients wander about, few of them able to socialize at all, while those who might are avoiding social contact that could become destabilized and prejudice their own behavioural track for getting out.

Richard walked with Dan up and down the long corridor and, later, about the surrounding grounds, through as much as possible of his ordeal until Dan was transferred back to the general hospital, where my sense of serious threat to Dan's well-being lessened somewhat. A colleague sent Richard a note that confirmed our perceptions and, for its sympathy, cheered us:

> Your note . . . resonates. Summer of 1969 I worked in [a] Psychiatric Hospital on the wards as an attendant. Bang on about chemicals and control. Have to suspect it is probably worse today given reduced dollars for health care—and lots of times then there weren't enough male staff for the problem times—and drugs had the wonderful function of reducing some of the violence against us. Bad shit, bad karma. Stories.

Dr. D.'s parting interview with me included self-congratulation "for having done the homework" as she proffered a "useful" literature review by a doctor intent on putting down any intervention for learning disabled children that wasn't part of the medical mainstream. Dr. D. denigrated The Listening Centre, saying that we "had taken a very sick boy to them asking for help and they had failed us," and warning us that taking Daniel back could cause serious damage. Although I stoically kept my mouth closed, the irony of the doctor's words pierced me: I thought her description much more accurately described what had happened to Dan under the hospital's care. I found the so-called "literature review" so shallow as to be useless. It contained erroneous statements about The Listening Centre's methodology. It demonstrated that the writer, whether or not he had an M.D., had given The Listening Centre only the most cursory attention in researching his "useful review."

During Dan's hospitalization I read two books on the function of the brain and parts of two others, one of which claimed that almost everything known about the brain had been discovered in the previous 10 years. Based on what those neurologists could and could not assert about their knowledge, it seemed to me that Dr. Tomatis's theory and method had discovered neurological explanations for behaviours and conditions that must impinge importantly on medical understandings of mental function, especially on such issues as the acquisition of language, neurosis, the personality, dyslexic syndrome and reactions to it, and possibly even diagnosed schizophrenia and varying conditions involving psychosis. But further investigation of the discoveries of the French auditory specialist had not yet been undertaken. In fact, the "useful literature review" demonstrated that the psychiatrists who had had control of Daniel were not interested in exploring such developments.

Several years later, I read Thelma Wheatley's book about her autistic son Julian and I learned of Dr. Huggins's personal research with Risperdal in respect to autistic symptoms, which was groundbreaking in terms of general medical practice and seems to have gotten into the mainstream.[13] Coincidentally, Wheatley, at one stage of her son's development, also had found some help for her son at The Listening Centre.[14] For such pioneers as Dr. Huggins, their innovative work entails courageously standing against strong currents of opposition in the medical, political, educational, and pharmaceutical establishments.

Dan was released from the Ontario hospital to the general hospital, this time with prescriptions for continuing medications and a follow-up appointment with a different psychiatrist and an appointment for a CaT (computer assisted tomography) scan. At the general hospital Dan was in restraints only briefly and had the run of the ward. He made friends, gained privileges, and was allowed to have his guitar to play in the common room. His "flat affect," that is, inability to show a range of emotion, was ensured by the drugs. His underlying personality was sometimes sad, but generally patient, and showed compassion for some of the age-mates he encountered.

After about two weeks, I received a call from Dr. B., the psychiatrist in charge of Dan's case at the general hospital, the elderly woman who had given me perhaps five minutes of her time spread over three interviews, two of them by phone. She said Daniel was being discharged and that I could come and get him. I asked her what her diagnosis was for Daniel. “Daniel is acutely schizophrenic,” she said.

“And what is the prognosis?” I asked her.

“Once a schizophrenic, always a schizophrenic,” she replied, and hung up the phone.

It struck me, even as I was trying to absorb the sickening force of her diagnosis, that what she was saying was at odds with the literature on schizophrenia provided in the waiting room on her psychiatric ward. Something was amiss; but where?

Returning Home

By the time Dan was out of the general hospital he seemed much more recognizably himself, although he still was showing flat affect and was emotionally scarred. As we started the drive home, I asked Dan if he would be OK in the car for a few minutes while I ran into the library. Once again, I couldn’t really afford the $30-per-year fee for non-residents, but I guessed that my friend Ruth, a librarian, would be disappointed when I talked to her later that day if I had not at least started to read up on Dan’s diagnosed disability. I figured that I would have the rest of my life to learn about his calamitous illness, but that I might as well start right away. Whatever our hopes for a normal life for Dan might have been, they had fallen away with the psychiatrist’s diagnosis. Dan said he could wait, and I hurried to find a starter course for the parent of a schizophrenic. I took out books on brain function, a textbook on schizophrenia, biographical accounts of schizophrenic illness, and books related to learning disability, which I felt was still my research specialty.

At home, I kept a close eye on Dan by continuing my watercolour painting in the room next to his rooms. I showed him a sketch that I had made at the Ontario hospital. Although the drawing was literal, as Dan knew from having been there, he immediately saw the the windswept half-tree as an image of himself. The tree missing half its branches resonated with that of the spruce he had planted in the yard that was halved twice, yet recovered. From my desk over the next couple of days, I noted a few of the things I overheard him say to family or friends:

“I can’t get over being locked in that room for three days. Why would they do that? I was in that little gown with my dick hanging out. I had no place to go to the bathroom. I had a urinal, but no place to shit. And they fed me hardly anything. Just egg salad sandwiches.”

“Why would they lock me in there like that? It was freezing.”

“I thought it was an old people’s home. It looked like it. Everything was damp and dreary and kind of depressing.”

"I was terribly upset that Eric wasn't in that little room with us."

On Dan's next visit to our family doctor, as Richard sat in Dr. A.'s chair while we were waiting for the doctor to arrive, Richard joked, "How can I help you?"

Dan responded, "You can enter into a death pact or suicide pact." His father misheard him and Dan explained, "You know, like Dr. Kevorkian." Dan had plummeted a long way from the optimism he had felt at the conclusion of his healing at The Listening Centre.

"What was it like in the hospital, Dan?" asked a friend.

"I had to lie to her to get out of there. I had to tell her my mom never listened to her." The implications of this remark were disturbing. Listened to her about what? The doctor had said nothing whatever prescriptive to me about how to treat Daniel at any time; I had not had any directions from her to follow. What was Dan told by the doctor about instructions she had never given me? And what bearing did that have on his being considered well enough to be released? I know now that Dan, even when psychotic and when heavily drugged for psychosis, still has an intermittently functional intellect. His comments were grammatical and rational, not "psychotic rambling." What sort of interlocution had Dan been subjected to so that his telling lies would have appeased his overseer?

The next day, I asked him how his first and second experiences in the quiet room compared. "The first time it was drug flashbacks, a purple and orange train dripping; one second seemed like an hour—weird optical stuff. The second time, when I wasn't drugged, I wanted to go to Sara's and find Mattie, which is kind of far-fetched."

A few days later, I asked Dan why he had not told me about his use of drugs. "I did it to myself. Being on Haldol was like being on LSD. Last year at Punkfest I did almost two grams of PCP. I was doing one-gram packages of PCP every weekend for a while."

Me: "Why did you do drugs?"

Dan: "It made me feel good. I did it for fun." "Ever since I did PCP one time my right index finger doesn't move properly. I know someone who has permanent neck spasms from the PCP they did."

For fun. To feel good. To escape from a reality too painful to bear?

7

Out of the Pit

Daniel's Diagnosis

> *While Dan was still in the hospital we were encouraged to attend as a family the "Life Group Therapy" session at the hospital, which Richard was not interested in doing because Dan was in no shape to be put through such a "social" event. Now that Dan was home, I pushed for us to return to the next session, on a Wednesday in mid-May. It proved to be a parody of what I imagined a "therapy" group should be. During the course of this get-together, I learned by asking for a show of hands that most of the "schizophrenics" there had had learning problems in school. Much of the meeting time was spent telling the patients in one way or another that they needed to stay on their medication even if they felt they didn't need it. Any questions asked by people present, family members of patients usually, were answered in the most banal and superficial way, as though we all were idiots. The main event, which took up more than three-quarters of the session, was all the patients lining up to hug the social worker, whose last session this would be. Hugs, it seems, were his forte. There was no effort of any kind to treat these people as individuals who might derive benefit from knowledge, sharing their experiences, or airing their difficulties.*

The Educational Environment for Male Children

What, I wondered, had moved so many of that roomful of people designated "schizophrenic" from learning problems in school to time on a hospital ward and the promise of a lifetime on medication? A number of authors have pointed out that our educational system, populated by women teachers for the last hundred years, has altered the expectations on male learners and is marginalizing boys with normal male learning patterns. This observation has been reinforced recently by Canadian educational sociologist Terry Wotherspoon.[1] Most IQ and psychological testing also has been "normalized," i.e., formulated or designed to eliminate sex differences, so that typical male learning strategies are obfuscated—covered up—by these assessment instruments. Thus, maleness, which for two thousand years of Western civilization defined the school environment and its measures of success, has become victimized in the co-educational setting. This state of affairs may please retaliatory feminists who point to the disadvantaged positioning of girls over the same time period, but neither justice nor democracy is served by ensuring either form of disadvantage along gendered lines. Schools, especially in the early grades, teach preferentially towards females, and the standard of success is defined by what comes easily to the female. My friend, the late Verna Hughes, who with her husband had raised six sons, was the only one of my children's female teachers who tried to offset these conditions. I know from things my mother said that she, too, had taught to reduce gendered disadvantage from her knowledge of her own large family of brothers and sisters and in her own way.

In most classrooms today, serious pressure is put on male children, whose brains and therefore learning strategies conform naturally to male models of learning that include the manipulation of concrete materials and more physical activity relative to most girls. Some children diagnosed as hyperactive are no more active than normal, but some classrooms and curriculums are no longer designed to accommodate their normal levels of energy and normally male modes of learning. Anita Griffiths insisted that the parents of dyslexic children brought to her make a commitment to good nutrition and strenuous exercise once a day for the children.[2] It is no mere accident that more boys than girls are diagnosed as dyslexic and that most of The Listening Centre clients then were (and perhaps still are) male. However, females also may experience listening modes that get them into trouble in the classroom. A friend, Nora, follows this configuration. She is learning disabled. She drifted into drugs, then lesbianism. She tried to kill herself, then worked her way through behavioural performance levels in a psychiatric hospital. She was released to live with family, then a partner. She had experiences both at home and at school that left her profoundly distrustful of "normal" society.

It seems that neither educators nor doctors/psychiatrists by and large really understand what happens when we try to shove certain people into our cultural pigeonholes. Then, when those shoved reach the breaking point emotionally and harm themselves, they are medicated without even an attempt at therapeutic conversation or useful research into the causes of their breakdown. Of course, not all so-called schizophrenia is the result of the process I am outlining, but some of what is being diagnosed as such in our part of the world surely must be.

> He himself still is questioning it [the Listening Centre intervention]: "It's like brain surgery," he says. But as I watch Daniel, I notice that

> he has retained the ability to stand straight, to listen and consider what we say to him, to weigh issues and come out with more balanced decisions, to write legibly. So the ear/brain listening intervention has survived his psychotic episodes and the anti-psychotic medication—at least so far. And Daniel now is a more reasonable mix of the old Daniel and the new Daniel. In other words, he is integrating the changes caused by the Listening Centre intervention into his sense of self. He does not any longer perceive himself as "flipping back and forth" between what he calls the two forms of perception, his dyslexic self and his non-dyslexic self. (Author's diary entry, May 1998)

Today, I would analyze Daniel's condition somewhat differently: Daniel's non-dyslexic condition, rather than having *survived* the psychotic episodes, had been *reclaimed*. And how would he have known how to reclaim healing if he had not experienced it at The Listening Centre?

Dan's mental condition was hard to discern because of the drug abuse factor, and Richard and I differed in our perception of how prescription drugs altered the picture. Certainly, what we were seeing of the way Dan's brain functioned overall had to have been a combination of those factors. Dr. A. said that he hoped that Daniel's condition was post-drug psychosis precipitated by the "shock" of his Listening Centre experience. In other words, he thought Dan's street drug experiences and his listening intervention both were factors in his having developed psychosis. I did not see how that could be so when my own listening experience was so positive; but neither did I do drugs and I had never been dyslexic. Another possibility was that Dan might fall into the category of people who experience one or two spontaneous "schizophrenic episodes" and never fall back into psychosis again. The remaining possibility, we were told, was that for whatever reasons, Dan's brain was "broken" and he might be looking at a lifetime on medication and severely attenuated hopes. Nothing was being said about Daniel's much more severe post-quiet-room psychosis and how much worse his underlying condition might have become due to Dr. D.'s decision not to medicate him immediately when he had been returned to the Ontario hospital.

The implications of schizophrenia for the family, according to the social worker at the Life group, are that schizophrenia is inheritable. The genetic pattern he described was: for a couple in which one parent is schizophrenic one-tenth of their offspring will be schizophrenic; where both parents are schizophrenic one-third of their offspring will be schizophrenic. I was struck by the fact that none of the doctors had questioned us at any length about our families' mental health histories. Who among Daniel's forebears were supposed to have been schizophrenic? Of course, genetic propensities for schizophrenia could reside somewhere other than the brain. We inherit our physiology, everything from the shape of our nose to our ability to metabolize the cholesterol in our blood stream to the configurarion of the middle and inner ear. The hospital's waiting room literature on schizophrenia had mentioned "the immune system."

For quite a long time, I wondered if a genetic pattern could be drawn for "learning disability." For example, I had been told by Eric's teacher that he and his wife both had trouble learning to read and that their daughter, one of four offspring, was having even more trouble learning to read than they had. Griffiths, writing in the late 1970s, reflected the commonly held belief that dyslexic syndrome has a genetic factor.[3] This condition—however the child comes by it—is what the Tomatis Method alters. Sound stimulation through the right middle and inner ear activates the neural pathway to the language centre in the left brain and facilitates the type of learning and lateral preferences, such as right-handedness, that our culture and most of the world's population consider normal. Is the Tomatis treatment overcoming prenatal and early childhood hearing/listening trauma, the effects of childhood ear infections, or even genetic tendencies towards malfunction in the vestibular–cochlear system? Can we inherit the shape and functionality of our middle and inner ears that predisposes us to laterality issues just as we can inherit a weakness in the nerves that predisposes the Islets of Langerhans in the pancreas to diabetes? It seems likely.

Unanswered Questions

Had Dan's doctors learned that Richard and I and Sarah had CFS and written all of us off as mentally defective, or as malingerers? A common misunderstanding among doctors and the public at large at that time[4] was that chronic fatigue syndrome likely was a psychosomatic condition. Certainly, it affected the immune system. Some hold the theory that schizophrenia is caused by a malfunction of the immune system.[5] At length I would understand why some might think such a thing, but at this point it was another one of those pieces of information that floated in my mind like icebergs in an Arctic sea. Why did some so closely resemble others? What were their origins? Where were the links?

The Emotional World of Dyslexic Syndrome Children

A friend who was nursing in psychiatry at a hospital near Toronto came to visit. How good it was to see her and catch up on the challenges each of us was facing in our lives! Among the psychiatrists she worked with was a woman who was of the opinion that some children using street drugs are trying to medicate themselves for underlying problems. For mental problems, I wondered, or instead for a simple exit from the unbearable stresses of their environments?

While Richard was supporting Daniel in daily visits to the hospital, I had been trying to understand better who Dan had always been that we had failed to understand. I re-read Paul Madaule's perspective on dyslexics in "The Dyslexified World," namely, that they are "immature" and that:

> even the most violent of dyslexics has, as a rule, only a pseudo character disorder and should not, under any circumstances, be confused with a psychopath. His violence is a counter reaction to the lack of tenderness in his life. His aggressiveness is an attempt to free himself from the suffering deeply rooted inside him.
>
> Rebellious, destructive, and even delinquent behaviour are the tools he has forged to break out of his cruel reality. They are just so many forms of escape.[6]

That the same sorts of rebellious and reactive behaviours may characterize female dyslexics, too, I had observed in women I had counselled. The pattern of Dan's reactions—the broken car window, the ignition he yanked from the car, the time he jumped from the hood of the car at Dad's house, the occasions on which he kicked his father in the shin, the obscenities spray-painted in the barn and on the road pavement in front of our driveway for all to see, the mess of junk in bedrooms, the ragged costumes of defiance and despair—were logical expressions of frustration towards a (however unintentionally) cruel social environment. None of this is to condone Daniel's behaviour, but it forces us to recognize that his extreme frustration was based in the failure of every single one of the adults around him to understand his severe disability and their persistence in penalizing him constantly for *their own* ignorance. However, my understanding of the neurological explanation for these tendencies towards extreme feelings and behaviours indicating a loss of self-control lay far in the future.

I re-examined Dan's drawings of the beast and the Christ-held baby and kept thinking of "The Marriage of Heaven and Hell." I was to have done my Master's thesis on Blake and Spenser—Spenser, the Renaissance poet of moral black and white, whom I understood only too well, and Blake, the artist-poet with whom I identified as an artist, but whose cosmology I found confusing and problematical. Then it occurred to me that "the beast" is not Satan, that is, evil per se, but the part of Dan that he hates, much of it merely "the flesh" that we may contend with but should come to love. I dug out my Blake and found some confirmation of what I suspected. I wondered if Blake experienced dyslexic syndrome, bipolarity, or episodes of schizophrenia. The problematical allocations of Dan's "Heaven" and "Hell" would reappear in further episodes of psychosis and more thoroughly answer those questions.

I remembered C.'s courtesy in asking Dan if he minded his mom's prying into his experience. Could Dan have arranged all that was transpiring to get away from my prying analysis? Had he found a place to retreat, even from my voice? Somewhere along the way I had forgotten the anguish of my own teen years, my need to be separated from my parents. I began to back off and felt that Dan improved in some ways. What teen wants his mom to do a thesis on him?

Reading Up on Schizophrenia and Bipolarity

A short way into the first textbook on schizophrenia I had borrowed, I began to feel that Daniel in some ways did not fit the typical profile for schizophrenia. First, there was no history of the disorder on either side of the family as far as I knew. However, indications of mood disorders show up on both sides of the family. My detailed genealogy goes back past 1066 A.D. on Dad's side and to the late 1600s on Mum's father's side. Richard's genealogy has been traced back at least four generations on his father's side and through some lines all the way back to the Pilgrims' *Mayflower* on his mother's side. No schizophrenia in those directions; and some of those families ran to 10 children. However, severe depression may have affected people on Richard's maternal side of the family and possibly depression or mood disorders on my paternal side.

The next dissonance to alert me was that Dan had never been socially withdrawn, quite the opposite; in fact, very strongly the opposite. This, too, was pointed out to the psychiatrists, so perhaps their disdain for my observations as a mere layperson and "Mom" caused them to overlook it. If they or the nursing staff had paid any attention to Daniel's interaction with other patients they could easily have spotted it in the friendships he quickly formed once he became stable. Even in his desperate condition at the Ontario hospital he communicated with and reached out to his fellow patients. He received phone calls from his friends, even when he could barely converse. He chatted with other patients, played his guitar in the lounge, went for cigarettes with others in the smoking room, and talked to people there. He came away from each hospitalization with names and phone numbers he sometimes followed up on. Moreover, dyslexic children are likely to make exceptionally strong peer connections because their distrust of the adult world has been reinforced by the failure of the adult world to understand their disability and its consequences. Daniel had transferred pretty well all of his hopes and interests to his social network (somewhat to his own detriment). Was this aspect of his personality of interest to his doctors?

Something I might have asked about at the beginning of Daniel's mental illness—because it pertained to a treatment I am old enough to remember as innovative—was lithium. Severe manic-depressive disorder (bipolar I) includes psychosis. I had forgotten the horror stories about incarcerated manic depressives who had been misdiagnosed as schizophrenics and who were returned to freedom once lithium therapy had been discovered and prescribed for them. Was lithium therapy relevant to Daniel? Perhaps Daniel's "normal" psychology had always tended to highs and lows. Or had his ups and downs emotionally corresponded to his being accepted or rejected by other people in relation to his dyslexic syndrome symptoms? Was he vacillating to a brain-chemical rhythm or to changes in his emotional environment? Dan had been exceptionally happy and even-tempered in his early years, although sensitive and emotive. That foundation had been eroded through misunderstanding and disappointment and despair. What was his baseline now? His first psychotic episode in March presented with extreme manic behaviour. His second psychotic episode in April presented with severely depressed affect: words like "disconsolate" and "heartbroken" and "frightened" seemed appropriate. Was this a result of his genetic inheritance, his "dyslexified world," the sudden change in that world resulting from the listening intervention, something in his home life, his abuse of drugs, the effects of medication (or of medication wearing off), his fear of the hospital, some hidden aspect of his psychology? Only some of these factors? Or all of the above? As far as I could tell, his doctors did not find these aspects of his mental function of interest.

The experiences of psychosis for the authors of the biographies I was reading were different from Daniel's. The onset of their illness was spontaneous and not related to drug use.[7] Anti-psychotic medication was not available to them and the treatments to which they were exposed—electric shock therapy, ice baths, and talking psychotherapy, for example—were not, as far as I knew, being used on Daniel.

The textbook authors taught me that the recommended procedure for diagnosis had not been carried out for Daniel. They suggested that the family should be interviewed extensively because the patient's family is the first and best site of observation of the patient—if the family is available and sufficiently functional to report observations. No one had interviewed us in depth in regard to Dan, although the nursing staff at the general hospital had listened politely to my description of the listening intervention. I had given them the notes I had taken during the listening intervention through to Dan's descent into psychosis (the notes had been returned to me without comment), but that would not reveal anything of family history. At no time had any doctor queried us about a family history of schizophrenia or other mental disorders or explained the illness to us.

The textbook authors explained that schizophrenia is a diagnosis of elimination; other identifiable conditions must be ruled out until the doctor is left with the residual possibility of schizophrenia. I was familiar with the process from the way our chronic fatigue syndrome had been diagnosed. How, I wondered, had Daniel's doctors had the opportunity to rule anything out? Besides schizophrenia, one of the other main causes of spontaneous psychotic behaviour is manic depression/bipolar disorder. Daniel may indeed have been demonstrating some of the symptoms of schizophrenia, but what about bipolar disorder, which was supposedly one of those disorders more easily identified that should first be ruled out? We had not observed *consistently* unusual behaviour in Dan for more than four days, during which only the last two would include behaviour classified as "extreme," before he was medicated, so my affirmations regarding his symptoms were tentative. The checklist of symptoms is on the following page.

Possible Genetic Factors

Richard's grandfather committed suicide nearly a century ago, in a fit of depression, by walking into the late-winter high waters of the Wapsipinicon River in Iowa just months after Richard's mother was born. His mother's oldest sister drowned in the bathtub. Although her bathtub drowning was officially deemed accidental, her sisters believed she feared herself infected with her father's mental illness. A family genealogist spotted recurring depression and, in the offspring of that family's marriages, two cases of diagnosed schizophrenia, including her own son (none of these, of course, blood relatives of Daniel). I wondered if a gene for depression or some other disturbance on each side of a person's family could mutate to schizophrenia? But the authors at hand were saying that the disease could be *clearly* identified generation to generation. So, there were some unanswered questions about our family and I would try to do further research through Richard's mother.[8]

What blazed out of the text for me as I continued with the authors' description of bipolar disorder was not that I was reading about Daniel, but that I was seeing clearly for the first time the patterns of behaviour that had

A Checklist of Manic-Depressive Symptoms[9]

Symptom	*Observed, but for less time, in Daniel*
Lasting two weeks:	
1. Extreme sadness	Yes
2. Loss of usual interests or pleasures	one week
3. Sleep disturbance	Yes
4. Weight or appetite disturbance	Yes
5. Difficulty concentrating	Yes
6. Intense feelings of guilt	Yes
7. Suicidal thoughts or behaviours	Yes
Other symptoms: these must cause problems for the person and last for at least a week:	
8. Extreme elation or irritability	Yes
9. Too much energy	Yes
10. Talking too much or too quickly	Yes
11. Having thoughts that are moving too quickly	Yes
12. Doing things that might put them or those close to them in danger	Yes
13. Having feelings of extreme confidence	Yes
14. Needing far less sleep than usual	Yes
15. Most people with mania also experience depression	Yes
Additional schizophrenia-like symptoms:	
16. Delusions, believe they are someone famous or have special powers	Yes
17. Very rapid speech sounding like the disorganized speech of schizophrenia	Yes
18. Hallucinations, usually related to how powerful they are	Yes
19. Depressed, lethargic, and without motivation	Yes
20. "The most important difference between schizophrenia and depression or manic depression is that people with the latter disorders usually fully return to their normal selves following episodes of depression or mania whereas people with schizophrenia rarely recover to this extent."	Too soon to tell.

characterized my father. My father's mood swings, talk marathons, tirades, deep depressions, illnesses, refusals to consult with anyone about his emotional unbalance, verbal abuse, self-concept as the expert in all things (his advanced degrees in what today is called engineering physics made him extremely knowledgeable on many subjects), and other personality extremes fell neatly into the bipolar pattern.

Dad sometimes had referred to his illness with scarlet fever when he was 16 as having initiated other health problems. As far as I knew, Dad had not experienced psychosis, but he certainly had lived a life of tumultuous ups and depressed downs for as long as I had known him. My father's life was marked by intervals of depression in which he would withdraw from conversation or any normal social activities and just lie on the couch between his arrival home from work and an early bedtime—evening after evening for weeks on end.

After another few weeks he would "feel better" and become the life of the party, the leader of church groups, the "best" at everything. After a few weeks of extreme activity he would crash and withdraw from normal socializing of any kind until he hit his next high. He could be "Prince Charming" or verbally abusive to my mother. He claimed to be insightful about people, but was anything but balanced in his treatment of others: he favoured me over my sister, scorned his parents, and considered himself the Renaissance man who always knew more than anyone else, whatever their field of expertise. He talked "at" people and lectured them for literally hours on end. He spent years isolating himself from family and writing in his office in the basement, always with vast goals, such as solving the world's economic crisis or "the problem with the Churches." After he started taking Valium his relationships with family, church, and other social groups became somewhat less extreme and remained so for the last 15 years of his life, and his writing continued. At his death in 1997 in his eighty-fourth year he had just completed a book-length manuscript providing a detailed scientific explanation of the Star of Bethlehem and was well along on another opus in which he intended to chronologize the first seconds of the Big Bang, precisely when it occurred, and God's place in that instant of the formation of the universe. There was nothing wrong with his intellect, but other aspects of his personality were definitely manic or depressed.

The uncle I have referred to (he was not a blood relative) had suffered from periods of mental illness until ameliorative medication had been found for him, and I now recognized my uncle's range of symptoms, too; his bipolarity had included more extreme behaviour, including psychosis.

I knew my husband struggled with depression, but he had experienced

10 years of CFS to cause that sort of neural problem and had some other "good reasons" for "feeling down," which was his term for finding it hard sometimes to be cheerful in our circumstances. Take any one of them: a seriously ill daughter; a daughter whom he loved deeply but could see only occasionally and who was being emotionally shredded in her environment; poverty and welfare despite his diligent work and Ph.D.; a devastating betrayal by friends to whom he had devoted a year of high-class intense effort; a wife whose activity had become more limited by CFS, also; and now a psychotic son. Not to mention having been the catalytic factor in the death of someone he had loved. Most people would consider him eligible for sainthood as he kept reaching out, soldiering on, even if he did sometimes feel discouraged and overwhelmed. In sum, I could not tell whether Richard had underlying depression or not.

Daniel's Symptoms

Some schizophrenic symptoms, such as flat affect (absence of normal emotional responses), appeared in Dan only after he had been taking anti-psychotic medication. That Dan's moods fluctuated despite the anti-psychotic medication was evident; but is that not the case for most "normal" persons? And might that not have something to do with the temporary "flipping back and forth" between bi-aural attentiveness and right-eared listening that the listening intervention had initiated?

In the context of what the textbook doctors taught about schizophrenia, Daniel also did not fit the profile for the disease because he was responsive to "talking therapy." Certainly, before the onset of his illness Dan had perspective and insight into himself. In fact, in the wake of his new abilities stimulated by the Listening Centre intervention Daniel's "maturity" and insight had grown rapidly. He was insecure with shaky self-esteem as a result of the unhappy experiences associated with his former listening disability, but he was already beginning to acknowledge these insecurities and come to terms with them. Even after suffering deep psychosis, Daniel, at least with my husband and me during the recovery period at the Ontario hospital, was trying to gain insight into what had happened to him. Furthermore, although he was still trying to process his experience at the Ontario hospital, it appeared to us that he had insight into others, i.e., the staff, because he said, "All they care about is behaviour; it doesn't matter to them what you think or how you feel about anything." On the other hand, it was not clear whether his responsiveness to talking therapy played a role in his recovery from the condition of psychosis.

We did not know if he had been exposed to talking therapy in either of his hospital settings. He had been in restraints at first at the general hospital, so it likely had not occurred there. He had been too afraid of Dr. D. to have been able to interact successfully even if that form of therapy had been tried, which seemed unlikely. We suspected that he had been exposed to some sort of talking therapy after his return to the general hospital, but had little at this stage on which to build conjecture.

Medications and Street Drugs

Some of Daniel's schizophrenic symptoms, such as continuing paranoia, i.e., thinking other people were putting thoughts in his head, and hearing voices, appeared in Daniel only *after* he had been taking anti-psychotic medication. He would spend long minutes staring closely at and studying the stamped identification on his pillowcase from the hospital laundry, seeking to discern a secret message from "the aliens." How many of his supposedly schizophrenic symptoms could be prescription-drug-induced? I looked again at the drug profiles for Dan's medications in my father's *Physician's Desk Reference*[10] andbetter recognized his symptoms and what was meant by certain side effects. Quite simply, any or all of Daniel's symptoms *could have been induced by his medications.* What did that mean in the context of Dr. C.'s warning that the sooner psychosis was medicated, the better? Even if the meds themselves produced other psychotic symptoms? If they produced psychotic symptoms, why did they? Was there a pharmaceutical optimum where the positive and negative effects of the medication reached equilibrium? And if "all they care about is behaviour"—as Daniel claimed—and if they are not interested in the particulars of individual patients' histories—as I had observed—how are "they" going to learn about pharmaceutical optimums for their patients?

I had insufficient information as to what role Daniel's use of street drugs played in his mental disturbance. What was the mental norm for kids who did dope? for kids who tried angel trumpet tea? for kids who got high a couple of times on LSD? for kids who did pot and PCP (phencyclidine; "angel dust") and "shrooms"? For Daniel, who had done all of them? I would be spending a couple of hours a day for weeks on the Internet trying to find the answers to these and related questions. We had only just begun to get some inkling of what Dan had been up to when he was skipping high school classes, and we were very poorly informed about the effects of what he gradually was disclosing to us. He had tried all those substances, plus a great deal of alcohol. It seemed the only available substances he had shunned were cocaine and heroin. He had avoided LSD, but had been "caught" twice inadvertently. He "didn't like mushrooms," but had succumbed to peer pressure and taken them, too, "a few times."

Psychological Trauma and Psychosis

The textbook authors asserted, "psychological traumas by themselves never cause schizophrenia."[11] At first, Richard and I had assumed that Dan's "descent into psychosis" was related to his feelings about the events in his life. In addition to (1) the intensely emotional experiences of his drug-induced psychotic states that he had written and talked about prior to the listening intervention, he had (2) seen the end of his first serious relationship with a girl; (3) discovered that all his perceptions were and always had been different from those of most of the people around him. We had thought that these events might in themselves have triggered his psychotic break. Now, I read that, even if they could have produced psychosis (which was highly unlikely), they would not define Dan's condition as schizophrenia. However, these emotional factors provided a foundation for psychological instability: low self-esteem and vulnerability to the "escape" offered by drugs. Drugs can cause psychosis. Drugs also can cause permanent psychosis.

These psychologically stressful events had been followed by the Listening Centre intervention, which, in his terms, stretched him between euphoria and despair, heaven and hell. I began to wonder how Daniel's drug experience *prior* to the Listening Centre intervention had heightened the emotional quality of these experiences *during* the intervention. Was the intensity of his reactions just a matter of his very acute hearing, as I had thought or did other factors weigh in?

The Onset of Schizophrenia

Some of the books I had picked up at the library that were "as told to" accounts by people suffering schizophrenia revealed that the sudden onset of their illnesses had nothing to do with drug abuse or emotional trauma, but some unexplained spontaneous alteration that affected the condition of the brain. That medical science had little understanding of the condition was brought to my attention through newspaper accounts of the work on behalf of schizophrenics by one of Canada's high-profile philanthropists, Hillary Weston, whose brother suffered from schizophrenia. I thought of the pamphlet in the waiting room and the disparity between its encouraging words and the flat statement of the elderly psychiatrist—and when and where had her views on schizophrenia developed?—"once a schizophrenic, always a schizophrenic." The onset of Daniel's symptoms had *not* been spontaneous.

How could Daniel's doctors have proposed a diagnosis of any kind until the effects of ingested street drugs and alcohol had worn off? *If* they wear off. They may have considered his weeks of hospitalization a "washing out" period as far as street drugs were concerned, but certainly not of the prescription medications they were giving him. And to what extent would the side effects of the prescription drugs permanently dispose him to psychosis? I began to realize Dan's diagnosis was suspect for a number of reasons. Whether or not Dan actually was schizophrenic (whatever that meant, since they couldn't agree on a definition or explain a cause) we would have to wait and see.

The Family Doctor

After Daniel had been released from the hospital, he again came under the care of our family physician, Dr. A., who (I assumed) thoroughly understood all of the issues I had been trying to sort through on my own. He expressed the hope that Daniel's problem was a drug-induced episode of psychosis. And that some people experienced spontaneous schizophrenia-like episodes that did not recur. The rule of thumb used in the medical profession was that a year—off medication and off street drugs—without a recurrence of psychosis probably reversed the diagnosis of schizophrenia or at least indicated that it merited further investigation. Meanwhile, Dan had been lined up for various tests and for an appointment with the next psychiatrist who would see him. Once we felt Dan was stable, perhaps this psychiatrist would recommend very slowly reducing his medication until he stopped taking it. A door had been cracked open to a different sort of hope. I could not formulate a plan, but I saw the possibility, at least, of further study and investigation of Daniel before we accepted as permanent the rendered diagnosis.

The Next Psychiatrist

Once again displaying a tendency to trust doctors, I wrote to the next psychiatrist, Dr. G. (the seventh, at least, who had seen Daniel), with my newly acquired information regarding other possible diagnoses for Daniel, explaining my family's history of mood swings, the depression on Richard's side of the family, and Dan's symptoms as he was entering his first psychotic break. I asked him to consider the possibility of a misdiagnosis.

For the first interview, Dr. G. had scheduled time with Dan alone, then time with Richard and me in his office with Dan. The doctor acknowledged that he had received my letter and gestured to it on his desk, but declined to discuss it with us. Dr. G. told us that some people enjoy psychosis and seek those experiences. I was disgusted that he would introduce Daniel to this idea. Dan, who certainly sought "highs," had not appeared to have enjoyed psychosis at all and had found his time in the hospital terrifying.

Then, Dr. G. asked me how I "felt" about Dan's illness. At first I tried to tell him what I thought. But that was not what he meant; he asked about my feelings. I tried to explain to him that the stress of the situation had reduced all of my feelings mainly to a kind of numbness. Also, that I tried to keep my emotional reactions in check so that I could deal rationally with Dan and his needs and the family's needs, and that my capacity for feeling had become deadened through stress before. (I had had the same experience following Michael's death. Emotional burnout for me had produced muted emotions, although I continued to function satisfactorily in work and other social relationships. I knew I could just carry on until my nervous system recovered.) The psychiatrist asked me if I didn't think that *not to feel guilt* at Dan's condition and *how I might have caused it* made me an inadequate mother? This question was levelled at me, not at Daniel's male parent. Nothing was said about the possibility that Dan had behaved irresponsibly or contrary to the teaching of both his parents. Dan and Richard sat there absorbing the doctor's view of my emotional state and culpability without uttering a word. We were there for Dan, so I guess I should not have expected a word of support from my husband. Nothing, of course, was said about the effects on Dan of having been locked up for several weeks under the control of an extremely powerful woman who kept him in a state of terror.

On the long drive home, I lay in the back seat of the car while Dan drove in front with his dad. I felt stricken. I was assaulted with feelings of guilt and inadequacy. I dared not cry for fear of upsetting the two men in the front seat. Why was I so vulnerable to Dr. G.'s criticism? Just because he had the magic word "psychiatrist" associated with his words? Then, it occurred to me that perhaps he was of the ego-destruction Freudian school that would see me as the cause of any one of our children's "maladjustments" and that he would be prepared, even eager, to tear down my ego or central self-concept in order to reconstruct it along his own lines of thought.

And where did his maleness and my femaleness fit into this picture of ego-domination and reconstruction? My reading of some of Freud's writing while I was in university had left me with a low opinion of his thinking, a view that contemporary analysis of his theories was confirming. Subsequent reading showed me how untrustworthy an analyst and confidant Freud had been. I began to feel better; I certainly could not be faulted for trying to be the best parent to our children that I could be. I had learned long before not to carry around guilt. My method was to face it, confess it, forgive myself, make amends if possible, and try to avoid making the same mistake again.

During my listening experience at The Listening Centre, one of the staff members had brought me a rather sentimental poem about God choosing some mothers for the special job of parenting a disabled child. I had thought of Daniel as a child with a problem, although that problem had taken on larger and larger proportions. For the first time I had grasped that Dan's problem was in the category of a disability—not something that could be isolated in an overall context of normality, but a way of being and perceiving, and of being misunderstood and mistreated—that was encompassing. I had wept and acknowledged my short-sightedness, even as I knew I was the family member who had fought her way through to at least a partial solution to that disabling condition in our son. Although I did not yet know what part Dan's Listening Centre intervention might or might not have played in his psychotic break, I knew that The Listening Centre was very knowledgeable in a highly specialized field and was performing important work in a way that was at least as ethical as—if, indeed, not more so than—that of the psychiatrists who were having some tremendously negative impacts on our family, apparently without qualms.

I could accept that I unwittingly had been the essential cause of Dan's disability before he was born and that I unintentionally had contributed to his frustration and despair during his childhood with unrealistic expectations of him. I could accept that I had sometimes failed miserably to meet my standards for myself in parenting. But I was one of two parents who did the same things; one of a larger social network, all of whom were making the same mistakes. At that point, I probably had a better idea than the psychiatrist of the nature of Daniel's dyslexified personality, but he had put me on the defensive so that we did not get as far as looking at Dan and how he might be helped. Furthermore, the destruction of my central self-concept was not something to which I would submit. Nor was I keen on the ideas a therapist of the Freudian school would be introducing to our son. On our second visit, I asked Dr. G. just what was his school of psychiatric thought and if he was intent on ego-destruction. He replied (uncomfortably, I thought) that he could function within a number of different paradigms. Daniel had decided this doctor was not helpful but he was unable to explain why, whether from reticence or short-term memory malfunction due to his medication or from mental confusion, we could not tell. In any case, we did not consider that it would be fair to Daniel or appropriate for us to pry into his interaction with the doctor.

Richard was finding the five-hour round trip to and from appointments, which seemed to him unhelpful, an incursion into his earning hours he would rather not make. I was very curious to know what else this doctor might be able to come up with that could be of help to Daniel or to us, but he certainly did not dangle any carrots before us. Were we to assume he had answers to what, why, and how Daniel had become mentally ill? Were we to assume he could reverse the situation? It seemed that the pursuit of this line of action was left entirely to our ignorance or assumptions. With no prospects from the doctor as to what he might effect in Daniel, Richard and I discussed the situation and decided to decline further appointments, at least until Daniel had resumed more normal mental function and could better assess the situation himself.

It never crossed my mind that Daniel's doctors might be as deeply—or more—in the dark than I was as to what caused a range of mental illnesses or how to cure them. They were the self-declared experts and were handing out the opinions and behaving as though their expertise was relevant.

The Learning Disabled in a Learning Society

The library books on reading problems brought me a few further insights into learning disabilities and a range of strategies—most were familiar to me—that had been found effective for teaching dyslexic syndrome sufferers to read, to attain better balance, to organize, and to feel self-esteem despite their differences from their normal peers. A Scottish educator, who spent much of her career in California, began her book[12] with the example of a Kalahari bushman's social environment. Why? How much more remote from our daily lives could an example be? Yet her example caught my attention forcibly: Dan had made the identical connection in his dream about the path that ended in a valley of mist—an image he attached to *The Gods Must Be Crazy*, the film about a Kalahari bushman (Sho) who walks to the edge of the world to destroy an object of modern civilization (a Coke bottle) that threatens his society because everyone covets it. This educator held in my gaze things to which I should have paid attention closer to home. She pointed out that for the bushman, the dominant life skill is that of using the bow with arrows to hunt game; tribesmen who are unable to learn that skill are in many ways disadvantaged and sidelined in their society, from the quality of food they consume to whether or not they will be permitted to marry and procreate. What, she asks, do we demand as the essential life skill in our society? The answer, of course, is "to read and write."[13]

I remembered the gentle courtesy of a man in our doctor's office in Arkansas who told the nurse behind the counter that he could not read or write and of the impatience and scorn with which she told him to mark his "X" on a form. I remembered the humiliated boys in each of my grade school classrooms who had trouble learning to read and how I shrank with a shared anguish and humiliation as some of them were strapped. I thought of my own climb through the grades and how advantaged I had been in learning to read because of the language-oriented and encouraging parents I happened to have. I remembered how confused and distressed I became when numbers did not behave like letters in my head and how I had been shamed and denigrated in even the slight disapproval of my parents—and how hurt I was when I realized I was viewed by them as less intelligent than they for having had that difficulty. And how much of my difficulty went back to the strapping I had witnessed during math class? The smarter the better; your father is a genius, your mother a schoolteacher. Language is power. Language is social status. Language is manipulation. Language is everything. In the beginning was the Word.

The Scottish educator reminded me that children will make heroic efforts to please. If you love them, children will love you in return and do anything for you. That relationship defines their vulnerability and our most solemn responsibility. She points out that the dyslexic syndrome child puts forth extremes of effort without winning recognition of that effort in the responses from teachers, parents, and other adults—or even from peers. At some point, which is unique to each individual, the child despairs, the central self-concept collapses, and the child becomes vulnerable to alternative sources of pleasure and ego satisfaction.[14] Now, sadly, I recalled our efforts to encourage Dan in tasks he could not perform and in ways that fell pathetically short of understanding—and therefore of appropriateness to—his needs. I had even taught Daniel that his brain was damaged and that he could not expect it to do some things like other children. I had helped to knit that network of misunderstandings and our inadequate efforts spread like connective tissue through life as he knew it.

So had others, albeit unintentionally. For example, Luke had organized a rocket hunt for his brothers during a visit. He had bought small rockets, and set them off in our orchard. The game was for his brothers to compete in retrieving the parts that parachuted into the adjoining fields. Dan's awkward gait over fences and stumbling into ditches was measured painfully against his brother's "Ahh, anyone can do this." Dan's inept and unhappy effort to participate played against Eric's plunge into the fun with his athletic co-ordination. Dan became increasingly frustrated, hurt, and angry—his damaged self-concept kept him from enjoying the competition at all. Eventually, competition of any positive kind became a situation Dan would avoid. Luke's hope of creating a joyful event for his brothers came to a very uneven success. His relationship with Dan was not enhanced. I would come to see that the subculture in our area was largely composed of dyslexic syndrome children like Daniel who had surrendered any hope of finding self-esteem in the dominant culture and had decided to make a virtue of necessity: they were determined to defend their differentness, emphasize it, and rationalize their ways of making it more extreme with street drugs.

The magnitude of the dyslexic syndrome problem is vast: multiply the failures in our family across dozens of children in our community, hundreds in our area, tens of thousands in our province, and millions around the world. Dyslexic syndrome is an extremely damaging disability both in the impact on our children and in the outcomes for our society, including most of our social institutions and the economy. The Tomatis Method brings hope for healing to that sector of suffering and despairing children and frustrated and hopeless parents. Whether or not Daniel's case represented a particular subset of problem cases, clear evidence was being gathered every day from 10,000 clients a year in about 200 Listening Centres around the world that the Tomatis Method brings improvement or normalcy to children of any language background who suffer from dyslexic syndrome, ADD, ADHD, and other ear-related disabilities, including autism.[15]

Respect for the Disabled

While this Scottish author recommended interventions less comprehensive in their effect than the Tomatis Method, she emphasized that showing respect and care for dyslexic syndrome children is the foundation on which remediation can take place. In her experience, all dyslexic syndrome people, including adults, can be taught to read if someone treats them with empathetic respect and takes the time and care to find the sensory avenue that will facilitate their learning. The *Language Patterns* series of readers I had helped to produce for Holt, Rinehart and Winston in the late 1960s had demonstrable success on the same principle, and the programs of Science Research Associates in the United States similarly proved the success of strictly positive reinforcement and a multi-sensory approach to the teaching of reading, even to children with the symptoms of dyslexic syndrome.

Plainly, Dan had progressed so well in the one-on-one sessions outside of the classroom when he was in elementary school because he was away from distractions; this also helped to explain why his success in school rollercoasted. While I still harboured some lingering questions about nature, that is, Daniel's genetic inheritance, questions about Daniel's nurture were being answered: he was a victim of the ignorance of those in his learning environment and of how our strategies had impeded his progress and self-esteem along his way. He had so much to forgive!

Daniel's Grade 8 teacher had bestowed on him in a kindly fashion the nickname of "Mountain Man," from the pile of books and papers on his desk he was unable to organize. When Luke brought from Virginia paintings he had made for each member of the family, the one he gave to Daniel was a scene of mountains like the ranges of Gondor in *The Lord of the Rings* that

Dan had drawn at The Listening Centre. Daniel would need Frodo's strength to climb those mountains for he already had been lured into a Shelob's lair. How would he overcome the neurological conditioning from the drugs our neighbour supplied that attacked his healing from dyslexic syndrome?

8

Mountain Man in Spider's Web

Addictions

Dan has overstayed the agreed-upon 12–1 curfew and Richard deems that we should dress and go to find him. Off to Spiderland at 1:45 a.m. We are walking through the borders of Mordor into Never-Never Land. We pass at least two dozen campfires scattered through the woods, proximate to a tortuous logging road. Stars are glimmering through the smoke. We walk a long way in the almost-frosty night, guided once or twice by Spider's Lost Boys: children in the coifs and warrior costumes of punk subculture, more and less drugged, reaching out with courtesy and good humour to the Mum and Dad looking for their prescription-balanced giant child. "Do you know Dan Tallman?"

"No . . . Nah . . . Yeah."

"Have you seen him?"

"Not for quite a while. Try the stage."

A deeper foray into the woods. Quiet tents. An outburst of "fuck yous" and "fucking" at one fire. The less and less distant amplified rhythm of a disorganized punk rock band. The hum of a generator. Two a.m. hangers-on at stage edge. The large fire beyond the shadow of the stage reveals the ravaged impresario himself, shrunken and quiet. A fresh-faced lad who says he knows Daniel lies on his back looking up at me and gestures towards the trail. But now Richard and I are flashlighting through unblazed bush on a hillside, our hopes slipping. At the next campfire no one knows Dan, but he hears my voice. "Mom?" from a yet-more-distant tent.

"Dan?" and I laugh as I did when he was born, lightly and over-brimmed with success. Richard's light catches him tumbling backwards out of a tent flap, like a surprised raccoon, mop of dark hair around bright brown eyes.

Caught between relief and embarrassment, he hesitates to make plans. I make them for him, and while his friend Adam joins us on the trail, he puts dark distance between us on the return hike through the woods. At length, I fall back from Richard's measured pace and light, and walk blindly wherever my foot finds a clear way. Suddenly a commotion at the road edge and something or someone breaks up onto the road. "Hello?" No answer, but a vague shape falls into step beside me. "Dan?" No answer, so it is Dan.

"I'm glad you came to get us."

"Me, too," I reply. As he runs ahead, a bright silver shape left of the path startles me. What could it be, in that black depth? Looking back and up I see the sinking moon—so perfect, pure, and luminous—has mirrored itself in a patch of swamp in this dark place of treachery. Youth, fire, drugs, the deep night, friends, the music of anguish and despair, the lost, the moon and stars, the found. Now the gate fire is in sight as I clear the trees.

The Listening Intervention

If not for my own experience at The Listening Centre, I might have blamed Dan's listening treatment for what had happened to his brain shortly afterwards, as at least one psychiatrist did and as Richard initially was inclined to. But I had been affected only very positively by the listening experience; I had regained my health and energy. Richard agreed that Daniel's drug abuse certainly was a factor. I was confident that drugs were the *only* factor in his psychosis both before and following his listening treatment. From the perspective of a dozen years, *I believe that for Daniel, too, the Listening Centre intervention was not only irrelevant as a cause of his psychotic break, but probably postponed the slide into psychosis that had begun before the intervention*, an episode of which he had described following the night of Kurt and Tonya's visit. The listening intervention, furthermore, had provided for him the neurological experiences that most people in the world define as "normal." That foundation of normal experience gave Dan confidence in himself; more specifically, confidence in the part of his brain that recognized the "normality" of what had happened to him as a result of that intervention. This inner sense provided a platform for his trust in those who cared most about his well-being. In those early months, however, I was not secure in my observations and "intuitions"; I was constantly observing, assessing, questioning, listening.

Without the neurological processes that are foundational to what most cultures in our age define as normal for learning and perception—and that were given to Daniel through the listening retraining—all of his subsequent periods of mental illness would have been much more severe and lacking in a concept of normalcy to give him hope for recovery and a reason, both for him and for those who supported him, to conceive of a brighter future. If we had known then what we now know, Dan would have been encouraged to maintain a daily program of listening and singing.

Dan Teaches Himself to Read Again

When Dan came home from the hospital after his first two back-to-back psychotic episodes, he had almost lost his ability to read. When he became stable enough to focus on a task, we started looking at books from his childhood that he particularly liked and we talked about his learning to read correctly this time, from the very beginning. I emphasized what he had gained from the Listening Centre intervention, not his losses from the psychosis and medications. Over the course of a month and gradually increasing the sessions up to an hour a day, Dan would read to me from a Grade 2- or 3-level storybook, then from the Bible he had carried around with him while in the hospital. At no point did he demonstrate the sort of confusion about letters or hesitancy in sounding out phonetic sequences as he had as a child.

In a couple of months Dan had progressed to the Grade 8 level and was stalled there for a while. Then, it occurred to me that the material was not age-appropriate (another nudge from Michaela Odone and *Lorenzo's Oil*). When we switched to more challenging material, Daniel soon was reading much better than he had prior to the listening intervention. Daniel could no longer be said to suffer from dyslexia defined as an inability to read. When he had reached a point where he could read a book at first-year college level, he stopped reading in a disciplined way, but would read when he felt drawn to something, such as an article in *National Geographic*.

I had heard from Dr. C., from a social worker, and from others that "Dan is 16, you can't be telling him what to do any longer." That advice might fit the existing social structures and accommodate common wisdom, but it was not suitable for a person with his mental experience and with his tenuous grip on how to get where he deeply wanted to be. If I had known then what I have learned since about the plasticity of the brain and learning plateaus in neural recovery, I would not have allowed him to stop there.

The Family

For me, the story becomes very complex at this point and I cannot begin to tell it fully here. Although his situation was critical, Daniel was not the only member of the family with severe needs. Eric was having serious trouble in school and already using marijuana. Sarah had made some gains, such as when she did a little gardening all spring and summer, and had experienced some losses by overdoing the activity. She had a better sense of what she could and could not do without harming herself and she was trying constantly to find ways of becoming well, in which I encouraged and supported her. Her losses, oddly similar to Daniel's, included normal life in high school, the academic background she would need for further education, diminished mental function, beneficial social contacts outside of the family, and incalculable other deficits of "social capital." Daniel's acute distress had not diminished my deep concern for Sarah's health and uncertain future. Luke and Kathleen, in Virginia, and April, as she moved from place to place, also were constants in my sphere of concern. Richard was not well and had special needs. My sister and other friends and relatives had places in my affections, and time, also, if I could find any. Richard and I found great solace in our mutual love and lovemaking, but most of what couples of our age (56 and 54 when Dan first was hospitalized) considered important for their relationship—weekends away, dinners out, a holiday without the children—were beyond our ability to afford and were precluded by caring for a teenager who hovered constantly at the verge of what the doctors described as "becoming a danger to himself and others."

Caring for Daniel

At first, we monitored Dan very closely to protect the community from him; he was never out of the supervision of one or, more usually, both of us. We saw ourselves as continuing the sort of surveillance provided by staff when he was in hospital. We lived in uncertainty about how or when another psychotic break with uncontrolled behaviour might occur and stopped attending functions of any kind, including church. It would be years before we learned enough to have confidence in our own ability to predict Daniel's behaviour and give ourselves more relaxation and freedom during the periods of Dan's psychosis. At length, we monitored him also to protect him from the community, that is, to protect him from what would happen if he were returned to the hospital by someone unable to judge his condition.

At first, I thought that Dan's essentially gentle nature and teaching in pacifism and his Christianity made the community much less at risk than might have been the case for some other mentally ill persons. I would learn that people with schizophrenic symptoms are more likely to self-destruct than to destroy and that the vast majority do neither—and for reasons of neurological function that will become clear in later chapters. This is not to say the belief system is not a factor, but it is not the only factor and it is not simplistic.

In many respects, Dan was in much more danger from others than anyone was from him. When Dan learned that some of his peers thought the murder by a local young man of his much younger girlfriend "was the sort of thing someone like Dan might do," he was profoundly shocked and hurt. For a while, the need he felt not to be viewed as a potentially violent person actually curbed his substance abuse. I would add that the sort of ignorance in a community, such as Dan experienced, makes the schizophrenic person vulnerable to persecution, or perhaps even unjust prosecution.

The following year was extremely difficult for Sarah and Eric with their brother in and out of hospital and their parents trying desperately to figure out what had happened to him and why. In case Dan should have another psychotic break (and we were uncertain about what had caused the first ones), we all slept with our bedroom doors locked. We understood that his drugged state was a form of chemical control that permitted him to be returned to the family and community in a condition the medical establishment considered reasonably safe—for others.

To protect Dan in our home I removed potentially dangerous substances from the bathroom and spice cabinet (mouthwash contains alcohol; prescriptions and over-the-counter medications contain substances for abuse; nutmeg and cloves are poisonous and mind-altering in large quantities; and so on), and we stopped having even an occasional glass of sherry or beer—the house became an alcohol-free zone. I removed the kitchen knives from the counter block and other sharp tools to where they would be out of sight in drawers. I did not know what possibly could happen, but the vivid memory of Dan's first psychotic break made me take the same sorts of precautions I would ensure if I were caring for an irresponsible child. Disclosures from both Dan and Eric years later revealed they had at least sampled every potentially mood-altering substance I had confiscated: Eric tried Dan's Risperdal, Dan used nutmeg, both boys drank mouthwash, and so on. Our household had become unusual, at least in terms of my experience. It was not so different from a great many others nearby where young people were "into the drug scene," except, perhaps, for the values and efforts of the parents to effect some kind of change in their kids.

As soon as Dan was mentally stable, we returned to family counselling. Through Dan's disclosures in that setting we gradually obtained a clearer idea of the extent of his drug and alcohol abuse. Sometimes Dan simply was too mentally ill to attend counselling sessions and they were discontinued until such time as he could participate.

Dan's illness wasn't easy for the children who no longer lived at home, either. The first time April returned when Dan was psychotic, she went into the living room to him, hugged him and said hello, then came back to the kitchen and fell into my arms weeping. When Luke and Kathleen next returned, Luke insisted on seeing our family doctor. Luke was hoping that some combination of talking therapies that had helped him when he had seen a counsellor would help his brother; but, as Dr. A. explained to him, until Daniel had regained his mental balance on medication he would not be able to think or remember well enough to benefit from talk therapy or counselling of any kind. When April had listened to my complaints about Dan's experimenting with drugs and what it had done to him, she offered the opinion that he would quit "when he found something better to do"—the implication being that we as parents still had not succeeded in that direction; or else, that the society at large eventually would offer Dan an enticing alternative to the substances messing with his head. Long years would pass before the older children, as adults, would be able to take a more generous view of what Richard and I continued to do as parents.

During the first months of Dan's illness, we took him to hospitals for tests and endured long waits for psychiatric appointments. However, as I began to educate myself about his diagnosis of schizophrenia through reading and through speaking with others who had interacted with medical support personnel for the mentally ill and drug-addicted, I discovered the local hospital was not following good procedures, their "teaching" was incorrect, and the diagnosis had insufficient basis to be more than tentative. Nothing had been said about or taught to Daniel or to us about addictions. In some ways the doctors and nurses had been very deceptive with us. Daniel's case was essentially like that of many, many other young people who had become addicted to street drugs; they could have said so. We could have been told that much, without their disclosing that many of the other patients on the ward with Dan were there for the same reasons. Knowing Dan's case was not rare, that is, identifying his illness as drug-induced schizophrenia (which has become common) rather than as spontaneous schizophrenia (which is rare) would have been helpful. Fragments of information gradually came from Daniel himself (although fierce loyalty to his subculture friends kept much information from us) and gradually were confirmed through people with similarly affected children in the community who were dealing with the same system. However, Dan's symptoms remained a problematical issue.

Daniel's medication did not protect him from what he came to call "torture." The suffering of addicts is viewed by some as just punishment for their unfortunate choices. Recognizing my own unintentional but undeniable contribution to Daniel's dyslexic syndrome, failure in school, reliance on friends, and disconnection from family, I shared in his suffering and did whatever I could to alleviate it without babying him. For those who recognize the role poverty plays in the despair of the impoverished, the comfortable and wealthy of our society might contemplate their complicity in such suffering, too.

Some nights Dan was so paranoid I slept on the floor of his room to reassure him each time he awoke that he was not about to be carried off by the drug dealer and that, no, his father was not in his closet with a prostitute. During one frightening night, Daniel had several episodes of shaking from head to toe. I realized only later that his heavy dose of medication probably had caused seizures—yet another circumstance for which the hospital staff had provided us with no instruction. For weeks Dan's paranoia persisted as a feeling that Spider, the Meister of Punkfest, was outside, prowling the yard, looking for an opportunity to pounce. I did not yet recognize his apprehension as a metaphor for his bondage to the easy "out" Spider proffered in an enticing setting.

When Dan caught a virus, he was particularly prone to vomiting. The colds that come and go barely noticed by most people took a heavy toll on the members of the household with CFS (Sarah and Richard, at this stage) and on Daniel.

Trying to Understand Dan

Half of Dan's muttered ramblings would be in support of the drug guru Carlos Castaneda's *Teachings of Don Juan*[1] and half of his ramblings would be in support of the power of Jesus to save. The tension between these two opposites seemed to be lodged in two separate compartments of his brain.

I had time, during the long night hours when I stayed awake near Daniel's room, to ponder his condition and the social environment and the culture into which he had been born. Why did he equate (not just associate) his dyslexic condition with hell and drugs and psychosis? Why did he equate Jesus with his "new normalcy" when his Christian attachments had been formed before the listening intervention? Where is "faith" located in the brain? How do the various parts of the brain function differently for men and women? How much of our social culture rests on patterns of ear-related thought—forms of brain function—peculiar to the male? Peculiar to the attentive male? When in the development of Western civilization did it become life-threatening to be insufficiently attentive? When did it become life-saving to adapt to a single system of listening and therefore of communication? Where did associations of left-handedness and "sinister," in the pejorative sense, begin in our culture? When did right-handedness become associated with "normalcy"? How much of our culture is imbued with a legalism that runs through the left side of the brain of both males and females and no other way? What, really, do we know about the way we socialize our children and the effect that has on their central nervous systems? Although these questions arose in my mind in the early stages of Dan's illness, I would wrestle for a decade with the effects of his problems before I would discover the answers to some of them.

We noticed that changes of environment stressed Daniel's nervous system when he was unstable. Driving in the car with a constantly changing scene streaming past the windows was especially difficult, and he would sleep if he could. But even the familiar setting of the cottage would destabilize him for the first 24 hours, as would the home environment when he returned. He could not be left alone at home or cottage; either he travelled with us or Richard or I would remain with him at home.

Struggling to Walk

walking on the shore
loose round rocks are as wide as
the length of our feet

Daniel had been on 6 milligrams of Risperdal along with other medications for several weeks, and we did not know to what extent the drug normalized his behaviour or impaired normal behaviour, or did both. Our family doctor intended to supervise his gradual withdrawal from the medication. Daniel spent the summer of 1997 easing off Risperdal (risperidone) and reclaiming his emotional security by returning to as much of his former lifestyle as he could get away with. We thought this behaviour might be more normal or "healthier" in some ways than the paragon we saw emerging immediately post-Listening Centre, a suggestion from a friend that I would now reject. Dan was adding gradually to his repertoire of responsible behaviour. He hoped to tackle a correspondence course or two at home so that he could qualify for a community college program.

By September Daniel had redefined his goals to doing school "on his own." He was not able to get the art class he wanted, so he capitulated to the school's assignment of Grade 12 science in addition to the music he wanted. I suggested that he was setting himself up for failure by taking on a difficult subject when his new mode of learning had not yet been tested. He said he wanted very much to conform to the school's suggestion. The school, of course, had no provision whatsoever for a student with Dan's difficulties.

As before, Dan spent almost all his time with his friends. I noticed that these were new friends who seemed needier than the boys I knew. He wanted to move into an apartment, but the plans he made with Chad fell through. By mid-October Daniel was missing most classes. He moved very rapidly from panic to apathy. He was obsessing about marijuana and using it again. Dan tried to dominate Eric, who developed severe behaviour problems again and I began home-schooling him. We found a quantity of marijuana in the barn clubhouse and told Daniel he could choose between counselling and facing the police. He opted to go back to the Beechgrove counsellor and the positive changes in his behaviour after one session were remarkable. As I wrote to Lauder Smith:

> Quite simply you gave him hope, lifted him out of despair. Suddenly, from the moment we left your office, he resumed normal, civil, caring behaviour with the rest of the family. That evening he shot baskets with Eric at the net in the yard for the first time in two years. We have seen him laughing, joking, singing boisterously around the house and outside—not the anguished grunge stuff of performance, but just hollering fun. . . . As I mentioned on the phone there have been some negative incidents as well, but he seems to be hanging onto new possibilities.

How different for Dan might the outcome have been if at this stage a group of teens similarly afflicted had counselling, support, and their own growth to keep them engaged? Isolated from successful peers and with no support group to relate to, Dan could not sustain his changed behaviour. He was addicted to alcohol, marijuana, to his friendships in the drug culture, and to a lifestyle that enshrined the laziness, bitterness, self-destruction, and rationalizations of blame that characterize despair. We discovered that both Daniel and Eric were stealing small things, such as scissors to cut up their pot, from my father's house. Dad was angry and said to me accusingly, "You have a serious problem on your hands." As if I did not know.

The End of Another Road

Just before his eighty-fourth birthday in November of 1997, my father died. For me, a very difficult relationship had come to an end and I was doubly grieved that my own life was so stressed and fraught with apparent failures that I had disappointed him at the end of his life. I loved and admired him for his many good qualities. The sincerity of his Christian beliefs and his fidelity to those beliefs were unswerving. However, in addition to his mood swings he was egocentric and a controlling, demanding husband and parent. He had "disowned" me when I became Catholic; he blamed my husband for our poverty; and he blamed us for Dan's worsening behaviour and mental instability. He also resented every penny of help my mother had offered us. Meanwhile, he maintained a sparkling public persona. I loved him very much, but life with him for Mum and for Lainey—who lived with our parents much of the time, perhaps partly to provide Mother with an ally—had been nerve-wracking. After Dad's death I found several executive-size file cartons filled with paper forms he had designed and printed detailing literally every waking moment of the last several years of his life, a memorial that easily matched the manic proliferation of notes festooned around John Forbes Nash's garage in the film adaptation[2] of the schizophrenic American mathematician's biography, *A Beautiful Mind.*

After Dad's funeral, Richard commented, "I believe he did the best he could." My husband's insight was not simply "charitable"; he realized that Dad's "impossible" behaviour from the point of view of the comfort of those around him had been more literally impossible in terms of his inability to understand and control his own behaviour.

My sister and I inherited his cottage, which we were in no position to afford to keep. The small house up the road, which we had inherited from our mother, was at our disposal.

Dan's Cognitive and Emotional Conditions

A normal consequence of the listening retraining experience for the client is to experience emotional ups and downs, since for a time the person is adjusting to fluctuations of perception between the new non-dyslexic perception and the former dyslexic state. Even I had experienced a certain amount of mood swing after my healing. I had felt elated; I had wept when one staff member showed me Dan's "I want to live" artwork and again when another revealed to me the extent of Dan's disability. I felt it was important to remember that after the Listening Centre intervention Daniel's symptoms of mania and depression appeared and retreated like the swing of a pendulum, possibly in response to or in some synchronization with his fluctuating "dyslexic" to "non-dyslexic" perceptions. Intervals of abnormality were interspersed with intervals of normality. Had he shown a gradual escalation in the intensity of his symptoms during the listening retraining that became more extreme after he returned to drug use the following week? *Or had the intervention prevented him from going "over the edge" sooner?* Or had his return to drug use sent him over an edge he might never have reached without those chemicals? None of the doctors mentioned, and I did not observe, the frequency and duration of oscillation of Daniel's changes in modes of perception. I discovered, over the years, that this was of pivotal importance.

Psychosis and Dreams

I thought that for Dan the drug abuse factor might have a couple of dimensions. First, drug-induced psychosis made up a part of Daniel's experience stored in memory. Dreams, which we assign to "the subconscious," also access memory. Dan's psychosis seemed to me to be dreamlike. I had been given the impression from Madaule's writing that Dan's *entire* experience prior to his listening intervention created memories *only* of "dyslexified" experience—what "normal" people consider at best a distorted perception of reality. However, all of us experience the psychotic state of consciousness in dreams; in that sense, all humans experience psychosis. So, I thought my best point of contact for understanding Daniel's perceptual world (in psychosis) might be through my awareness of my own dreams and about which I had learned through my education and my religion to become acutely introspective.

Dreams may be composed of a uselessly bizarre mélange of elements experienced in waking reality, or may produce a "psychologically significant" version of reality (the sort of thing Freud was digging for) often through symbolizing, or may convey images some consider "supernatural," like the dreams of Joseph affecting the early years of the life of Jesus. Following Dan's listening intervention, I thought, he had begun to build a store of *non-dyslexified*, normal memories. However, I imagined his newly awakened ability to introspect and analyze his thoughts and memories perhaps made it difficult and stressful for him to distinguish (among his memories) his two different modes of perception, especially since his brain was not yet consistently operating in one mode. He seemed to me to express thoughts and feelings the way I thought and felt while emerging from the dream state.

Dan's Insight

I had been amazed at how clearly Dan had written about his mental functioning in history class before his first full psychotic break. How had he been able to figure out: What is "real"? What is "real," but filtered through drugs? What is imagination dwelling on the variously warped and twisted perceptions of my dyslexic—or drugged-dyslexic—forms of perception? Did psychosis occur when the effort of putting all these sorts of awareness into a rational framework had become more stressful than he could continue to grapple with?[3] His artwork at The Listening Centre had shown a dark whirlpool of "The Confusion" that was comprised of large aspects of his dyslexified experience and included all the temptations and cravings associated with street drugs that had flourished because of it.

Like most of those who try to understand consciousness without understanding brain function, I was contemplating my own mental processes and trying to use that awareness to rationalize the forms of thinking I had observed in Dan. I had made the dream/psychosis equation fairly well, but I had made the dyslexified perceptual situation seem much more complex than I would discover it to be.

I wondered whether PCP, "magic" mushrooms, and other hallucinogenic drugs, including his medications, had damaged his former ability to withstand emotional stress, which then predisposed him to psychosis? This notion was based originally on my "ignorance" in connecting emotional trauma with psychosis, but it also was true that when Dan was psychotic, he was emotionally "labile," i.e., super-sensitive emotionally, so that he was likely to become more confused cognitively if something touched him emotionally (made him angry or jealous or happy) or stressed him physically (such as hunger, needing the bathroom, having a cold). The kindling into rage and other extreme displays of emotion, which are familiar to me in bipolar individuals I have known, Daniel sometimes displayed during his periods of mental instability with an accompanying intensification of his psychotic symptoms. I now realize I was seeing Dan display bipolarity with psychosis (bipolar I); but I did not recognize that as a phase of his illness because I had been taught by the text on schizophrenia to think of bipolarity with psychosis as a separate malady.

Since Daniel was not so mentally damaged that he had lost recognizable features of his personality, it was reasonable to ask what other psychological pressures he might be trying to deal with. For a start, following the intervention, every step he took towards actualizing the "new Dan" that was available to him simultaneously made him feel he was losing his essential peer relationships. For the dyslexic syndrome child, peer relationships may be stronger ties than any others. Peer relationships for a teenager are becoming dominant over family ties in any case. Furthermore, an impoverished teen is sure to have self-concept challenges.

During Dan's third psychotic episode (below) I felt he might have given himself permission to behave bizarrely so that he could find refuge in the hospital from his emotional conflicts. He had moaned to the hospital attendants calling him "Dan" and the babyish "Danny" at the time of his third admission to the hospital, "I have no name." That turned out to be a precursor to greater losses of identity; I would learn that his self-naming provided an indicator of the severity of his mental condition.

The Wastelands

In the second year of Dan's illness he had abandoned the idea of schooling and was constantly on the prowl for people who would supply and share drugs. Through every kind of weather in all seasons and around the clock I was prepared to follow his trail, finding him at various places within a 5–

10-mile radius and bringing him back home in various states of drugged or drunken insanity. Occasionally, one of his friends would call to alert me to his condition. One day my neighbour Wendy noticed Dan in another village and called me. I found him walking along a country road without a coat, gloves, or boots, and stoned, drunk. A spot on his face had frozen. Sometimes at Eric's reporting, I would set out and find him somewhere in town. Or at someone's house.

The young people in the drug culture came to know that Daniel's mental condition was more fragile than theirs (some of them) and that his parents were trying to save his life. This did not improve Dan's sense of manly independence, but it frequently interrupted a degenerative process, occasionally saved his life, and protected his brain to some degree. The drug culture had become extensive and pervasive in our part of Ontario. For young people, it was the affordable alternative to all the camping, travelling, entertainment venues, socializing scenes, and sports activities available to their more advantaged peers. That's stiff competition for parents short on resources. And we were about to lose our share in our highest trump, the cottage.

Daniel's Third Psychotic Break

Two or three times during the first couple of summers Richard and I went out with flashlights searching for Dan through the dozens of stoned and drunken bodies at Punkfest. During the second summer following Dan's diagnosis, a few weeks after Dan returned to friendships involving alcohol and drugs, he had another psychotic break and went back into the psychiatric ward at the general hospital again, with a police escort and by ambulance.

This was not a gentle slide like the second instance that had led us to take him to the hospital that returned him to the Ontario hospital. The family was present for "the break" and it was, briefly, frightening. Eric and Sarah locked themselves in their bedrooms while Richard called the police and paramedics, and I tried to reason with the cataract of irrationality spouting from Daniel in the upstairs hall, where he wanted to break into Eric's room. The explosion lasted a couple of minutes and was not so extravagantly expressed as his first psychotic break had been; then Daniel went and sat calmly, but frightened by what had happened, on the deck at the back of the house. Richard sat with him. Observing the scene, I was surprised to see how rational Daniel had become. Had it been a mistake to have called the police and to be sending him to the hospital? He was rational when the police came and talked with him, handcuffing him, but kindly. Dan went to the hospital by ambulance; I followed in the car. Dan was able to describe to me his conversation with the ambulance attendants. During the admitting process one of the nurses scolded me for not taking him to an even more distant city to an appointment with a psychiatrist. I wondered if she had had our experiences with psychiatrists (including those in her own hospital) she would have held the same opinion.

This time, I could see more clearly Dan's progression from the semi-rational state in which he arrived at the hospital to the homogeneously psychotic state he quickly achieved on large doses of Risperdal and other drugs before he became more uniformly rational a few weeks later. He was not removed to the Ontario hospital. I also noticed a number of peculiar practices on the ward. Here came a nurse with a drug cart, "Does anyone here need anything for pain?" These were not post-partum or post-op patients—these were psychiatric patients and drug addicts. Dan brightened at the offer and I told him bluntly he was not to accept any but prescribed medications, including painkillers; but I wasn't going to be there when the cart came around again, was I?

Another evening I heard a nurse reading questions to a recently admitted patient, of about Dan's age. She asked the questions, then rephrased them to presuppose the young man's answers: "Have you ever hurt a pet or an animal? You wouldn't hurt a pet or an animal, would you?" And his barely audible reply, "What was the question? No, of course not." And so on down her list. The questionnaire was meaningless—that item alone told me the questionnaire had no cross-cultural or rural sensitivity of any kind. Our perfectly sane neighbours regularly hunt, slaughter, and dispose of animals. And what are the chances of a spaced-out teen answering honestly that he has tossed cats out of barn lofts in tiny, inadequate parachutes? One former neighbour reportedly taught her children to shoot rather than drown unwanted kittens. I have decapitated and dressed chickens. As repellant as I find some behaviours, these are cultural issues, not matters of insanity. Also, the nurse's mode of interrogation would have rendered even a useful questionnaire meaningless. Why would the staff engage in such a charade?

Daniel was more mentally alert during this hospital stay and reported that he was exposed to a kind of "talking therapy" in the hospital that taught him I was responsible for his mental illness—a pseudo-Freudian theory our family physician declared to be a half-century out of date and without adequate foundation in the first instance. This approach by the hospital reflected the line of questioning to which Dr. G. had subjected me months before. I revised my original notion that Daniel was getting anything like "up-to-date" psychiatric care. The only reason I could see for the staff teaching vulnerable patients such false and damaging propaganda would be to drive a wedge between the patient and the parental home, which was still the primary point of daily care for the patient—a strategy for ego destruction and control and a livelier trade in outpatients. During his off-ward privileges towards the end of this hospital stay—Dan revealed months later—he had been able to exchange little items from the ward for pot in the hospital parking lot. As he left the hospital that time, the nurse said to him sweetly, "Now don't forget, Danny, we're here any time you need us." Great, I thought. Encourage him to think he can do drugs, go psychotic, and come here for three weeks' vacation whenever he feels like it. Her attitude was the final bit of code that allowed me to decipher that Daniel was no better off in medical custody during a psychotic episode than he would be at home.

Dan began to gain weight, one of the potential side effects of Risperdal. He gained one hundred pounds before he had tapered off the medication enough to regain his normal metabolism and his self-control in regard to eating. He was very upset with what he deemed his unattractiveness to young women and probably was motivated by that as much as by any other reason to try Alcoholics Anonymous (AA).

The Eighth Psychiatrist

Our family doctor again thought Dan should see a psychiatrist and after having sorted through the list of doctors he had met professionally, contacted another woman doctor for an appointment. Again, we drove to a city over two hours distant. Eric was with us because Richard needed to work rather than to son-sit. We arrived slightly before Dan's appointment and the doctor did not appear for another hour-and-a-half. Keeping Eric occupied in the sparsely appointed hospital waiting room was not a lot of fun for either of us. At length an apparition minced down the corridor in heels that would have challenged Tina Turner. Dramatically coiffed and elegantly clad in an exquisitely tailored suit, the woman's appearance was striking. I would learn 20 minutes later she was the doctor we had come to consult.

Dr. H. thanked us for coming such a distance, which mollified me somewhat, as I anxiously wondered what Eric might be up to alone in the waiting room. She glanced at Dan's file and noted his dyslexic syndrome and proceeded to tell us briefly what "dyslexic" means. Dr. H. then told Dan he should choose science and math courses in high school since he would find the study of literature difficult due to his lack of speed in reading. I was agape at this woman's ignorance of the problems my son presented—for example, Dan had been unable to learn the multiplication tables—and that she seemed to have no interest in discovering anything more about him than what might reside in her file folder. She inquired little from him about his experience. Not surprisingly, Dan sat there unimpressed. Dr. H. spent the rest of our 15-minute interview kindly lecturing Dan on his clothing, telling him earnestly that *her* teenaged children told her that the way *they* separated themselves from kids who didn't do well in school and took drugs was to dress nicely. She said she expected to see Dan again in a month and would be available to talk to him on the phone whenever he felt like calling her. Dan had no inclination to keep his second appointment or to call her on the phone when he was upset about something, although I had kept my opinion of her to myself and encouraged Dan several times to pick up the phone and see what she might contribute to his current dilemma.

Talking Therapy

When Dan returned from the hospital this time, and so heavily medicated that he was functioning at a very low level and required minimal care or supervision (i.e., he was incapable of running off somewhere and getting into some kind of trouble), I was able to resume editing university textbooks to ease somewhat Richard's financial burden. Some of these manuscripts proved very helpful in broadening my knowledge of Canadian society and how it was changing; the problems inherent in the neo-liberal policies that were eroding the social advances of the previous few decades; and the mounting ecological crisis. Getting back into academe, even peripherally, helped to keep me objective, analytical, hopeful, and seeking. I was making significant changes to some important publications, which was a satisfaction that buoyed me up in the face of what in our home sometimes felt like a losing battle.

As Dan began to reduce the amount of medication and adjusted to greater mental capacity I resumed a larger role in conversation with him about his lifestyle. The task of confronting, teaching, reassuring, and counselling Dan was a constant for years, for both Richard and me—although I bore the brunt of the input during the first few years of his illness while my husband kept us treading water financially. The cornerstone of our opposition to Daniel's self-destructive activity in the drug subculture was his importance to the Creator, who had a purpose for his life; his value to the family who loved him; and his innate ability to recover from whatever had happened or was happening in his life that was harming his well-being. Daniel knew he was struggling with the mountains of his addictions already, although I but dimly understood the metaphor. As I saw it, a battle for his imagination and future was being waged against the lures of the drug culture. Some of "the teachings of Don Juan" of Carlos Castaneda had an insidious grip on his imagination. Castaneda's books had been in Richard's library when we married. After I had read one, I wanted to destroy it. We talked about it and Richard thought I was being fanatical. I had not felt I had the right without his permission to obey my impulse. How very much I wished that I had been bolder. Dan took some kind of approbation for drug use from those books having been his father's. Associations can have a profound influence; ideas can warp and kill.

Alcoholics Anonymous

An aunt and uncle, who had been especially close to my parents, had a son addicted to alcohol and marijuana. They used to visit with Mum and Dad en route to relatives in Ottawa. After Dad died, they came to us instead. Not only did Stevie and Bill mourn with us our mutual losses, they gently and persistently shared their experiences of AA and what they had learned in their son's recovery process. Should Daniel's diagnosis be, not schizophrenic, but addict? Or both? I felt it was too soon to reject the psychiatrist's diagnosis entirely, but I began to ask whether the motivation that drove Daniel towards self-destruction was not just the wounding from the misunderstanding by others about his dyslexic syndrome, but also the neural pathways of craving created by addiction. Those readers familiar with drug and alcohol addiction will wonder why it took me so long to ask those questions. The answers are (1) sheer ignorance and (2) a tendency to trust the medical authorities, who hadn't raised those questions.

We had just gotten on-line with our editorial work. I turned to the Internet and began to educate myself about addictions. The location of my desk in the hallway outside Dan's room made this activity tricky, with his wanting to lean over my shoulder and luxuriate in information about the various plant and chemical sources of his terrible illness. I quickly shifted my inquiries to times when he could not look in on my educative process. A few people, most of them unknown to me by their real names, at an AA chat site helped me to understand Dan's symptoms in the context of their own experiences. I spent time most days for almost a year with those anonymous friends. They were very supportive of me, and helped me to understand how much of the recovery process Dan could take responsibility for and how best I might and might not help. I began to grasp the multiplicity of forces that drove Daniel: not only factors in his psychology, but his addictions that drove his physiology. The courtesy and kindness and good humour of those AA people helped me to understand myself better, too; I will always be grateful to them.

Through Daniel's uninhibited talk and behaviour I came to the opinion that his cravings for the substances he abused were entwined with his basic drives, such as thirst, hunger, and the needs for sexual attachment and companionship, that are said to "originate" in the brain stem. My friendship with M. had prompted me to explore what was known about obsessive-compulsive disorder, a disabling mental condition she suffered and that was responding to new neural-transmitter medications that, I was told, facilitated the neural messages between the frontal lobe and the brain stem. Her sharing about her suffering affirmed my observation that the compulsions under which Dan floundered also must be related to his brain-stem function not being controlled by the judgement centres of his frontal lobe. Two years later, I found a breakthrough discovery had established the role of the brain stem in addiction. It was summarized in *Harvard Magazine*;[4] that knowledge has since become commonplace.[5]

The *Harvard Magazine* article confirmed another of my observations: as the addict brings one addictive behaviour under control he or she tends to shift addictive behaviour to another arena of expression. Thus, the ex-alcoholic may switch to gambling; the ex-alcoholic ex-gambler may switch to pornography; the ex-alcoholic, ex-gambling ex-pornographer may switch to sport; and so on. Another advantage of AA, of course, is that the alcoholic or drug addict switches to an addiction to sobriety, which replaces harmful behaviour with behaviour useful both to the alcoholic and to others who want to stop drinking and/or other addictions. This information tempered my critical attitude to some of the behaviour of other AA people that so upset Dan. Eventually, it also would become pertinent to his struggles.

Dan was interested in the on-line AA group but could not type quickly enough to participate in on-line "meetings." To the degree that he tried, new hope grew in him that he might start to build a different kind of life. I used the Yellow Pages of the telephone book to find out when AA met in the village and found a woman there who graciously shared her experience and the wisdom she had gained the hard way. She explained the difference between closed and open meetings and arranged for me to attend the one she was missing by talking to me, although it usually was a closed meeting. I was encouraged to think Dan might find some kind of help with these people, although his diagnosis and hospitalizations were beyond her experience.

As Dan's girth expanded, he agreed to try AA. He couldn't drive and our rural location meant someone had to, so I did. Another motivation for going to AA, certainly, was Dan's desire to find enough sobriety to get a driver's licence. Since many of the meetings were open, I attended and soaked up messages of hope; for me the meetings were as helpful as going to church, or more so. When it became too evident that Dan was there with his mom, I started going to Al Anon meetings, which were poorly attended, run very differently, and evolving far away from their roots. The 12 Steps were not a lifeline for these women in the same way that they were for their addicted partners—possibly because society had become supportive of separation, divorce, and "trial" relationships for solving the problem of an addicted partner in a way the social milieu of the founders had not—the social pressures on the early Al Anon women to save their marriages were much greater—and New Age thinking was supplanting the Christian ideals of the wife of Bill W. and her companions.

During his first year with AA, Daniel made real progress in dealing with his addictions, especially alcohol. He worked at integrating the 12 Steps into his thinking and behaviour, although he did not get all the way through the steps to #12. He also became extremely controlling about food and lost more than the 100 pounds he had gained on Risperdal. At the same time, he encountered some very serious problems in the groups he was attending.

The first was that he was by far the youngest member in the meetings. In the closed meeting he was at least 25 or 30 years younger than the others. The one man who cared about Dan may have been mentally challenged, yet in many ways he showed himself staunchly faithful to the teachings of AA. In the larger, open meetings further away, the average age of the membership was 40-something, many were elderly, and Dan was 17. Of the two or three men he turned to, none was interested in "sponsoring" or in forming a friendship with Daniel. The one person who tried to sponsor Daniel lived at too great a distance to do the sort of travelling that would be involved in providing support outside of meetings. He came to the house once and in his genuine kindness provided the sort of companionship Dan longed for. But the limited financial resources of both men prevented that sort of event from happening again. This conundrum was, of course, exacerbated by Dan's dyslexified world in which peer relationships had become of supreme importance and by the fact that the men he related to were family men with those interests and responsibilities. Thus, the lack of anything like peers in AA was a huge stumbling block for Dan.

The second problem with the local AA was partially cultural, partially a failure of the group to move forward in their own programs of recovery. Early in his association with AA Dan attended a weekend retreat. The men's sharing of sexual exploits that demeaned women shocked and alienated Dan, who was not entirely sexually inexperienced, but was idealistic and reticent by comparison. He came away from the weekend disgusted and disappointed in the people to whom he had looked for guidance and inspiration. There was no one he wanted to choose as a sponsor to confide in, although he tried several times to make that kind of contact.

The third problem had to do with the local culture, also, and to the impoverishment of learning and the anti-intellectuality of the men he knew. Dan might not be continuing with formal education, but his superior intelligence was soaking up learning in his environment and he still wanted very much to find his way back into formal learning. He found no support for these hopes among his contacts. The one younger man I hoped might mentor him shortly returned to college. There were no others.

A further problem arose from the way the membership was handling the combination of alcohol and drug addiction that younger people might have been bringing to AA. The old-timers didn't know much about drugs and ignored the fact that AAs everywhere (as I knew from my on-line contacts) were adapting to the double addictions of younger people. The many older people who simply were alcoholics had been sober for a long time. The patterns of their thinking and their interests were not attuned to the newcomers. During one meeting, a man in his thirties stood up and objected vehemently to how the meetings were being run. Although I didn't know all that lay behind his objections, it was evident that the members tended to be "cliquish." Once Dan had achieved a high level of control over alcohol, little of the verbalizing at meetings related specifically to his addiction to drugs or to the settings in which he did drugs or to the people with whom he did drugs—and had nothing whatever to contribute to his need for sober peer relationships.

Among the men who considered themselves part of AA but who rarely went to meetings was Mike, a computer genius with health problems that he had learned to manage without alcohol. Dan related to Mike in some ways despite their age differences and decided that Mike was able to maintain his sobriety because he had a lovely wife, Sherrie. I became friends with Sherrie and could see why Dan would think so. He also was attracted to the way Mike had integrated his beliefs and sobriety and seemed to be able to maintain them without much attendance at meetings. One day I heard Mike out in our yard with another AA man, barking orders at Dan like drill sergeants while Dan trundled wheelbarrow loads of earth and rock past the big locust tree to the edge of the east field. When I stepped outside to investigate what sounded like abusive treatment, Mike explained that Dan had asked them to treat him as though he were in the army so that he could get over his phobia against working. Dan had found his own method of reversing his childhood declaration of rebellion and somehow had engineered that process to occur on virtually the same spot on earth! Even during several further episodes of psychosis, Daniel was cheerfully responsive to requests for help with physical labour and would happily choose such tasks when he was capable of doing them.

What alternative for support might exist for Dan? No Narcotics Anonymous meetings were held within reasonable distance of our home and, for Dan, NA would have put him into close contact with peers who were users of even more dangerous street drugs than the ones he was abusing. As Dan was sliding into his most recent episode of psychosis one of the things he cried out was, "And if I go to NA I'll meet people who do worse drugs than I do and I'll be dead of a heroin overdose at age 40!" He knew himself that well, and struggled still to maintain certain barriers in his choices. The travel distances, our lack of money for gas, and the needs of other family members made getting to several meetings a week stressful for the whole family. Nevertheless, Dan and I made the AA trips several times a week for a year at first and for spates of months at other times. Richard occasionally stepped in to drive or to attend meetings. AA people in our area had very little money, for the most part. Only very rarely did anyone volunteer to pick Dan up or to drive him home, and I understood why. The sponsor who drove from the city to spend time with Dan had made an extremely generous gesture, which Dan probably lacked the adult experience to appreciate.

The 12 Steps program is spiritual; that is, it engages the belief core of rationality and experience. The brain-stem activity tied to compulsions is counteracted through AA teaching both by the reasoning and judgement and control capacity of the frontal lobe in the left hemisphere of the brain and by the more generalized capacity for God-experience. Where the principles of AA are fully practised, they draw notions of the Divine—the belief system—into self-control over addictions. That aspect of AA was something to which Dan did relate. The glimpse we had of AA functioning "by the Big Book" came during a trip we made to the Midwest to visit Luke and Kathleen and Sarah. Dan wanted to go to AA one night en route so we found a meeting at a fire hall in the Wisconsin town where we were spending the night. The reception of those people was so warm, the resources of mind and spirit they shared so marvellous, that the experience turned Dan away entirely from the groups he knew back home. He could not imagine himself, the youngest person in attendance at local meetings, becoming a force within AA to bring about that sort of change. Frankly, I couldn't, either.

Dan and I had explored the provincial and community programs for support in the city nearest to us and investigated independent living arrangements for him there. The prospects were bleak. In any case, he had his next psychotic break before his name came up on the long waiting lists for that sort of assistance.

Dan was the one of our children who claimed to be a Christian and despite the agony and failure in his addictive behaviour, his mindset was evident—perhaps not more so than in the forgiveness he offered endlessly to the members of his own family. Not only when he was well, but when he slid into trouble mentally he continued to try to treat people in terms that expressed his Christian belief. In deep psychosis, he did not have control of his verbalizing or expressions of mood, however. Peers were as absent from local churches as from local AA meetings. One factor probably had to do with the overall fall in church attendance and the diminishing of church programs for young people throughout the country. Another was that the trials and tribulations of most young people had not become great enough for them to be seeking for answers beyond what they could provide for themselves. The churches were no better equipped than I had been to deal with the massive shift in social problems that had arisen, including and especially the new challenges posed by drugs. The memberships of both kinds of social group—AA and the churches—were too ingrown and exclusive and aging to offer young people camaraderie or insight into their problems.

The See-Saw

Over the next few years, Daniel would become stable on medication, more or less stop using alcohol and pot, be weaned off the medication, and then perhaps return to AA. We boarded one AA friend one winter and supported that man's return to work, which gave Dan support for sobriety and gave me valuable insights. After he left, Dan returned to substance abuse, developed psychotic symptoms, and climbed "on the wagon" of sobriety again. A serious problem for Daniel was that his motivation towards sobriety was so often via the incapacity brought about by psychosis. A psychotic episode is a profoundly ambivalent form of reinforcement, as I will show in Chapter 14. Depending on how one defines Dan's wavering mental condition, this happened five or six times in a major way and several other times when he came to the brink of psychosis and pulled himself back by going off alcohol or pot or both for a while or by going back on medication. Each psychotic episode included dreadful losses—physical, emotional, or mental; yet, with each episode he learned some new and important lesson.

During the entire summer we were refurbishing the cottage to prepare it for sale he vomited at least once a day, but usually several times during the day, as a side effect of his medication. Many people would have found that experience in itself sufficient motivation to get off and stay off the substances that necessitated medication. This kind of sickness did not in itself prevent Dan from returning to alcohol abuse, but combined with other factors it reduced the attractiveness of the psychotic episode. In his third year of recovery Dan told me he had figured out that alcohol could cause psychosis fastest, but that he was more profoundly addicted to marijuana. As other unhappy experiences negatively reinforced his damaging behaviour, he gradually assumed more and better control over his drinking, stopping for intervals that became longer and longer until these could be measured in months and years.

After one spring and summer on the couch semi-conscious, Dan decided he hated being insane and that during the intervals when he had to go back on medication he lost something he valued. At last he saw himself as "like a vegetable," and perhaps realized that if this happened much more often he would be profoundly schizophrenic or not have a brain left at all. Or perhaps he would die of the side effects of a medication or of a street drug. For a while that awareness motivated his attendance at AA. Of course, the fact that Dan was insightful and was trying to reorganize his life after each of these episodes was a fair indication of misdiagnosis—at least in terms of the definitions of schizophrenia available to me at that time. On the other hand, Dan never stayed off pot, alcohol, or medication long enough to fit the criterion of "no psychotic breaks for a year substance- and medication-free," which also left his diagnosis unresolved.

Lower Dosages of Risperdal

During the first couple of years (1997–9) of Dan's episodes of mental illness he was seen frequently by Dr. A. and all of his medication was prescribed in a more-or-less conventional way. Dr. A. seemed to think that the levels of Risperdal Dan was getting in hospital were excessive because he weaned him to around the 2-milligram level shortly after Dan was released from hospital on each occasion. The dosages of Risperdal in the hospital were at least 6 mg, and he apparently had been given several other drugs that were discontinued after he was discharged.

I started to notice the changes in Daniel's behaviour during the drug phasing-down process. He displayed incremental changes in certain kinds of behaviour: his interest in food, his level of alertness, his manual dexterity and co-ordination, his ability to remember, his ability to focus on a task, his capacity for sociability, the complexity of his language. I've lost count of how many times this occurred because his intermittent or continuing substance abuse had the effect several times of obliterating his improvement on reduced amounts of medication and muddying the waters for observing the effects of his prescription mind-altering substance. It was felt necessary by the doctor—and capitulated to by us—that his medication level be raised again to deal with the reappearance of lesser psychotic symptoms.

My initial obedience to the doctors' suggestions about dosages sometimes resulted in Dan's being destabilized for weeks or months. After about a year and a half of his see-sawing between street and prescription drugs I noticed that Daniel's behaviour and cognition improved with as small a *decrease* as 0.25 mg of Risperdal. This perception set me on a long, slow process of trial and error—trying to give smaller dosages of medication when he became psychotic and trying to get him off the medication at the fastest rate that would not destabilize him. "Stepping down" sometimes took weeks or months. I was beginning to notice how Dan reacted to the medication when he went back on it, when he was in the process of weaning himself off it, and what he was like in the intervals before he began using pot (and/or other substances) again, and then how his behaviour changed when he started using pot, etc., in the run-up to another episode of mental illness. I did not at first realize how long the titration period (time from the initiation of medication until it reaches its full effect) for his medication should be. Risperdal was still being tested in those years and the medical authorities didn't really know, either. Daniel's level of co-operation, in terms of his abstaining from substance abuse, varied tremendously, so the process was sometimes painfully frustrating. I poured my anguish into my journal and the ears of a few friends and kept to the path that was teaching me something about his well-being.

At some point in 1999 I explained to Dr. A. that I was observing positive changes in Dan's behaviour in response to the reduction of as little as 0.25 milligram of Risperdal. The first few times I pointed this out, the doctor more or less ignored me. As I proceeded with the protocol I describe above, and could show that Dan improved cognitively on less medication, the doctor paid attention. When Dan required medication again, the doctor prescribed less of the drug, which proved to be as effective as higher doses of the drug had been in alleviating his symptoms. In other words, Dan recovered as well—or better—from psychotic symptoms on very small amounts of Risperdal as he had on higher dosages. By then, I had also learned to wait for the "titration periods" that could last from three to six weeks before Dan's behaviour would become consistent at that functional level.

During one phase, Richard primarily was relating to Dr. A. with Dan, on the assumption that Dan's self-esteem would fare better with his father in attendance than with me. Both Dr. A. and Richard knew I would object to Dan's going back on a higher level of Risperdal than had been effective earlier; they even talked about it, as Richard reported when he returned home. Nonetheless, together they had agreed to a 2-milligram level of dosage. Dan did no better on the higher level, eventually phased down from it, eventually recovered his sanity, and eventually returned to substance abuse—although he was learning to control aspects of his addictions by then.

About a year after this incident, I read a news item on the Internet that two doctors in Vancouver had published their finding that psychotic patients should on no condition be started on more than one-quarter of a milligram of Risperdal and that the drug should be titrated for at least 6–8 weeks before being increased, if necessary; and that a maximum dosage in the most severe cases of psychosis should be considered 2 milligrams.[6] Dan was surprisingly co-operative about the lower dosages once he had been weaned to that level. But, he nevertheless was walking around drugged 24 hours a day, seven days a week. From that time, I took a very hard line on the minimal doses of Risperdal that had proven adequate for Daniel's recovery process. At least one of us had scaled another mountain.

Sarah Starts to Sing

When Sarah no longer had the ability to focus mentally on her school work during the winter of 1995, I had found books that might amuse her and hold her interest. Limited television reception offered little to teach or entertain, but she could watch movies. When it became clear that she could no longer continue in high school, I thought I might be able to create a home learning environment of some sort. I scraped up the money to add a few art books to our library, most of which was too challenging for a young person whose mind felt perpetually foggy and whose ability to remember from one moment to the next had become impaired.

I was playing the piano and singing when Sarah joined in. I was struck by her ability to reach notes I could not and, as I asked her to try a few scales, we discovered she had an amazing coloratura soprano range. Her interest in classical and operatic music grew as she played records and sang alongwith them. She found vocal exercises among Mum's sheet music and started to teach herself to sing. We did not realize she was doing the best thing possible to counteract the damaging effects of the virus that initially had harmed her, and the virus that had injured her more seriously two years before. We found a teacher for her about an hour's drive away and she began singing lessons, building on and correcting her own first efforts. Some weeks she was not well enough for the long drive and her lesson, but she pursued that activity for most of a year. Soon, I could no longer tell whether I was hearing Sarah or Maria Callas or some other diva. She had a prodigious talent. When she prepared to sing for the Christmas Eve service, it was her first community involvement in three years. She sang beautifully. That evening was a giant step into hope for Sarah.

In the second winter of her illness Sarah had planned a garden and ordered seeds; in February she started plants indoors. By spring she was able to spend short intervals out of doors. With some help in the digging, she planted a beautiful garden. While not healthy, she certainly was not as dreadfully ill as she had been. Not energetic enough to study in a consistent way, she nevertheless was reading quite a lot. She kept assessing her resources and testing their limits in an ongoing effort to reclaim her health. As Sarah regained a degree of health, her influence on her brothers expanded again. She made artworks for them—paintings in watercolour and paper sculptures. They noticed her reading. They helped her to move from the small corner bedroom to the large room over the kitchen that she turned into a bed-sitting room. She explored oil painting and clay sculpture, and work in other media. Working from a couple of my theatrical costume books, she designed dresses and sewed them.

Sarah Goes to The Listening Centre

For a year following Daniel's treatment at The Listening Centre, Sarah continued in wavering states of health and strength. As she contemplated the obvious changes in my health, pondered the changes in Dan, and evaluated her father's doubts, she came to a decision in 1998 to try a few days of listening retraining at The Listening Centre. *She lost all of her chronic fatigue syndrome symptoms in two days*, although she continued for the five days of the intervention that had been proposed. It took her about six months to regain the full use of muscles that had not functioned properly for at least four years. She had learned to be "in touch" with her body in such a way that she did not push herself past her limits, but built up her strength gradually. On the other hand, Sarah had no idea of what she needed to do to *maintain* her healing, or even that a maintenance program would be essential to her health.

During the intervention Sarah had wanted me to go home and leave her in Toronto with friends while she finished her five-day course of treatment. Was that a mistake? I think now that it was, because she was very vulnerable to advice, very suggestible. A teen normally develops independence gradually through increasing autonomy in high school activities. Sarah had been away from that socializing environment for two years. When she returned home after the listening intervention she quickly became extremely aloof. An unnatural silence came between us with her growing rejection.

Even worse, she stopped singing. A long time would pass before I came to understand why a much broader program of teaching and support should be available for a young person making the sort of mental adjustments that are involved in the listening effect—especially for someone following a long illness that interrupts social development; or for someone like Dan who has never before known full left-hemispheric dominance and is skirting quicksand in his peer relationships. Ironically, the very means of their healing would separate them from the inclination to recover that methodology (because I upheld it and they naturally were trying to grow separate from me) when they again needed it.

Eric and Luke and Kathleen and Sarah

All of our sons, as December babies, were among the youngest in their grades starting school, a situation created by our inflexible educational system that creates disadvantage in many ways, especially for males. Eric also had started high school at a disadvantage. The previous year, in Grade 8, the innovative principal at the junior school had set the bar lower for him and offered him rewards, such as being allowed to go on the Quebec ski trip, if he would meet those standards. By the end of that year, a young man with a great deal of insecurity was making strides.

When September came, there was no money to buy the new clothes that most young people would be wearing to school for the first couple of days at least. I had to explain to Eric that we simply had no money at all, but that when our next paycheque came a few days later, some new jeans and shirts for him were the first priority; I knew he was hurt by that situation. Eric had left for school the first day, nevertheless, full of excitement about being a high school student, and with high hopes. We learned a few days later that after lunch on that day (when his self-esteem was almost as fragile as it had been nine years before when he was starting kindergarten and was abused by his teacher) the French teacher yelled at him before he even got into class—because he was carrying a can of pop? because his clothes were shabby?—and told him he didn't belong in her class, to go to the guidance office and change his program. He was shocked and devastated. Already slipping towards the drug culture, his tenuous attitude to school collapsed.

A new principal in the school must have decided that the way to improve the standing of one of the least prestigious schools in the province was simply to eliminate any student who was struggling. Our contact with some other families over the following months confirmed our suspicions. Repeated meetings with the principal and new vice-principal did nothing to enhance our opinion of the administration and of a school with very little in the way of supportive academic programs for students who might for various reasons be having difficulties adapting to high school. By January we were partially home-schooling Eric again in an effort to keep him away from the same slippery slope Daniel had taken.

That fall Luke and Kathleen had returned to Canada with the intention of staying. They wanted to see if they could live at the Muskoka cottage through the four seasons and moved there. When Eric was set up for partial home-schooling with assignments from high school teachers, Luke and Kathleen offered to take him with them to the cottage for a few weeks. He made significant progress with his school work. He loved being taught by Kathleen, and Luke saw him through withdrawal from pot. Eric completed the assignments given by his English teacher and turned them in when he returned from Muskoka. When phone messages to the teacher at the high school were not answered for several weeks, I drove to the school only to learn the teacher had died of cancer. We were told subsequently that Eric had no English marks, and that the assignments he had been given covered only half the course. What happened next to Eric had mostly to do with failures of the school system, teachers, and principals who exacerbated his difficulties. He finished all of his geography assignments and Richard personally carried them to the high school and quite literally put them into the hands of the head of the department. Later, when Eric did not receive his mark and we asked why, the school "could find no record of his work."

Eric had returned to an inhospitable school and a surrounding community with a rich supply of marijuana. In fact, this part of Ontario was fast becoming a premier growing locale and some people in the area were making tidy incomes while others were paying the price for their entrepreneurial skills in jail. We became aware of a battle that was tearing families apart, destroying sanity, killing aspiration, and widening the gaps between the various social groups in the community. Alcohol was an ancient and accustomed problem; drugs were not. Because of their illegality and the nature of their long-lasting effect on the brain, social traditions and strategies had not yet developed for coping with the effects of drugs.

Luke was asked to do a small building project for C.'s Muskoka cottage, which he accomplished to the satisfaction of his clients, and this raised the prospect of further work from them. In December, Luke and Kathleen found the cottage was insufficiently insulated to be habitable and they came back (with their two dogs) to live in Lainey's house in the partially finished basement living space. (The upstairs was inhabited by a tenant.) Luke tried to get Eric started on building a cabin at the top of the hill but even when Luke worked with him, his younger brother was reluctant. He was torn between the non-work ethic of the subculture sucking him in and by the enthusiasm of his older brother to be creative with logs. The cabin-building became very cold work as January got its icy grip on the hill, and was set aside.

Luke came to me one day in February and said that he thought he and Kathleen and Sarah could renovate our dreary old kitchen—except for appliances—for about $500, if we could find that amount. I begged for a day to think about the proposal. Over the 15 years of tending linoleum that peeled in strips as I cleaned it, a ceiling that crumbled a little more with each yearly cleaning (and sometimes spontaneously in between), wainscotting that blistered 100-years' layering of paint when wiped, and open shelving that I had built in desperation when we'd moved in, I had composed some lovely images of the Italianate kitchen I would someday, somehow afford. I set those dreams aside; I knew the older children had made me an offer I could not refuse.

Since the project would be spread over an estimated six weeks and I also had an editing job, we thought we could eke out the money. Daniel was heavily medicated and recovering at home from his fourth psychotic break and was too fragile to be in a house with that sort of activity going on, so I took him to the basement room at Lainey's house during the day and played movies that he could barely follow while I perched on an armless chair and edited.

The older children gutted the kitchen in a day. The first night I returned to the house to discover with a shock how comprehensive their plans were. The worn-out appliances were in the yard. The plaster ceiling had been taken down and bagged. The old, chipped porcelain sink had been taken out and the newer sink—in the cabinet Dad and I had put together 15 years before—likewise was gone. The plywood flooring with its barely discernible pattern of linoleum was stacked in chunks on the lawn. As the days passed in a dream for Dan, Luke and Kathleen and Sarah performed miracles and even Eric helped and had some brilliant suggestions. The floor was reinforced in the basement with logs, and cement was laid to flatten the wavy old pine floor to provide a foundation for ceramic tiles.

Luke's intention was primarily to recycle materials from the old barn, which he did in a number of ways. Boards from the barn's exterior were used to build the cabinetry, which Sarah and Kathleen washed, sanded, scrubbed with wire brushes, and then painted with clear varnish. This was an arduous undertaking and proved the efficacy of Sarah's healing from chronic fatigue syndrome. Sarah found a set of handles in the barn, sanded them until they resembled pewter, and later discovered there were enough of them to put matching handles on all the cabinet doors. The ceiling was finished in cleaned (but not sanded or brushed) barnboard and Eric had the idea of adding corbels in the corners as finishing touches. Two huge cedar planks from one of the barn stalls were planed and sanded to become a pass-through windowsill between the kitchen and enclosed front porch, and a bench under a new window. Other cedar planks from the barn stalls were planed and joined to form a harvest table, with a large flat stone set in the middle to accept pans hot from the oven; the table legs were sections of a timber found in the hayloft—the one Daniel had nearly dropped on Eric.

The new window was one Luke had found left over from one of my father's building projects, and once he'd installed it Sarah remembered some pieces of "gingerbread" trim from the old porch that she had seen stored in the driveshed. From two triangular cut-outs she fashioned a Gothic arch over the window. In the barn, she also found an old oak medicine cabinet a neighbour had left. After ensuring that the neighbour had no further use for it, Sarah sanded it, made little shelves, and put new brass handles on the doors that she glazed and repaired. She had produced a charming spice cupboard for a kitchen blessed with an herb garden and a mother who enjoyed using herbs and spices liberally in her cooking.

Kathleen repainted the old walls a deep cream and the ancient wainscotting a cheerful apricot. The look of Mediterranean adobe complemented the warm woods. Luke was pondering what he might do for a countertop when he looked down at my father's workbench of solid birch, which he had inherited, and decided that it was too fine for its purpose. He built himself a cheaper workbench while he turned the birch bench into a countertop, which he sanded and oiled until it glowed like satin. The original front door with twin lights was sanded and installed with barnboard trim around the door frame. For the outside door of the enclosed porch Luke built a new "Dutch" door with a small window in the upper section from thick pine planks he finished in a dark stain.

As the kitchen project drew near to completion, I rolled slabs of clay and asked each family member to imprint the tiles with found objects. These were fired in our kilns. Sarah glazed them in greens, blues, white, and beige, fired them again, and built the sink countertop from them.

When all the bills were tallied, Luke's guesstimate had run over by less than $100. Richard's mother heard of the project and sent $1,500 for the floor tiles. By the time we were ready to buy tiles, Luke and Kathleen were back in Muskoka, building an addition to C.'s cottage. Our friends Ian and Rachel came for a weekend, as Ian had some expertise with tiles. He knew exactly what was needed and had come with some useful tools. With help from Eric and his friend Andy, the job was finished under Ian's direction over two days. By then, Dan was well enough to lay some tiles.

Over the next year, since I was working again, we were able to replace the tiny RV-size fridge—which had "made do" for almost a year for a family of seven—with a regular fridge, to replace the rusty old range with a new one, and to add a dishwasher. A year later, we had the chimney repaired and installed a new wood stove. During the next four stressful years, the children's labour-of-love kitchen was the constant reminder of their creative genius and their ability to work together. It kept me hopeful for the restoration of sanity and for the healing of hearts in our family.

North and west five hours distant by car, Luke and Kathleen renovated our cottage to prepare it for sale—a task done splendidly, but in sorrow. All of the children were deeply attached to that place somewhat separated from the ups and downs of their home, woven with happy memories and linked to history through my father and his family's pioneering in that location. Lainey and I simply could not maintain that property physically or financially. Luke was hoping to work in the area, but finished the renovation too late in the season to find employment. Kathleen had her sights set on art school in Wisconsin. So Luke and Kathleen said goodbye to the cottage and to the family and headed west. Lainey and I sold the family cottage and took back a mortgage, which gave us a small but important quarterly source of cash. She bought out my share of our parents' home, which allowed us to pay the debts accumulated from our ongoing support of an ailing family.

Following the kitchen renovation, Sarah and Kathleen had used my ceramics studio in the enclosed front porch to make pottery and to weave baskets from grape vines. Kathleen, with two years of college to her credit, had been exploring the options for her art education. The result of their sharing of art and ideas was tht Sarah decided that she could study in the U.S. and perhaps do something similar. (Both she and Luke, having been born in Arkansas, could exercise their U.S. citizenship.) A few months after Luke and Kathleen had established themselves in Wisconsin, Sarah followed them. The huge duffel bag she shouldered, along with a hefty suitcase and other parcels, as she trudged unsteadily into Union Station in Toronto, seemed a perfect symbol of the unresolved burdens she was carrying into her new life. It was time for her to move on, but surely not in that frame of mind. My cheeriness hid a heavy heart as we said goodbye.

Luke and Kathleen sent us news of Sarah. Luke helped her to find work. We heard that she had a boyfriend and had taken an apartment across the street from them and had decorated it beautifully. She studied successfully for the first time since she had become ill, and earned her American high school equivalency.

Eric

Eric had lost all of his sane and sober siblings. He was more and more often missing school or in some kind of trouble in a school that suddenly had become much "stricter." At 13, he, too, began to detach from the family. A neighbour girl, also in serious emotional trouble, became his constant friend and they aided and abetted one another with pot and other aspects of the pot subculture. When her family moved to the city a couple of years later, he would hitchhike to see her. Winter made that increasingly difficult and he shifted to spending most of his time with his desk-mate from Grade 7. For months, Eric slept at home only once or twice a week, yet to his father he remarked that he continued to think of this place as home and us as family. It didn't look like it to us; we felt as though we had lost all of our children. The nest had not so much been flown, as plundered.

Eric adopted a husky-lab puppy with one brown and one blue eye, which he named Tequila; he loved her and she made a significant difference in his depressed outlook. I got to know the school friend and his mother a little better and came to see that the woman's essential kindliness towards Eric perhaps had given him a better option in his dark night of despair than the one Dan had found, although marijuana remained in the picture.

Eric was too young to leave the school system officially and while the administration was happy to see him go, some kind of home-schooling arrangement had to be maintained. His school work with Luke and Kathleen had gone very well; but by the end of the school year a lack of support and co-operation from the school had undermined Eric's hope in that sort of arrangement. In September, a social worker was assigned to Eric to check on his home-schooling and she had some wonderful suggestions. She revived Eric's interest in earning school credits. The school system did have a program of off-grounds co-operative work in the community. Some students were working at gas stations and others were clerking in stores, working on farms, assisting in an art studio, and so on. Since the school's facilities in business and the industrial arts were extremely limited, these non-paying jobs gave the teens some idea of the sorts of work available to them in the community if and when they were graduated and developed some skills.

As the social worker got to know Eric and interviewed the family at home, her attention was drawn to the small log cabin Luke had helped Eric to start at the top of the hill. She asked how much time he had put into the work, how much was contemplated, and then explained how the co-op system worked. We made those calculations and Luke estimated how much more time would be required to finish it. Luke, as a local contractor, was Eric's work supervisor and we arranged for him to follow his brother's progress even though he and Kathleen were primarily at the cottage finishing the renovations there. The social worker's instructions to us were meticulous, including our keeping copies of her signed reports in case the school "lost" these documents as well. A year later, I found her concern to be precisely warranted and had to make special inquiries and keep up considerable pressure for Eric to obtain his well-deserved credit for over a hundred and fifty hours' work on the cabin. In the process, I pressed for the geography mark and someone "discovered" that grade, also. However, by then irreparable damage had been done to Eric's relationship with formal education.

After Eric had fulfilled the requirements for his co-op assignment, the family (mostly Richard) helped him to finish the hand-numbing job of cutting wire mesh for the remaining chinking. Eric had completed a local landmark: the cabin sits on the top of the highest hill for miles around. It immediately provided a place of escape, not always of the right kind, but a place that was his alone.

A Family Reunion

Luke and Kathleen had moved to Wisconsin and made friends, and Kathleen was studying while Luke worked. They were ready to formalize their partnership. We could not attend Luke and Kathleen's wedding, but Richard assured them right away that we would come to see them as soon as Dan became well enough to travel. A few weeks later, we set out and reaped the wisdom of Richard's decision. Our drive across the top of Lake Michigan was a welcome relief from most of the kinds of trouble we had been dealing with for such a long time, even though Dan and Eric were present with all of their issues. Dunes that swept and towered and shape-shifted for a day of our journey soothed my soul and stirred my imagination. Although Dan came away from the Wisconsin AA meeting depressed, I took strength and insight from the way those people shared their experiences that lasted for a long, long time.

Luke and Kathleen welcomed us warmly. We were uncertain whether Sarah would see us at all. The second day she appeared, but barely spoke to us, and would not invite us to her apartment or introduce us to her boyfriend. From her manner, the few fragments of conversation, and her artwork, it was not difficult to see that she held us, particularly me, responsible for the shattering of her former personality. She welcomed her brothers, however. Eric's new puppy Tequila was an ice-breaker and as Luke and Kathleen enthusiastically toured the family around local sites, the relationships among the siblings were reaffirmed. During the time we were together, Sarah allowed me once or twice to reach out to her.

I had the feeling that Luke and Kathleen soon would be starting a family, and, indeed, that joyous news came shortly after our return home. Although he had registered in university, with the prospect of a family to support, Luke abandoned his ideas of going further in school.

The trip for Eric was both joy and agony, for he would have given anything to stay with Luke and Kathleen. After I had shed tears off and on for a couple of days—which must have been terribly hard on Eric, who was sharing the back seat with me and his dog—I wrote a sequence of poems for Sarah that brought me a sense of peace and revived my hope for hers. Eric, who wore his long, fair hair in "dreads" of which he was extremely protective, allowed me to untangle some of them and braid them in a gesture of his sympathy for my grief and mine for his. Thus, a new bond grew between Eric and me. In the following weeks he began to treat me with more respect and every so often would play on his guitar the Credence Clearwater Revival song that he knew I liked, "Have you ever seen the rain?"

Listening Retraining Alternatives

My healing at The Listening Centre (February 1997) proved to be vulnerable to viral infections. About a year afterward, I caught a virus that caused intense dizziness, so I knew it was affecting my inner ear. My CFS symptoms came back and there was no money to undertake "listening retraining," so I just plugged along with CFS symptoms for another year. I thought I might start to sing and hum on my own, wishing I knew how to sing Gregorian chant. About six weeks after I had made this undisciplined effort, I was watching TV one evening (winter 1999) and an ad struck me so funny that I burst into laughter. In that instant, the same sort of "click" in my brain that I had experienced at The Listening Centre was accompanied by the immediate return of my normal level of energy. This time, I realized that the reawakening of my stapedius muscle with laughter must have ensured the firing of neurons and the reconnection of a synapse, completing the chemical mechanism of healing between the middle and inner ear and the level of neurotransmitters in the brain. The benefits of laughter to healing were famously explored by Norman Cousins[7] and a branch of yoga emphasizes laughter. The connection between immune system function and laughter in Cousins's experience and through laughter yoga probably can be explained more specifically as reactivating the muscles of the middle ear in transmitting sound energy through the vestibular–cochlear system to the brain.

After my "laughter healing" I remained free of CFS symptoms for longer, about two years. By 2001, when we first realized that Richard's mother had moderately advanced Alzheimer's disease and I started to look after her in Louisiana, I had become unusually tired. Again, I caught a virus and was extremely dizzy for several days. CFS had struck again and what could I do? Jean, who also was recovering from a broken ankle, needed my constant attention. I was helping my husband to assume all of her business matters, some of which were in disarray. I was continuing to edit via the computer

I'd brought along. Richard had to return to Canada. Since my care-giving would be remunerated, I could afford the Tomatis Method of treatment, but not conceivably the time to spend a week in Toronto away from my responsibilities. So, I purchased headphones and some Mozart CDs, and listened to music on my computer when I could, during the evenings or when Jean was resting. This inner-ear exercise would give me spurts of energy, but not the sustained healing that I had experienced twice before. Clearly, the Electronic Ear® and the Tomatis Method provided a more refined treatment than I could devise. Nevertheless, the focused listening allowed me to carry a considerable workload.

A Change of Season

I am not sure when I dropped all expectations of and efforts towards my children returning to "mainstream" life and just concentrated on loving them as well as I could wherever they found themselves. Eric and Dan, despite their troubles, were treating us and each other more kindly. The gross destruction of property had stopped; the stealing and loss of valuables (guitars, amplifiers, CD players, clothing, etc.) that had been given to the children as gifts diminished, although we knew that any pocket change or allowance they received was going straight to our neighbour, the master of Punkfest.

That Daniel was growing in insight was obvious from some of the things he typed into our old computer that had been passed on to him. He wrote creatively: a story for children. He wrote to Luke, who was in college, about his own academic aspirations, his spiritual life, his disappointment in the loss of the family cottage. He also wrote for himself and expressed his thinking about his creativity and emotional life in these ways:

> The creativity part of me is buried deep in the middle of my mind and trying to get that information and that creativity and that intelligence that I can't penetrate and all of the nerve pathways around that part trying to get out the useful information [I] keep hitting the nonsense and the crap that isn't me and running into little clicks and popping sounds and things that I don't mean to think or say.
>
> . . .
>
> The thoughts from my childhood come and go at random. Most of memories start when I started school. But one memory comes to mind from before school; it's kind of funny because it has to do with school.
>
> My brother and sister were starting school and [I] had just gotten this stuffed animal of a duck from my grandfather, that is, my dad's

> dad. I had spent my entire childhood to that point loving and looking up to the two of them so that when they left I kicked and screamed for them to stay and then I sat in the bookcase in the living room squeezing and crying into my little ducky, while listening to my favourite record of the ugly little duckling. I remember my mom coming to comfort me but making her leave and feeling rather satisfied in my self-pity that she had left me to my crying.
> (*early summer, 2000, a few spelling mistakes have been normalized*)

As I worried about Eric following the same route as Daniel, I began to appreciate their differences. While most often I had felt it was important to go out to find Daniel, I now noticed that Eric almost always found his own way home. I learned to turn from the losses of the past to the possibilities for the future. Perhaps Eric became hopeful by sensing my hope or by no longer hearing my anxiety expressed, however subtly. We began to see that the boys' behaviour was improving. Ironically, Eric had entirely stopped stealing shortly before he was arrested and tried for possession of stolen property and of possession of pot with intent to traffic. He was acquitted of both charges. We were not alone in the porticos of the law. Standing with Eric in the halls of justice meant, mostly, sitting in the waiting room outside the court with the parents of many other local children we knew, most of them with parents who struggled to make a living (and not always on the sunny side of the law), and sometimes with parents who had addictions to alcohol, pot, or much more damaging substances. Richard took the larger role in supporting Eric through that ordeal and their relationship was strengthened. Eric found his experience with the law a surprisingly educative one and almost enjoyed it as a form of class trip, so to speak.

9

Long, Muddy Road

Addictions and Mental Illness

When Richard and I met, we walked in April darkness through the silent grounds of my university while he wept and our conversation explored the issues of death in life that each of us had encountered. We met again in June sunlight on the Toronto Islands, immersed in the music he knew so well, listening and singing along with the folk at Mariposa, Ontario's summer music festival. We knew unusual difficulties lay ahead and as we talked late into the night with music we cared about flowing from the record player, the words of "Woyaya" sung by Art Garfunkel on his Angel Clare *album seemed to capture our sense of the road we would be walking together: "It will be hard we know/And the road will be muddy and rough,/But we'll get there, heaven knows how we will get there, /We know we will."*[1]

We sang love songs to one another, lullabies to the children when they were babies, then, playing songs and working songs, hymns, and the bedtime songs interspersed with Jack tales and invented yarns that drew widely on Richard's folk repertoire. It is hard to know whether chronic fatigue syndrome silenced the songs or whether the waning of the bedtime ritual as the children became older gave illness a toehold. Still, the songs sing themselves in our minds, rise in snatches to our lips, unite us in outlook, renewing our faith that somehow we will achieve an unimagined destination at the end of the long, muddy road.

New Learning from Psychotic Breaks, 2003; 2005

From the time we took over the care of Richard's mother, at the end of 2001, Dan was entering into a new phase of his recovery from his addictions. He had weaned himself off Risperdal early in 2001 and had remained free of psychotic symptoms. He had very good control over alcohol, and apparently avoided other drugs, but continued to use marijuana. During Dan's previous episode of psychosis he had been able to take his medication responsibly—that is, regularly and without bartering it for street drugs—under more relaxed supervision. He had control of his anti-establishment attitudes and had managed to stay out of trouble with the law. He knew that we considered him far more trustworthy than he had been and for that reason were leaving him and Eric to look after the house, the pets, and themselves while we returned to Louisiana in the fall with Richard's mom to prepare her condo for sale and complete her move to Ontario. While we were away, Luke would be coming to make renovations to our house to accommodate her; we expected the younger brothers to become involved in that work.

Dan didn't accomplish much during the renovations; friends who used pot were guests of Eric and I suspected they and the younger boys were doing dope every day. Eric, nonetheless, began to take a serious interest in carpentry under Luke's direction. Shortly before we returned with Richard's mother to Canada, Dan ran our van into a tree, putting an end to his perfect record vis-à-vis the police. Over the phone he reassured me that he had not been hurt, although the police could not understand how he could not have been. I said with a sigh, "Oh, Dan; why did you go out again and bring this upon yourself?"

"Because peer pressure is real?" he asked rhetorically. Once again, we had the dubious privilege of accompanying a son through the intricacies and costs of a court case. Once again, we watched a son rise to the occasion as the details of involvement in the court system provided a learning experience spiced with tension. I suspected that for members of the subculture such events constituted rites of passage.

Luke and Kathleen moved back to Virginia and we were able to visit them en route from Louisiana with my mother-in-law (with all her possessions) at the end of January 2003. From Virginia, Luke joined our entourage to help with the U-Haul truck. This meant Richard needed to take Luke back to Virginia after the moving-in at our place had been completed and on the spur of the moment he decided to treat Eric, Dan, and their friends to the trip. I was concerned about Dan's mental stability before they left, but at that stage we were encouraging him to take responsibility for his own mental health and we did not insist that he stay home or take medication.

An hour after the happy group returned, I was in the upstairs hallway expressing to Eric my concern for Daniel, because he was talking incessantly, was preoccupied with himself, and had turned night into day. Of course, these signs of increasing mental instability seemed like normal life with Dan to Eric, who was maintaining that his brother was just fine. At that moment, Dan rushed into the hall and said, "I'm a fucking alcoholic and drug addict and I've just taken a quarter of a milligram of Risperdal." I don't suppose Eric was any more taken aback than I was. It seemed to me that Dan not only had recognized his own increasing mental instability for the first time, but actually had taken responsibility for his own mental health in a new way. Furthermore, he had affirmed the very facts Eric either could not see or was trying to deny. Daniel became stable on less Risperdal than he had taken that evening. I immediately suggested he reduce his dose to 1/8 mg, which was the amount on which he had recovered during his previous episode of psychosis.

The Snowboarding Accident

Daniel's sporting activities had not been enhanced by his continuing use of pot, which together with the Risperdal may have affected his vestibular canals in the inner ear that provide the sense of balance. His "pothead slouch" (which I failed to recognize as identical to his dyslexic inability to stand up straight because Eric showed the same symptom) and posturing under the influence of antipsychotic medication both indicated alterations to normal vestibular action. In mid-February, Dan was snowboarding on the hill in the west field and misjudged the jump, hitting the back of his head on a corner of the half-pipe when he fell. Because we were into the "new phase" of Dan's taking responsibility for his own health, I more or less ignored him when he mentioned he had hit his head. It never occurred to me he might have sustained a serious concussion. I got him ice and aspirin, advised him to rest, and let myself be otherwise engaged with his grandmother, who required a similar sort of care as Daniel had needed when he was severely psychotic. Dan complained very little, although he took it easy for a few days. He remained on the 1/8 mg of Risperdal he'd been taking since the end of January. During the next several weeks he seemed more lethargic, complained of headaches for which I advised aspirin, and greatly reduced his use of pot until he was off it entirely. I ascribed his lack of verbalizing that produced a quiet, subdued manner to his having quit using marijuana.

Stabilized

On my first trip to Dr. A. with Richard's mother in 2003, he first took me aside and apologized for "almost" having thought of me as a prejudiced mum who just didn't want her son suffering the side effects of a medication. He had seen the study of the two Vancouver doctors in regard to Risperdal.[2] Thus, Dr. A. had come—at least to some extent—to trust my observations of when Dan was in need of medication and how much medication to use. However, Dan soon would be in a situation that confounded both the doctor's and my sense of how to medicate his psychotic symptoms.

I do not recall noticing prior to that point that Dan "flipped" between cognitive states, apart from his condition for about three or four days the first and third times he entered hospital, the third being when he calmed down before being taken to the hospital by ambulance with police escort. Perhaps after a while I simply wasn't looking for that symptom. One of the amazing features of his mental roller coaster over the years, I thought, had been that his healing from dyslexic syndrome seemed to have remained more-or-less intact. He wasn't what I would describe as an avid reader, but I had noticed that whenever he recovered from an episode of psychosis he showed no signs of dyslexic syndrome. Dan had retained his ability to read (although he did not read much), to write legibly (although he did not write much), and until the snowboarding accident he had enjoyed a normal level of physical co-ordination that had allowed him to develop his athletic abilities in such sports as skateboarding and snowboarding that require a good sense of balance.

Dan reconnected with AA. Once he was attending meetings again, Dan would leave his brother and friends when they smoked pot, but he surely picked up a certain amount of second-hand smoke. He talked almost incessantly during this period, a clearly manic symptom. We saw less and less of the friends, which provided a tapering-off period for Dan in terms of his desperate attachment to peer friendships. Dan kept going to AA. Around mid-March he started having so-called "dry drunks"—as some of the people from AA and Al-Anon called his almost daily rages. Usually in the afternoon, he would burst into arguing and shouting, usually at me, sometimes at Eric. Today, I would call that behaviour bipolarity II (without psychosis). Next, he flatly refused to have anything to do with AA or any AA person. After about six weeks on 1/8 mg Risperdal per day—around 21 March—as he continued with these scenes, I laid down an ultimatum that if he could not stop the verbal abuse he would have to leave or go on more medication. He decided, with Richard's support, that he would go on an additional 1/8 mg every second day. Dan's behaviour became better controlled almost immediately—within about three days. I became less stressed and had more energy. Eric's behaviour improved. The friends left and stayed away—perhaps Dan's changed behaviour and constant preaching at them had something to do with their plans. Dan resumed going to AA.

Dan started to talk about studying again and I spoke to a friend who is a teacher about tutoring him. While we three were discussing the tutoring scenario in our kitchen, Dan suddenly began to demonstrate serious mental problems: wandering thoughts and somewhat bizarre verbal responses to our friend's conversation. That certainly terminated our plan for tutoring. I think that she did not realize how suddenly this change had come upon Daniel.

Medication Withdrawn

Daniel became insomniac and his manic symptoms increased: paranoia, a desperate desire to sleep, racing or tumbling thoughts, religious fears. He decided to burn the Dungeons & Dragons books I had noticed in the guest room. On the fourth night, he went to bed around 10, then, came to us at midnight saying he had been crying and had only dozed off for perhaps an hour. Around three in the morning, when Dan had slept for perhaps a couple of hours, Eric did something he'd never tried before—he got up and made a chocolate cake with marijuana in it. The stench permeated the whole house. Dan got up very distressed. I was upset, too, and talked to Eric to little effect. It was Good Friday and I thought Eric's persecution of Dan resonated. Both Richard and I were terribly tired. I gave Dan an aspirin; it was nearly 5 a.m. and Dan tried again to sleep. After 5½ hours of sleep, he felt well, rested, and "on top of things" in his head.

It did not occur to me until much later that Eric's way of demonstrating his own upset over Dan's deteriorating condition—and fear for himself—was to act out in a symbolic way that showed his own desperate need mingled with defiance and an assertion of his own idea of the solution to his problems. This incident demonstrates how the crises Dan's condition often produced (sometimes for days, weeks, or months) could deflect attention away from another family member in need. Even when I learned how to watch for such "Catch-22s," I was not always insightful or energetic enough to respond adequately to the second crisis.

For the next few days Dan wavered between a sort of weary normality or drowsiness and a state where he was hard to talk to because his attention would drift, the way it had when he was a dyslexic syndrome child. One had to speak his name several times to call him back to reality. He wanted to play dominoes with his grandmother and the rest of us, but he was doing everything in slow motion. Obviously, his neurological condition was seriously abnormal. Yet, when he did attend, he sounded quite normal and in control, just very tired. He displayed typical symptoms of mania: smoking more and drinking juice, milk, pop, much more than usual ("polydypsia" is "excessive thirst" and as sodium levels in the blood are compromised can become life-threatening). He was eating well. He was trying "to do good things," but could hardly focus. He had serious difficulty with a simple game of dominoes and had to be "called back" from his distracted condition in order to play. His cognitive processes were correct regarding play, but very slow. However, he did not seem to be upset by his mental changes, as I would have expected.

We wondered if the extra Risperdal had built up in Dan's system to destabilize him and decided to reduce the Risperdal back to 1/8 mg per day. Dan stayed awake all day and seemed insightful and analytical about his symptoms. By afternoon he again was becoming lethargic and distant. He felt "pestered" by thoughts and sounds in his imagination. He was noticing what he calls "visuals," which he described as the coloured trails from objects as he moved his head, and he tended to move his eyes to follow them; he said they were attention-catching and distracting. He seemed quite insightful and clear about what images in his mind constituted imagination and he had no problem distinguishing reality from the unpleasant sounds and images in his mind.

We were greatly relieved when the doctor's office reopened after the long weekend; Dr. A. saw Dan and suggested that perhaps he had developed a sensitivity reaction to the Risperdal and recommended phasing him off it as rapidly as possible. Daniel retained a fair level of normalcy and self-control during the phasing-down period, which encouraged us to agree that Risperdal was the cause of Dan's increased mental instability.

During May and June, all day long, Dan continued to drift back and forth frequently between periods of confusion and periods of clarity. For the most part he was able to continue on the path he had set for himself of painting, working in the garden, reading, and walking. His behaviour was not normal, but he seemed to be in better control, except for one disturbing outburst of temper when he kicked the bottom panel out of the back storm door. Friends were expecting him to work at their marina and the teacher remained willing to coach him in a high school course over the summer.

The Psychotic Interval

Several weeks passed before it struck me that not only was Dan's mental condition deteriorating overall, his cognitive condition also altered constantly and fairly rapidly in the course of a day. In some of these phases his thinking was far more lucid than in others. I remembered having noticed near the end of Dan's last period of illness that he had alternated between two distinguishable levels of abnormality. I began to wonder whether *all* of Dan's previous psychotic breaks had involved a moving in and out of psychosis that had been masked by medication, but that disappeared when he recovered his mental balance. I did not have the perspective to connect this mental see-saw with the fluctuating from dyslexic to non-dyslexic modes of perception following his listening intervention.

I was less comfortable than Richard was with the doctor's complete withdrawal of medication from Daniel. Because the doctor's authority informed that decision, Richard was especially reluctant for Dan to use the medication at all. Richard also seemed less distressed than I by Dan's deterioration. I wondered if he found Dan easier to deal with when he was more severely ill than when Dan was functioning better mentally but having trouble finding his feet in society more generally. Dan's condition had been deteriorating for months. Richard was also watching his mother lose her formerly very sharp wits and formidable self-control incrementally day by day; he was coping with two forms of unusual stress around him and without relief from his chronic fatigue syndrome. And he was the chief provider of income.

I recalled Dan's normal and abnormal thinking that had "flipped in and out" during the first three or four days of his first hospitalization. I had asked Dr. A. in those early days if bipolarity could create mood swings so rapid that they would occur many times during a day and he said, no, that bipolarity moods (mania and depression) swing over periods of days or weeks. Nevertheless, I wondered how Dan's alternating forms of cognition affected his mood. I had assumed that the overall situation of "flipping" became wearisome and eventually caused "low mood" from which he would swing back. But, what if his mood swung up and down with the alternations in his cognitive state? Which came first, the mood swing or the alteration in cognitive state? Or were they consonant?

I was getting a better perspective on differences in Dan's cognitive behaviour over the long haul. As I thought about it, I realized that during Dan's first stay in the hospital his alternating cognitive states had subsided into an almost unrelieved psychotic state only a few days after his being medicated, when he was on 48 to 96 times the amount of Risperdal that I had learned was effective in treating his psychosis. I began to suspect that his uniform psychosis had been induced by the effect of the drug on the normal part of his brain. Risperdal (sometimes prescribed along with other drugs) had obliterated the normalcy of the "non-psychotic" part of Dan's brain, while acting much more gradually on the supposedly "damaged" part of his brain to reduce his psychosis. My reasoning about Dan's changing cognitive conditions was applied within the "broken brain" paradigm I had learned in the psychiatric text.

One day in August, Dan surprised me by asking if he could do some kind of school work. The first idea that came to me was English grammar because it is simple and I would not need special materials; I found paper and pencils and we began to parse sentences. As I worked with Dan and noticed that his efforts to focus on the task waxed and waned, I thought there might be a *measurable* beat, like the "moment" of a pendulum, of a couple of minutes to his periods of confusion and to his periods of clarity. His ability to maintain his focus on the grammar was so often frustrated by his mental switches that we agreed to set the activity aside.

Upledger's Discovery

As I observed the regularity of Dan's switching from focus to inability to focus, a fragment of learning from a very different context suddenly occurred to me. Following an automobile accident in 1995, I had seen a massage therapist who sometimes used Upledger techniques. In response to my curiosity she had loaned me his book; and I read of John Upledger's observations of fluctuating contractions of the meninges (the membrane that encloses the brain and spinal cord) and the theory and practice of healing he developed based on his observations. Dr. Upledger describes the hydraulic system under the meninges, thus:

> It possesses its own physiological rhythmic activity. It has all the characteristics of a semi-closed hydraulic system. Functionally, it is intimately related to the central nervous system, the autonomic nervous system, the neuromusculoskeletal system, and the endocrine system. Its boundaries are formed by the meningeal membranes, most specifically the dura mater. . . .
>
> This hydraulic system includes the brain, which is more compressible than the cerebrospinal fluid. Severe pressure upon the external boundary of the system is transmitted through the relatively incompressible cerebrospinal fluid to the more compressible brain substance.
>
> The craniosacral system proceeds through cyclical flexion and extension at a rate of approximately 6 to 12 cycles per minute under normal circumstances. Flexion is an extreme range of motion during which the head becomes wider transversely and shorter in its anterior posterior dimension. During flexion, the whole body externally rotates and widens. After flexion, the physiological motion passes through a neutral or idling zone on its way into extension, during which phase the head narrows and elongates. The whole body internally rotates slightly. A complete cycle from flexion through neutral into extension, back through neutral and again into flexion requires about 6 seconds.[3]

Here was a description of something happening to the brain that fluctuated about every 2.5–5 seconds to produce a cycle of 5–10 seconds between 12 and 6 times each minute; or between 24 and 12 times over two minutes. How might that impact Daniel's brain to produce another cycle of longer periodicity? I could not answer that question, but Upledger's observation was an idea that hovered in the back of my mind as I continued to try to fathom the swings in Daniel's cognition.

Upledger teaches the extremely gentle—far less force than chiropractic adjustment, for example—manipulation of the cranial (head) bones and sacrum to effect balance in this hydraulic system. The object of the practitioner is to sense manually the "direction of ease" and "direction of restriction" during the flexion cycle and to exert very gentle but exceedingly steady pressure so that the cycling will come to a "still point" from which it alters to correct the restriction. "The still point may extend from a few seconds to a few minutes."[4]

Dr. Upledger's descriptions of the positive responses of patients to craniosacral therapy include children with a variety of disorders involving the brain, including dyslexic syndrome, behavioural and learning disorders (some of which I might subsume under dyslexic syndrome), cerebral palsy, and autism.[5] Of particular interest was the case of the teenaged boy whose dyslexic perception was altered "after four treatments and correction of internal rotation of the right temporal bone."

Concussion

I almost never go to the LCBO (the liquor store) since we gave up even the occasional "treat" of an alcoholic beverage once we understood Dan's addiction, but Richard had forgotten to obtain wine for his mother's self-prescribed daily glass. A brochure caught my attention and I plucked it in passing. It turned out to be about brain injury and described nine possible symptoms of brain injury. *Daniel had seven of them.* All at once, the car accident in January that the police thought surely would have injured Dan and the snowboarding concussion in February seemed pertinent.

Dan was rationalizing his periods of confusion to himself in strange ways. We took Dan to Dr. A., along with my Venn diagrams analyzing his behaviour and different possible explanations for it, and requested a CaT scan, which was set up for late September. As Dan's condition steadily deteriorated, that seemed too long to wait. Dr. A. wanted him to see a particular psychiatrist, Dr. H., who I knew by reputation and who was one of the psychiatrists who had seen Daniel on the general hospital psychiatric ward. We were not happy about that idea; we thought Dan needed a neurologist. Dan was growing more lethargic, losing more and more of the ability of the "better" cognitive state to remain normal. Watching Dan slide back towards unremitting psychosis, losing more and more of what he had so often regained, was sad and distressing for all of us. He was behaving as beautifully as he could, fighting however he might to maintain his sanity. The alternation between two kinds of mental condition continued, even as the "normal" intervals became less and less normal.

We took Dan to New York State, for a CaT scan at our own expense. Richard joked ruefully about this outing being his birthday present, his sixtieth having occurred a few days before. Dan seemed more relaxed and happy that "something is being done about my head." His symptoms, however, continued to worsen. At that point Dan was able to behave well enough in public so that his mental state was not immediately obvious to strangers; in conversation he sometimes was able to focus on the speaker and reply, at other times he was hesitant, sometimes confused. He managed social situations overall by staying very quiet and sometimes by letting his father or me interface for him.

Two Blanks

We had to wait until late August for the CaT report, which found nothing abnormal about Daniel's brain. Meanwhile, Dan's periods of confusion had become clearly psychotic and we were concerned that he was deteriorating more rapidly. Dan became insistent that his grandmother was his aunt, for instance, and became quite angry if anyone tried to reason with him, demonstrating deteriorating self-control. He became even more lethargic and spent much of his day on the couch, watching videos or just resting. Occasionally, he tried to hit a golf ball around the yard. He was not unco-operative, but was too tired and confused to be able to handle even simple tasks. He socialized poorly; he was very edgy and irritable. This was especially distressing to me in relation to the impolite way he treated his grandmother. Of course, it didn't bother her (she could not recall an incident five minutes later) and she wasn't particularly polite to anyone, either. Dr. A. wanted Dan to go back on Risperdal. We questioned him about the psychiatric route he had suggested weeks before (Dr. J. had not called) and requested a neurologist. We still were worried about Dan's possible reaction to Risperdal, given what had happened in the spring.

Minuscule Medication

For a few days we talked about Dan's going back on Risperdal, but a very, very small amount. The smallest possible dosage of Risperdal would be a quarter of a quarter milligram (¼ of a 0.25mg) pill, or approximately 1/16 mg every second day. Richard agreed reluctantly that we might try this for a week, but thought consistency over a 24-hour period was important, i.e., that medication taken every day would be better. The smallest pills of Risperdal available were the 0.25 mg-size that I routinely had cut into halves. Now, I divided them into quarters and gave Dan one fragment every evening. Richard and I agreed to continue the trial for at least a week before altering the dose. We could not discern any improvement for the first seven days and were strongly tempted to increase the dosage, but resisted. On the eighth day I wrote to my friend Ruth:

> I just had to tell you that Dan has come to the end of a day (Saturday the 13th) still being polite to Jean and referring to her as "grandmother." After he became agitated with me on the way into town, he apologized on the way back home. His weird sideways looks and grimacing have not been in evidence today. His periods of clarity are clearer. . . . for the first time since Easter Dan has shown some kind of improvement rather than some kind of decline over 24 hours.

Dan was less aggravated by his grandmother's odd behaviours; his ability to socialize with all others improved slightly. Towards the end of the week he asserted "I am sick, but I'm not mental." In fact, he appeared to have a virus. He did not sleep as well; he had one very severe incident he told us about only the next day where he nearly passed out in the kitchen. For a couple of days his symptoms of confusion again worsened; he addressed his grandmother as "aunt" once or twice. Again, I updated Ruth:

> Dan's symptoms are improving as though we were seeing a video of the last three months in rewind. However, the absolutely regular fluctuations in cognition continue (every two minutes approximately) between his ability to think clearly or to struggle with his mental process. This looks a whole lot like when he was dyslexic and hearing with both ears. He has serious trouble "hearing" when you speak to him during these intervals, i.e., he cannot attend or focus. That there's nothing wrong with his hearing is evident because he stores what he has heard in his short-term memory and responds as quickly as he can when his thinking clears. I so desperately want to find a neurologist or a craniosacral therapist who will recognize this phenomenon and have some idea of how to restore Dan's full mental functioning. If he has relapsed into dyslexic syndrome, I could take him to The Listening Centre. But that treatment supposedly causes mood swings, which I fear could exacerbate his symptoms; and in any case they will not treat him until he has been symptom-free (psychosis-free) for a year.

During Dan's third week on the tiny amount of medication, definite improvements in his mental condition were evident. The periods of confusion continued to alternate with periods of clarity, but the confusion phase, if he was not stressed, was less severely psychotic. For example:

- He was no longer speaking of himself in Messianic terms.
- He was no longer paranoid about some unnamed conspiracy between Richard and me.
- He was less easily upset.
- He had a little more energy and was trying to order his day.
- He began to get his own breakfast without assistance, and sometimes fixed his own lunch.
- He socialized better.
- He started to read a novel.
- He watched TV less.
- He offered to help with cooking, wood gathering, and other household tasks.
- His affect improved: for example, he could smile or laugh.
- His sociability returned: he could show concern for others and he reinstated good-night hugs.

At the end of his first month on the very low dosage of Risperdal, Dan's improved energy level brought with it an intensity during his confusion cycle that was very hard to live with. He became upset with me because I would not tell him where his machete knife had been stored. Later, he apologized. On other occasions that week he became unreasonably angry, upset, accusatory, etc. and later apologized. It did not occur to me that, once again, we were seeing Daniel in a bipolar phase of behaviour.

Sometimes Dan would mutter to himself, but not actually attempt to communicate what he was saying. He would become upset easily, but apologized later for his outbursts. Dan was preparing his own breakfast and usually his lunch, though I had to remind him to get a meal. He was socializing well enough to spend a short time with his and Eric's friend Andy and brief times with Eric and Eric's girlfriend Robin. He found it hard to continue reading because his focus fluctuated every couple of minutes. By early October, I recorded his improvement in one cognitive phase and problematical behaviour in the confused phase, thus:

> Dan was up when I came downstairs at 8 a.m. He was outside smoking and greeted me in a friendly, cheerful way. He came in and told me he didn't like the black velour shirt I am wearing (something he had told me a year ago that he liked—Daniel rarely comments on my clothes). He objected to the teapot in which I was making tea, so I made tea in two pots. He gave me a long, silent glare when we both arrived at the cupboard for tea cups. Then, he cheerfully made his own

> breakfast. Later, he admired the new curtains I had hung in the living room.
>
> I objected to the TV program he turned on to watch while he was eating, but he persevered. Once Jean was up I objected again (he was eating at the table) and under that stress he became coldly resistant, rude, telling me to "fuck off, you are not my mother." But he turned off the TV and found a book. His cold nastiness was extremely out of character, and I am still frightened and distressed. I asked Richard to speak to Dan about his attitude to me and treatment of me, but he did not.

A few days later, Dan seemed not as well, and was edgy and confused. Later in the day he made comments about me as being dead and his father knowing better than I about things, muttering to himself, and singing a song very loudly accompanied by weird gesticulations. The weather had turned to Indian summer, a sudden spate of warm days. Beautiful; but I was uncomfortable adjusting to the temperature changes. Was it hard for Dan, too? Was his busyness that morning too much for him? Was yesterday's fragment of medication too little? Was this deterioration in rationality because Andy had been at the house with Eric all day providing an increased level of emotional stress? Was this because Lainey was at our house for quite a while and kindling towards a meltdown? Was Dan picking up on the other family concerns Richard and Eric and I were discussing? I was beginning to see that the accumulation of irritants that weary anyone took their toll on Daniel, too, exacerbating his symptoms. This difficult day marked the end of Dan's fifth week on 1/16 mg per day. We were committed to at least one more and perhaps three more weeks at that level of medication before altering it.

Dan tried on headphones to listen to the chamber group[6] playing R.E.M. tunes for a couple of minutes and said it "helped the pressure in his head." The following day, Dan was a little irritable and vomited in his room and later vomited breakfast. Could his vomiting be related to a minuscule increase in medication? Or to his smoking? Or to some virus making the rounds?

Measuring Daniel's Cognitive States

I decided to find out if I could measure Dan's fluctuating cognitive intervals. He co-operated in a timing of his ability to focus while counting from 1 to 100. I set his counting task to a 10-minute interval because I felt that more than 10 minutes of focused activity might become too stressful for him. At first he seemed a little unclear about the directions; possibly I gave them while he was partly in his unfocused mode. He didn't like the term "confusion" so we agreed to use the words "focused" and "unfocused" instead. The time intervals available on the clock did not show seconds, so breaks in mode of attention-span time intervals were a bit rough. (The full chart is available as Appendix B.)

Counting Activity, October 2003

The following outlines the 10-minute activity as Dan carried it out:

9:22 a.m.–9:23 a.m. False starts.
9:23, 9:24 *focused*
9:25 *unfocused*; focused with great effort
9:26 *unfocused*, slightly focused
A rapid vacillation or "fibrillation" occurs in the next two-minute interval:
9: 27, 9:28 *very focused*
9:29 *unfocused*, then *fairly focused*
9:30 starts *unfocused, regains focus*
9:31 *unfocused*
9:32 *focused,* then *unfocused*
9:33 *unfocused*, then *focused*
Dan: "Very good, now what?"
[Me: "Do you want to start again at '1'?"]
Dan fairly focused [long pause] "1, 2, 3,"
9:34 *focused* "4 [continues to:] 31."
Time ends.

This activity demonstrates Daniel's fluctuating cognitive modes: about two minutes of clarity or focus followed by about two minutes of confusion. Notice the "fibrillation" effect, probably due to fatigue (at 9:28–9:29), about halfway through the test.

A few days later, Dan seemed quite a bit better. He still was complaining from time to time about "a sense of pressure" in his head. He no longer jumped manically out of bed, but got up after lying in bed awake for half an hour. One afternoon he talked to me about his theory regarding "marrying someone who has died before." He was edgy and defensive, but not extremely upset the way he sometimes had been when this notion had occurred to him in the past. He was trying to rationalize the notion as part of his belief system, perhaps because he could not imagine meeting in real life "the woman of his dreams." In general, he was more comfortable with being helpful and co-operative; performing small tasks was less stressful for him and he could do more of them in a shorter period of time. Later that day he gathered some kindling for the stove fire. He had made his own meals except for dinner, and in the evening his comments about some TV programs indicated more selectivity and judgement.

On the afternoon of 17 October, when Dan was walking on our road to the bridge and back, two dogs ran after him and he was so frightened that he sped home, losing his slippers (!) in the process. He associated images of war with the incident, probably from having watched *The Last Emperor*[7] for half the afternoon. He was calmer in the evening; I think a brief (five minutes or less) time of listening to Gregorian chant had a deeply calming effect on him. He expressed an outburst of frustration: "I am like the Chinese emperor imprisoned in the Eternal City" and it came to me a while later that, as a child, he had expressed a longing to be a hermit working with leather. When I reminded him of that childhood whimsy, suggesting that perhaps a part of his wish had come true in the isolation his illness has brought about, he seemed to think it was a "calming idea." This incident was one of dozens that reflected Dan's responsiveness to talking therapy even while in psychosis.

The following day he improved further. He had fewer episodes of garbled talk. Instead, during the intervals of confusion he was aware that he had to wait for his head to clear before continuing to speak, and he more often had access to short-term memory to carry him over the confused interval so that he could pick up on what he had been trying to say after it passed.

I noted a couple of days later that Dan was expressing emotion more normally. He would occasionally ask for a hug. He would smile and sometimes laugh at an appropriate time. He had stopped the giggling-over-nothing that had been prevalent in the evenings for a few weeks.

Dan's mental focus continued to drift. He would forget why he had gone to the barn freezer—for hamburger—until he was back in the kitchen holding ice cream. Back and forth, up and down.

Then Dan caught a virus, as I recorded in my journal:

> Dan got sick again a couple of days ago. He has been vomiting in the mornings for a couple of weeks, then developed a fever. Two nights ago he hardly slept. BUT, he did not go into a decline. . . . Dan's thought processes are improving, regaining normal complexity and nuance. He is a bit fearful, but no longer paranoid. The level of confusion in those regular intervals is something he can deal with or dismiss or even ignore. Not perfection, but progress.

> Last night I noticed from his conversation that his paranoia has segued from irrational suspicions to what one would think of as normal anxiety about his health and mental condition. His use of language is much more complex; his capacity for introspection has increased greatly over the last couple of weeks. He has better self-control and is calmer so that one is less conscious of his needing to exercise self-control.

I was still on tenterhooks as to how a physical illness might impede Dan's recovery process. The next day Dan was feeling worse and vomited early in the morning, but he was able to keep food down later. As soon as Dan got up the following day, I was treated to a somewhat confused lecture on the way some people actually "mind-meld" with the past through reading, study, stress, extremes of various kinds (drugs?) to become "beside themselves" with Napoleon or some other major political figure. His display of mental confusion tinged with megalomania was not reassuring. Dan got his own brunch, which included doing some cooking, and he seemed energetic, but not quite manic. He did not chatter—the way he had done all the previous day—about how awful he felt; but he sighed deeply from time to time.

Over the next two months Daniel regained his mental abilities at a rate that made an upward curve in the mirror image of the downward curve of his spring-and-summer's rate of deterioration. His behaviour became more controlled. He became more polite, then thoughtful and considerate. He stopped calling his grandmother "aunt." His irrational musings about Napoleon and his "dead girlfriend" stopped. His irritability ceased. He resumed taking care of himself, doing little tasks for others, and taking an interest in his environment. But the steady vacillation in levels of cognitive awareness that recurred at two-minute intervals persisted.

Confession and Healing

While the vacillation in Dan's cognitive states continued and his mental balance remained very fragile, during the confusion cycle his problematical thinking was much more highly rationalized, less "psychotic," and closer to the range of "normal" misapprehension. For example, his paranoid feelings transmuted to more normal fears, his megalomaniacal impressions of hero-identity sounded more like ideas that get into print than like the ravings of a madman.

By then, I had learned to recognize the symptoms that indicated the progression and regression of Dan's deteriorating and improving mental condition as he had displayed them over the previous nine months. I could discern the phases within these descents and ascents. I knew that his drinking and use of marijuana were the underlying problems that had pushed him to some sort of tipping point. I had seen that the car accident and snowboarding injuries had him teetering on the brink of psychosis. Then, too much medication had nudged him over the edge into the long psychotic descent. And "too much medication" was, in fact, a small fraction of what in the past he had been able to recover from. He continued to improve on the 1/16 mg.

Luke phoned one night to say that he and Kathleen had discussed the situation—Dan had been quite psychotic the last time they had seen him—and they were nervous about how Dan would be with their new baby Jemmy, born in September. So, when he became well enough they would love to have us visit. Which could be two weeks or two months . . . . And it hurt so much not to be able just to pick up and go. Our grandparenthood was a fact that we had little opportunity to express and on that account we also had feelings of loss and sorrow.

Near the end of October, Dan wakened us at 3 a.m. to let us know he was having trouble sleeping—usually a signal of danger. I was pretty sure the situation was not serious, but I got up to see what I could do to help. While milk was heating on the stove he told me, "I think I'm finally beginning to forgive myself for what happened."

"Forgiven yourself? For what, Dan?"

"For going out [drinking and drugging] again," he answered.

"What do you mean?"

"When I went out drinking last winter, I could see myself; it was like watching myself from a distance. I can't ever do that again or I'll be locked up. I'll be a vegetable, a human vegetable being cared for, for the rest of my life."

"Well, there's no point in beating yourself up for that. It just slows down the healing process, doesn't it? God forgives you."

"I know. Good night."

The next morning Dan got up about 10 and without so much as a greeting started ranting to me about how "confession isn't important and shouldn't be a part of the church and . . ." He spouted on and on, hardly stopping for breath as he followed me upstairs while I was carrying laundry. After about five minutes of his barrage I realized he had been coherent right through or past the predictable interval for the "flipping" process. The "dyslexified" condition of his brain had healed. Once again, however, I failed to make the connection between the cessation of the condition of dyslexia and the end of the episode of mental illness that had included psychosis. This failure in my thinking left me with the impression that Dan had reverted to a foundational condition of normality—that which had been produced by the Listening Centre intervention and which is *my* condition of foundational normality. Instead, Daniel had achieved a condition of mental health that was *not* his norm and made him profoundly uncomfortable. So I had a "Whew! Now he's OK" attitude to his condition, rather than, "OK, he's come this far; now how can we help him to solidify this plateau so he can start cautiously to recoup his losses."

It was important for me to know that Dan was having trouble forgiving himself. I can counsel and teach to that trouble. And it was important for him to have someone hear his confession or disclosure. He had been going to AA whenever he could that year, but the people he knew apparently were not able to counsel or to take the confessional role, which demonstrably is essential to the 12-Step program. During the week that followed, Dan reclaimed a normal level of complexity in his cognitive activity and was becoming introspective again. It was reassuring to see that those aspects of his brain function were intact and coming back into use. By the end of October, four days after Dan had come to terms with his conscience, he had improved to the point where I would forget to make journal entries about his progress. These are the sorts of changes that spelled normality:

- When he woke me at 1:30 a.m. on Hallowe'en because he was feeling paranoid, talking to him briefly easily persuaded him to go back to bed. I thought it strange that he could be so far recovered cognitively and yet actually be feeling paranoia late at night that someone from the drug culture was lurking at the gate wanting to shoot him. I began to see the similarities between the states of consciousness in sleep, in dreams, in the condition of psychosis, and in the twilight condition of transitional un/consciousness as one awakens.
- He described at some length how his paranoia comes and goes and how his vision simultaneously is affected. He talked about formerly thinking he was hearing people say things when they weren't and we discussed that situation—he had not realized that we were aware of his having that experience while it was happening to him.
- He was getting restless about what he should be doing next.
- He was much more even-tempered, less volatile, more polite and considerate, and quite normally verbal in his responses to chiding or teasing.
- He made his own lunch, but asked me to make cream of wheat at one point. I stopped worrying about whether or not he had eaten a meal, because he was mentally stable enough to miss one or to prepare it himself without being reminded.

- He commented that he was smoking too much, but actually was smoking somewhat less.
- He became responsive to the beauty of the landscape and showed joy and amusement as we toured through Eric and Robin's little fixer-upper.
- He commented on their boots, his and hers, on opposite sides of the dresser without jealousy and with delighted appreciation.
- He came to town, met an AA friend at the grocery store, and participated in birthday shopping for his grandmother (although he found the store environment stifling). For the first time I was not apprehensive about how he might behave in the store.
- He got along well with Lainey.

Some vital reconnection had been made in Daniel's head, apparently unrelated to the medication because that had not been altered, and I wondered how and why and just where. It had seemed to have something to do with his belief system. How, by forgiving himself, could healing from psychosis occur?

Dan's rate of recovery quickened. He was much less hesitant to respond in conversation and handled the continuity of conversations better, although he sometimes would drift. He was about the same as or better than he had been at the beginning of July. I had forgotten that the stapedius muscle can be shut down for so-called "psychological reasons," that is, unconsciously or semi-consciously but with purpose; that it sometimes can be shut down—or revived—at will. I lacked other important facts and I did not recognize the fragility of Dan's new-found rationality. And I lacked the knowledge to teach him how his brain worked and how he might best take care of it.

By November, there were further indications of his restored mental balance:

- His conversation was both introspective and insightful; but it would become a little weird in the religion category, especially late in the day.
- He could respond to humour and make jokes.
- He smiled and laughed at appropriate times.
- His voice had nuanced tonalities.
- He was remembering details of his past both before and during his illness and was integrating them into his conversation.
- One could see him tiring, for example, finding the restaurant atmosphere for Jean's birthday dinner a little stressful, but still coping like a quite normal person.

- He joked by asking me if he could drive to the city; but I could see that he almost felt ready to get behind the wheel again. I didn't know whether his reaction times would be fast enough while he still was taking medication, but we would have to see sometime soon if he could do that. I thought that in a couple of weeks he would be well enough to travel with us to see Luke and Kathleen and their babies.

By mid-November Dan had recovered to the point of losing interest in advice and asserting his own preferences in making decisions. He had some difficulty controlling his emotions. When Eric came to the house, Dan would pester his brother unpleasantly. He kept looking for entertainment in directions that would not be satisfying. He almost skipped the Friday AA meeting. He cooked special dishes a couple of times. He was restless and did not know how to channel his energy. Wood-cutting for fuel was the obvious chore that needed to be done, but he did not do much. The weather was exceptionally mild—perfect for outdoor work. But he seemed not ready or able to let go of the lethargy of illness and embrace health. He had decided to stop taking Risperdal and began the process of reducing the medication by skipping it every second day.

The Honour of Work

At the end of December, a neighbour who has a small furniture refinishing and building business talked to me at length about Daniel, whose illness he knew about through his son of about the same age. The son at that point was living away from home. Artie needed a helper in his shop. We worked out an arrangement with Artie somewhat like an apprenticeship, knowing his flow of business was sporadic. Dan would refinish some pieces of our old furniture and we would pay Artie for it, which would allow him to pay Dan. He hired Daniel with the understanding that the work flow was uneven and the situation could end at any time. Dan wasn't thrilled about the idea of becoming a refinisher of antique furniture. On one hand, the "work sucks" ethos of the subculture tempted him not to work at all; on the other hand, the idea of being able to earn money was appealing.

On 20 January 2004, after having cut back his Risperdal to every third night for the previous couple of weeks, Daniel said, "I've just taken the last ever Risperdal." We could only hope.

Dan was living at Lainey's house again. The furniture shop is minutes away and Artie picked Dan up in the morning. Dan, who had paid little attention to normal daylight hours for years, would be up and ready to go on time. Dan continued to come to our house for a main meal and I fixed his lunches to take to work with him. He looked after his own breakfast. He was off for a couple of days with the flu, but had no scary mental symptoms, although the first day or two he felt dizzy. I was astonished to see how important to his self-esteem his employment was. Artie was pleased with his work and told me so; and Dan responded to his approval with diligence and care on the job.

A couple of months later, Dan asked for time off work and accompanied us to Florida for a couple of weeks, following Jean's death in February of 2004. Work had slacked off for Richard in early March, enabling us to spend time with his brother and sister-in-law. Dan and I also took a side trip to see people his grandmother had been close to in her condo. Dan had a string of good experiences, too: walking for miles on the beach, eating out in restaurants, touring local sights, doing quite a lot of the driving, returning some of Jean's ashes to the condo golf course in an informal ceremony with family and friends, and dining with his great aunt.

Dan's return to work was less than enthusiastic. Then, the request to Lainey and me for foreclosure on our cottage mortgage put money into my hands that allowed us to help our children in various ways. Dan's gift was a used car to take himself to work and to AA. For almost six months Daniel gained an apprenticeship in fine furniture refinishing and did excellent work. For the first time in his adult life, he experienced pride in accomplishment outside of his home that graced a number of other homes and offices. As a result of Daniel's work, our house took on a more comfortable and civilized atmosphere. With the advice of his friend Andy, who was studying art in college, Dan took photos of his own work with great satisfaction.

In May, Artie ran out of steady work, although he called Dan back a few times. Only then did Dan mention that Artie's son, a young man as seriously addicted as Dan, had returned home and was enjoying sharing his pot during Dan's work breaks. So, Artie may also have been watching an impossible situation arise between his son and ours in the arrangement that he had made. Once Dan no longer had employment, he also lacked the motivation to find another position—a side effect, partly, of the marijuana. As his motivation eroded, the downward spiral was set in motion.

In June, Dan got a registered chocolate Labrador retriever, to the delight of the entire family. However, I was concerned to see Dan was letting others take over more of Garm's care.[8] Also troubling was Dan's refusal to take into his own living space any of the pieces of furniture he had restored. Where lay the conflict? Was this a matter of taste? Was he rejecting the product of his involvement with Artie because the work situation had been terminated? Or had he just reverted to the mindset that characterizes despair and perpetuates addiction? Or had something else affected his brain?

The conflicts that had arisen in Dan's workplace were profound. The peer connection Dan was so vulnerable to is a natural one: like animals that run in packs, young men are especially predisposed to male bonding. Against that bonding was set a work opportunity that was unpredictable, without a future, and intellectually unfulfilling. The community offered no similar venues for the skills Dan had developed. I had thought often about April's pronouncement years before that Dan would quit drugs "when he found something better to do." It is almost impossible for parents to provide an alternative that will outclass male bonding in a workplace (or anywhere else, for that matter). The community held no prospects. We talked about setting Dan up in his own woodworking business and made some purchases of equipment in that direction. But by then he was losing the battle with his addiction.

Same Old Story

In May, Daniel resumed using pot, but little or no alcohol. By midsummer, Dan's self-control and judgement had deteriorated to the point where we would not let him drive alone except to AA. Then, not at all. Eric commented in one argument he had with Dan in mid-August that "Dan uses pot uncontrollably." Dan immediately started trying to spread out his use of the drug—a clear indication of the positive potential in peer pressure of the right kind. From Dan's comments, I think he was using every day. I summarized Dan's situation to my friend Ruth that summer:

- He may not be going to AA and it is clear that, if he is going, it is only to regain his freedom with the car and not to start "working the program."
- Dan has lied so often that our relationship of trust has eroded.
- Dan has developed mood swings. He is irritable, particularly towards me.
- Dan's eating and sleeping patterns have changed. He more frequently will go for most of a day or longer without eating. He has turned night into day, despite the needs of his new puppy.
- Dan refuses to go back on medication. Dr. A., whom Dan refuses to see, recommended it and supplied the prescription. My psychiatric nurse friend also suggested he go back on medication.
- Dan denies that he has any symptoms. Typically, Dan recognizes his symptoms only when they involve psychosis, and sometimes not very clearly then.

That Dan actually *did* recognize some of his symptoms sooner, but viewed them as intrinsically *valuable* within his subculture (symptoms not unlike some religious mysticism), was a fact I failed to appreciate for another few years.

I had a slate of questions that continued to haunt me as I saw Dan sliding once more towards the incapacity of psychosis: I wondered why Dan would not comply with what seemed like good advice when he did not seem to enjoy being psychotic? Had he an extremely strong psychological mechanism of denial? Or did the judgement capacity of his brain become disconnected from his ability to introspect and analyze? Was he embarrassed to admit the pattern of using had caught up with him, again? Did admitting to his problem interfere with his practice of his addiction? I knew that when he recovered from an episode of psychosis he could not remember much of what it felt like to be fully psychotic; and, once he became ill, with the inevitable egocentricity of the mentally ill, he was unable to understand what the family was suffering on his account. And what *causes* "the inevitable egocentricity of the mentally ill"? And why *did* his memory cease to function normally once he became ill? On this muddy road, Richard and I certainly had not arrived at anything that fit "We'll know we're there." We would just have to keep walking.

10

Listening for the Light

Tomatis Reclaimed

Eric returned from a visit to Luke full of enthusiasm for the building of timber-frame structures like those his brother was constructing. That trip coincided with the breakup from his partner of the previous two years. He soon went back to continue learning from Luke. When he came home again, he began methodically to rescue the timbers and other materials from our old barn, which had been taken down the year before. He viewed with more critical eyes the rough woodshed he had built over the deck outside the living room the previous year. Casting about for other ways to exercise his new skills, Eric tore the old stoop away from the back door and built a comfortable deck out of salvaged materials from the old deck, which he dismantled after pulling down the woodshed he had built.

Impressed by this display of initiative and skill, we gave him free rein to build a sunroom where the scorned woodshed had been. Figuring out how to manoeuvre the timbers for posts and sills safely took a long time. Daniel felt less enthusiasm for the work but could usually be talked into helping. Eric had a special sensitivity for Dan's battered self-esteem and kept engaging his interest and skills while trying to forge ahead on his own when Dan could not be cajoled. Eric was developing important leadership skills. The process was slow and painful for both young men; the financial rewards were few (but appeared occasionally, thanks again to my sister) and the temptation strong to run off in other directions. The work would languish for weeks while other issues dominated their lives. There were no particular deadlines apart from the changing seasons and the only pressure was the need for a more-or-less weatherproof and insulated structure by winter, since the sunroom adjoined the living room through a double doorway. My sister helped to buy second-hand windows, a collection of ends-of-line tiles we found at a recycling store, and some of the other materials needed. Daniel helped to dig the holes and pour the cement, then he built the stone piers to support the 10' x 20' room. He helped to build the frame for the decking, install the windows, raise the rafters, deck them, and shingle the roof. He insulated the interior, put up wallboard, and finished the wallboard joins.

Despite all the interruptions, a year and a half later the room needed only finishing details, such as some wiring and a tile surface for the exposed portion of the European-style stove in the adjoining living room. For the latter, I would simply have used the leftover floor tiles, but Dan thought a mosaic should be made from them and he designed a lighthouse, "because you want your house to be a beacon of hope."

Burned Out and Aware

During the two years following Jean's Alzheimer's diagnosis, while I cared for her, Dan relapsed into drug-induced psychotic illness again in the middle of that period (2003), as described in Chapter 9. Thus, for about eight months we were looking after two mentally disabled individuals. Jean was sufficiently aware of Daniel's disability to ask often "what will become of him?" She lived to see Daniel recover from psychosis, cheered his going off to work every morning, and experienced the kindness that he was able to show towards her when he was well again.

For me, Jean's death in February of 2004 was somewhat like losing a child. While I felt an undeniable relief as the burden of caregiving ended, I also mourned the loss of a friendship that had ended in unavoidable dependency. I approached my sixty-third birthday feeling fragile and mortal. Years of caregiving and stress seemed to have eroded my optimism. Waves of fatigue would wash over me and after a couple of fairly good days I would have a couple of desperately tired ones. Richard was quite unwell, too, more so than I, which made me suspect we were prey to the most recently circulating virus. I wished I knew how to fix my middle ear, which possibly once again was the cause of my fatigue. I renewed my promise to myself to listen to Mozart with my headphones every day.

I tried to speak words of wisdom and encouragement to Daniel, but he was no longer listening. I learned from Artie that his son was in the Ontario hospital. Without question, Dan had been working in a challenging environment. Within days of my effort to reach out to him, Daniel had gone to another Punkfest, used ecstasy along with the usual pot and a few beers. Dan was back on Risperdal at the former low dosage.

I poured my grief and anger into my journal and shared it with my friends. Then Rinske, who I had not seen since Dan's infancy, invited me to an Anglican women's AWARE conference in October. Mary Hutton, who had encouraged my singing decades before, would be leading the music. The speaker for the weekend was the remarkable Diane Lang, whose son Jason was murdered at his high school in Taber, Manitoba, in 1999. Diane's forgiveness for the deranged young man who had killed her son showed me clearly that my lack of forgiveness for those who had hurt Dan was blocking my own optimism. During the weekend I confronted my anger, especially towards Dan's chief tormenters and tempters, and forgave them. As when I first had met her, Mary revived my freedom to sing.

Talking with other women at the conference had lifted me out of my sense of imprisonment in the place where I had been working so hard for so long. While I had been isolated in a social and economic backwater, I learned that I had been no more challenged in my peculiar circumstances than were many of them in theirs. I saw how my experience with Daniel's illness was relevant to the struggles of other parents in similar circumstances. The idea of writing to their particular needs sent me home with fresh ideas.

The Listening Centre Helps Again

A week after the conference, I returned to The Listening Centre to use their LiFT® program. Seeing some of the staff members again and reporting to them a little of the continuing benefits of the listening interventions in our family made me feel Daniel's and Sarah's benefits from their listening interventions were solid achievements. I still had not noticed that those benefits had tended to diminish, then to be reclaimed through their subsequent singing and listening experiences. My personal story of healing was requested for their website.

From The Listening Centre I drove to a motel near London for the night and went on the next day to Michigan to visit my friends Minka and Franklin, who I had not seen for many years. I was refreshed in many ways by their hospitality. I began the LiFT® listening intervention at their house and had used it for about four or five hours before my return home. I did not experience the "click" in my brain, but felt a dramatic increase in energy that was demonstrated as I drove back home. The distance from my house to theirs I had needed to break up into three stages, including an overnight at a motel. I managed that distance on my return in one eight-hour stretch of driving. As I continued with the LiFT® intervention during the following week, my normal energy once again was restored. I became stronger, healthier, happier, more confident, more assertive, calmer in general but also capable of sudden and intense emotion (which I found a little uncomfortable to readjust to). I experienced less pain from arthritis and found the breathing and humming exercises on the Earobics Series #1 and #2 tapes I had obtained from The Listening Centre improved my lung capacity and muscle function. I had the energy to reconnect with family and friends, made new friends, and started to describe from my own point of view what the journey of Daniel's disability and recovery had entailed. The discipline of listening and singing revived other creative urges: I copied one of my paintings for Ruth, completed several craft projects, and gradually picked up watercolour painting again. And, I was still editing.

For the next year I continued to sing along with Mary's CD and to use The Listening Centre's tapes frequently. At last I had learned that *singing and listening disciplines must be exercised faithfully* to maintain my healing.

During my time with Minka I had come to think I should step back from Dan's problems and allow him to wrestle with them without my input. At the same time, I gained a clearer sense of where I stood in relation to the rest of the family and determined to prioritize my relationship with my husband. My journal a few weeks later reflected this alteration:

> Dan has "gone out again" after we trusted him with his own automobile. For the same old reason—"because peer pressure is real." . . . [I]t is time to allow Daniel to go his own way.

When Dan decided to trade the car we had given him for a beat-up van belonging to a friend deeply alcoholic and drug-addicted and dishonest, with the rationale that he could use the van for transporting furniture he hoped to finish in the business he would start on his own, we kept out of the deal. Within a month, someone had t-boned his van at an intersection (not really his fault). Within weeks of this accident, the battered van broke down and was abandoned at Spider's second property 80 miles away. I drove there to have it towed to a local wrecker. Dan was back on Risperdal at the former dosage: ¼ of 0.25 mg every day.

Stepping Aside

Although Dan was psychotic and on medication during the winter of 2004–5, I no longer gave him continuing verbal positive reinforcement or did much for him. I spoke expectantly of his healing to others and stayed alert to what new intervention might help him, but took a decidedly hands-off role. Dan had moved into Lainey's house after Sarah and her husband Tom had returned to England in January 2004, where Tom was in art school. When they returned the next year to stay, they chose to create a private flat in part of the second floor of our house, partly because Dan continued to be based at Lainey's. The attention they gave to Dan during his illness seemed an amazing gift. They included Dan in their meals (they did their own cooking and eating separately from us) and invited him to their apartment to participate in activities they carefully planned to accommodate his limitations. Although Dan slept at Lainey's house, where he kept all his things, he spent the days at our house, mostly in front of the TV. I was too busy to find the change in the way we related to one another difficult. I took on more editing, helped Sarah and Tom to become involved in some aspects of our editorial services, and started attending the local Anglican church on a regular basis.

Richard Tries a Listening Program

Richard, who had suffered with his chronic fatigue syndrome without remission for 17 years, decided to try the LiFT® program from The Listening Centre and was feeling day by day that it was helping him (an improvement very obvious to me). During the last session of his program of listening I heard him shout. He said that a loud sound—which he thought was related to the sound equipment—upset him very much. I had never seen him angrier or more distressed and he expressed some of this over the phone to the Listening Centre staff in a manner that surprised me. The staff explained to me later that they had carefully checked the equipment and nothing was found to be malfunctioning. I felt something eerie in that juxtaposition of a healing process that was working for Richard—as it did for most people—and an unusual, sudden, distressing event that intruded on the healing process. In retrospect, I think he had momentarily experienced what Dan had described as being "like brain surgery." However, Richard's listening experience, despite or perhaps because of the unpleasant incident, also may have prepared the way for further healing that took place a few months later. Richard remained aloof to my listening intervention and ambivalent about how the treatment had helped him. Nevertheless, in the years since then he has regularly listened with his own headphones to music of his own choice to improve his energy level, mood, and immune response, and with significant success.

Early in September 2005 Richard had said about our parenting, "I realize now that you knew what you were doing from the git-go, and I was ad-hocing it all the way along." Certainly, he didn't mean I was a perfect parent; but he was acknowledging my solid foundation both in learning (that was scientific and rational as well as artistic and creative) and in belief (broadly Christian), and that I had been building on it for a long time. He didn't mean he was making bad decisions all the time, either; but that he was constantly breaking new ground and finding it harder to learn as he went. Unfortunately, it also created a tension of suspicion towards me (and therefore towards my thinking and my beliefs) and of insecurity towards him (and therefore towards his thinking and his beliefs) on the part of the children. We were learning to do a better job of "getting onto the same page," although it took effort to be sure we were in agreement.

Richard rarely affirmed his faith verbally; however, his experiences at an Anglican men's FLAME conference, which included his frequent enthusiastic singing with the 200 men present, healed his chronic fatigue syndrome—he came home from the conference with an identifiable cold, his first such positive immune reaction to a virus in 17 years. His remission lasted over a year and led him to become more involved with the children. He also apologized for his sometimes unkinds treatment of Lainey over the years, an apology she graciously accepted. The change in their relationship impressed the children and allowed my sister and me to become closer again.

Daniel Asks for Listening Exercises

After Dan's medication was reduced by one-third (all of October 2005) his periods of drifting had a goofy quality. Frequently during these periods he would be amused and laugh to himself. When I would question him, he would avoid telling me the details of what struck him funny, explaining, "You know. The sort of thing you don't want to hear about"; in other words, what the medical people call "strange thoughts"—a moderate level of psychosis. I wondered if the goofy periods were encroaching on the rational periods. Dan would become apologetic, often when apology was not needed, as if he were trying to excuse himself for his private world of weirdness and confusion and, perhaps, intrusive thoughts.

Early in November 2005 I began taking notes on Dan's condition again. He had been tapering off his prescription and had been on half-meds (1/4 of 0.25 mg Risperdal every second day) for about four days, behaving little differently from the way he had all summer. However, the deep lethargy that made me feel he was overmedicated did lift when he dropped to half-meds and he became restless, not to the point of manic but like a normal person who doesn't know what to be doing. He expressed this by taking long walks to the bridge and back, or up over the hill—although we were nervous about walks on the property with so many bear sightings. Dan himself had seen bears three times during the fall and there were many reports in the local newspaper. All during this time his "two minutes of focus, two minutes of confusion or drifting away from normal consciousness" was clearly evident.

After a return of my chronic fatigue syndrome symptoms, when I realized

I had let my discipline slip, I returned to a program of focused listening. This was just after Dan's birthday in mid-December. My immediate improvement prompted me to ask Dan if he thought he might have dyslexic syndrome again and he said, "Of course."

"Of course"? His words struck like a splitting maul shearing a pine log. How could I have missed the symptoms?

He said he wouldn't mind trying "listening exercises" again. He was referring to breathing and humming exercises we had done with the Earobics Series #1 and #2 tapes on a few days during the summer. Before I had time to find the tapes and organize a listening station, Dan noticed the listening arrangement I had for myself: the headphones attached to my computer so I could listen to violin music without inundating the household. He picked up my headphones and began listening to a Mozart CD and liked what he heard.

He was a little irritable in the afternoon—listening does cause some mood swings. But I could see, too, that he had become more fully present—more "awake," more aware. He could feel the irritation of a rough fingernail on his right hand and was complaining about it. His pace in speaking accelerated. At the grocery store he was so much more reactive to the store and the things in it that I was afraid he might behave oddly. (For example, earlier during this episode, when Richard would sometimes take him to town when he was going to the store, Dan would buy chips or a drink or candy bar with his own money and invariably refuse the change proffered to him by the store clerk with a mumbled "I don't want it" or "That's yours—I don't need it.") He noticed a pretty young woman in our aisle—he hadn't shown that sort of awareness for months. He wasn't rude, but he gave her an appreciative glance. He noticed her again as we proceeded through the store—once more, as though he were awakening.

He showed small signs of motivated activity—for example, he offered to make dinner, but couldn't sustain his motivation while the meat was thawing in the microwave. While I was cooking, Dan had put on a Mozart CD for ambient listening, so the house was full of music for a change. I tuned the music to Gregorian chant before we sat down to the meal and during dinner Richard complained, "Doesn't that music just make you want to scream?"

"Absolutely not," I said as I saw how thoroughly I had internalized my husband's tastes and tolerances for music over the years and resolved to make some changes. After dinner, Dan cleared the dinner dishes with the flair of a waiter—with much more grace and nuance in his thought processes and physical co-ordination than we had seen for nearly a year.

I made a couple of suggestions to Dan about what he might be thinking of doing next, but he seemed too lethargic to imagine activity other than walking and performing the lifting and carrying tasks sometimes asked of him that required careful supervision due to his seriously deficient short-term memory. He did suddenly decide to sand a wooden bowl that I had given him during the summer, a project he'd not been able to attempt.

Whether because of the reduction in Dan's medication or because he had started to listen to Mozart with headphones, or a combination of those factors, Daniel's brain function was changing. The next night, he was able to help make a meal. Two nights after that he noticed a pie shell on the kitchen counter and said that he wanted to make lemon filling for it. With my doing the prep steps and giving him constant direction he was able to stay on task and was very pleased with himself. It turned out to be one of the best lemon pies he has ever made, so his joy was reinforced by each slice he ate over the next couple of days.

The second time Daniel donned headphones he also noticed displayed on a shelf the small painting that he had done years before of the two trees and the valley filled with mist "like in *The Gods Must Be Crazy*"—his vision of his future. Because I had returned to doing craft and art projects, the dining room happened to be equipped very much like the Listening Centre rooms where art materials are available for use while listening. Wearing the headphones, Dan helped himself to the art supplies and made two paintings, the first being a recapitulation of his picture on the shelf, but differing in some details, such as the recurring motif of a "grinning sun," which he said represented himself. I had the impression that the music was allowing him to be much more alert mentally. That he would even consider picking up a brush after so many months away from making any art at all was an indication of some sort of mental improvement. When Richard got up and slammed his office door "because of the noise we were making," Dan didn't react much, remaining calm and focused on what he was doing. This surprised me, because Dan might just as easily have become hurt and angry and unable to do anything in response to his father's outburst—we were making no obtrusive noise and were in the downstairs, whereas Richard's office is above the adjacent room, the living room. Sound does travel through the grate in the living room ceiling, but most of the music sounds were coming through Dan's headphones, which I could hardly hear sitting next to him. At the time I found Richard's outburst inexplicable; he later claimed he was "trudging through the prolix prose of a less than elegant writer and any sound was too much distraction."

Watching Dan over the next few days was like seeing a person awaken ever so slowly. He was walking more during the day and I wondered if he had become over-stimulated by the music, although he didn't show any other more severe manic symptoms; evidently the music simply had energized him. On a Wednesday afternoon I bought some cheap furniture and did some gift shopping. He spent quite a while helping me put together the furniture—mostly screwing legs onto the table tops. I noticed he had trouble staying on task, especially during the phase of construction that took place at Lainey's house. He stayed at it, but with periods of confusion that were causing him distress by the time we were finished. This assembly task and the bit of cookery were the first manual activities Dan had done since he had spent time cutting up baskets of apples for freezing three months earlier, in September. These activities also were the first in which I had participated; I really had remained on the sidelines during this episode of illness. On the Wednesday night Dan listened to Mozart for about 20 minutes and then said he'd had enough. He seemed restless.

Thursday, I tried to organize his help in building a small wooden platform on casters that would raise one of the end tables he'd assembled to a comfortable level for my computer keyboard. Again, he was able to stay on task only with constant direction from me. "Do this. Now do that. Be careful of such-and-such." Occasionally, he would protest that he knew what he was supposed to be doing, but most of the time his short-term memory functioned so poorly that he needed to be "pulled into reality" by being reminded of what should happen next. His procedural memory was good for some tasks: he remembered how to operate the drill press to put holes in a slab of oak. I thought his hands showed the slight tremor I'd observed during the summer, which is a typical side effect of Risperdal, but also a typical symptom of dyslexic syndrome. He was able to hand-saw the rough ends off two small posts and was more verbal in reflecting small problems that had to be solved within that task, although he became frustrated and irritable at the way the saw kept sticking and the posts kept moving. He did quite a lot of wandering around looking for tools that Eric must have taken with him to Virginia for Luke to use in his work.

Dan discovered the little Christmas gifts I had shopped. I dropped the idea of trying to keep any of them as surprises, because his reawakened ability to show an interest in anything seemed an important behaviour to support. First, he expressed his appreciation for the drill-bit set I had purchased for him. Later, he expressed his liking for a star-shaped candle lantern. From someone who had had such flat affect for so long, these reactions were outstanding. In the evening, almost as soon as he started to listen to the Mozart concerto, his face became normally animated—a startling change. His eyes were "alight," his features mobile, smiling. I thought of The Listening Centre environment again and found some paper and pens for him. He drew enthusiastically, commenting that his work was like Andy's (the friend in art school). I took only sidelong glances as he was working at the table beside me, but I could see he was sketching in a style typical of his work several years before: many fine lines building up the borders of figures. Long ago he had explained his technique as a way of camouflaging his dyslexic inability to put a single line where he wanted it. I was astonished at the fine-motor control he had regained, displayed where he had dotted intersecting points of a squared lattice. I wondered if he had lost the tremor.

Near the end of his listening session, which probably extended beyond an hour because he had moved from finishing one CD to playing another, he came over to the computer; I said something to him and as he responded he appeared to me from a momentary rolling of his eyes as though he had hit that next "phase of confusion" in mid-sentence and yet had been able to continue his sentence right on through the barrier. He stood near my work station, still wearing the headphones, and swayed, nodded, smiled, hummed, and moved a bit with the rhythm of the music. This kind of spontaneity and reactivity had not been possible for Daniel during the previous six months. In sum, for the first time since May I had seen:

- normal animation in Dan's face
- normal expressions of pleasure in the things around him
- normal stability in relation to superficial aggravations
- the reawakening of his problem-solving ability
- improvement in his short-term memory
- greater physical energy
- better fine-motor control
- a range of spontaneous responses to music
- motivation to take on more complex activities.

Dan dragged through the Christmas season. At the end of the first week in January 2006, he was showing fairly mild symptoms of mania—pacing, regular intervals of confused thought, wanting to eat or drink very frequently, predilection for religious discussion, etc. I wondered what relation these symptoms might have to the fluctuating forms of cognition I had been observing throughout this episode of his mental disturbance. Pulling a book from a set on the living-room shelves, I asked him to read from *Robinson Crusoe* in the place where I had opened it randomly. I timed his periods of reading or of attempting to read, recording results that closely matched the pattern I had observed in October 2003 (compare Appendices B and C).

Integrated Healing

A few days later, Sarah and Tom had the idea of Dan's colouring some mandalas, as an activity within his capabilities. Tom found a website from which I could print out some of the circular patterns. Sarah insisted on buying coloured markers herself. They gave him the mandalas and I watched from my computer station in the dining room as he worked. When I went into the living room and complimented him on his work, he looked up and said with child-like gratitude: "Sarah and Tom gave these to me."

The next day I suggested he try another of the patterns, but he felt too tired. The following day, he picked one up and worked on it and I noticed he was using his hands with greater control, was approaching the task with more imagination, and was using the markers to make more complex patterns.

On Sunday after church, the priest asked me how Dan was and I expressed concern. "He's not worse?" she asked.

"No, not worse," I admitted. I felt Dan had taken a couple of tentative steps towards recovery. That afternoon I watched Dan "run up the rest of the steps." Following the pattern of the recovery of his mental functioning that had taken place over a period of 6–8 weeks at the end his previous episode of psychosis, he regained full mental function, but within 6–8 hours—a marvel to observe. Apparently, the reduction to an even smaller amount of medication, the sound stimulation with music inspired by my knowledge of the Tomatis Method, and the integrative task of colouring circular designs had combined to restore the function of his brain to the range of normal.

The next day, Sarah and Tom invited Daniel to dinner in their apartment. Two hours later, Daniel came downstairs deeply distressed. He could hardly bear to talk to me, not wanting to betray confidences, but neither could he handle an unexpected emotional burden. "I know my addictions are my own fault. You and Dad have been the best parents you could be. They're sicker than I am," Dan said with tears in his eyes. Dan had encountered again the "blame the parents" motif, the limping Freudianism repackaged in postmodernism and a spurious structuralism by whoever had influenced the thinking of our daughter and her husband. His disclosure caught my breath—such an emotional assault by two people who had endeared themselves to him while he was most vulnerable and who now threatened to withdraw that support while he was still so fragile, so recently restored to normal brain function. I recognized, of course, that this approach is simply how people evangelize, whatever the platform of their belief; they reach out compassionately with their most cherished beliefs to those in obvious need. It also is my experience that people tend to deal as they have been dealt with. Sarah and Tom doubtless had been offered the beliefs that now bulwarked their hopes and desires at some equally vulnerable point in their lives. That much I knew, without the further knowledge of brain function that would come to me a few years later. I recalled, in particular, Sarah's sudden turnaround in her thinking and feeling towards the other members of the family after her experiences at The Listening Centre and of her severance of our friendship. My reading of Tomatis's link with psychoanalytic schools of thought prejudicial to the female parent still lay in the future, but I felt them in every fibre as the family relationships that had been tenuously rebuilt buckled under this influence.

While Daniel would not side with his sister, what he did next closely followed her pattern of behaviour. He separated himself from the family in general and from Sarah and Tom specifically. Eric stepped into the breach and allied himself with Dan. The end of Eric's partnering with Robin gave him the leisure to become closer to Dan. But Eric brought his addiction to marijuana into that fraternity. At first Dan resisted, but before long he was drawn back into the familiar patterns. I continued to remain on the sidelines because I believed that Dan had to come to terms with whatever life was throwing at him, including whatever was dished out from the people in his own family.

Sarah, whose conversations with me had tended to be fraught with tension, stopped speaking to me almost entirely—a behaviour valiantly supported by Tom as he guarded his wife in the beliefs on which she, or they, had staked her well-being—the beginning of a process whereby she wrapped herself more generally in a blanket of sound deprivation. Over the next two years she would relapse into severe chronic fatigue syndrome. I suspected I was observing such a process as Madaule describes:

> Many people suffering from depression unwillingly [*sic:* unwittingly] put themselves into a state of sensory deprivation by cutting off all sources of sensory energy. This, in turn, weakens their ability to act and react, further increasing the depression and thereby creating a spiral effect.[1]

Sidetracked by the grinding necessity for work (we were entirely supporting three of the young people in the house and augmenting Dan's disability stipend) and by my own emotional pain, I had forgotten the revelation implied in Dan's "Of course" remark about his dyslexified self. I failed to notice whether Daniel's dyslexic syndrome had vanished along with his psychosis. Nor did I ask myself *how* focused listening to high-frequency violin music while engaged in art work had tipped the balance for Dan's recovery from psychosis.

Daniel on His Own

Dan maintained his mental function for most of the next two years despite his increasingly heavy use of marijuana and occasional use of alcohol and hallucinogenic mushrooms. He pushed his limits in several directions, many of them unhealthy. However, his ability to take responsibility for himself was vastly improved:

- He worked with Eric on several projects, including the timber- frame addition of the sunroom.
- He lived independently in my sister's house for most of that period.
- He survived on his disability allowance by talking my sister into a low rate of rent and by feeding himself sparingly. These ploys gave him more money for clothing of good quality in counter-culture style.
- He shared one meal a week at our house.
- He contributed to the family's wood supply, although he was not planning to live in the house where it would be used; in other words, he gave generously to all of us living here—Richard and me, Eric, and Sarah and Tom—by felling trees, hauling logs, splitting logs, and stacking the burnable fuel.
- He planted our long-neglected garden and tended it sporadically, yet brought in a small, useful harvest: enough tomatoes for 25 quarts of spaghetti sauce in addition to seasonal-use potatoes, beans, and herbs.
- He provided an additional small supply of wood for my sister's house and tended his own fires in November and December.
- He maintained a friendly demeanour with his father and me that did not include closeness or an interest in advice or support. In other words, he had an adult relationship with us.
- He developed his friendships within the subculture and apparently could be counted on to help Spider when asked, going off to cut wood for him, to make stone steps at his more northern property, to shovel snow from his driveway, and so on.

On the down side, my sister's house was deteriorating with Dan's neglect and carelessness. She and I became closer as we tried sporadically to deal with the mess Dan and Eric, when he visited, made of it. She admitted on one occasion that her own tendencies to neglect her property were merely "writ large" in Dan's malaise. I explained to her what I had learned about our father's bipolarity. Our cousin shared how the steroids she'd been given during chemotherapy for breast cancer accentuated her own bipolar tendencies to the point of disability. Although my sister had an opportunity to see herself on a continuum of mood fluctuations that had made her life so tumultuous, she continued to hold other people responsible for the ups and downs she experienced—she viewed my family as the source of the tumult in her own life. It might be countered that only when she read this manuscript did she learn of the breadth and depth of the challenges we had faced over the years. During most of our difficulties she had lived and worked separated from us by distance and by long intervals between minimal communications. Happily, that situation has changed.

A business venture with Eric and a friend brought a renewal of hope and self-esteem to both brothers during the early part of the summer in 2007. "Something better to do"—April's sagacity shared so long ago—galvanized the three young men into a tremendous amount of activity—logging, transporting timber, and building— with deep bonding. Their plans crashed when Dan and his friend became drunk and narrowly missed a run-in with the police: Eric felt betrayed and called off the collaboration. Dan became increasingly angry and bitter, refusing to recognize his own part in the fiasco by his personal use of marijuana and alcohol with his vulnerable associate. He slipped into stronger expressions of his allegiance to punk culture in dress, music, and frequent trips with Spider to Montreal. It was no secret that Daniel hoped to find a girlfriend through these excursions. As his mental balance started to slip during the autumn, he began showing up at our house for one meal a day and seeking the kind of involvement in art and music that had helped him to recover from his previous episode of psychosis. But his turnaround had come too late. Just before Christmas he attended a punk-rock concert by "The Misfits" in Montreal; I had specifically asked him if he had earplugs with him and he said he had. He did not use them. Just after Christmas, at which he had given thoughtfully chosen but extravagant gifts to each member of the family—and two years following his previous recovery from psychosis—Dan developed serious manic symptoms. By the first week of 2008 he was back in his doctor's office seeking a prescription for medication, which was initiated at the same level used at the end of his previous recovery: 1/4 of 0.25 mg Risperdal every second day.

Luke and Kathleen finished their house renovations in Virginia and sold their property, heading for Canada via a visit with April in Ohio, where she now lived. April's mother was there, helping to care for April's two-year-old twins Star (the girl) and North (the boy), which gave other members of the family a first opportunity to become acquainted. Kathleen talked about her surprised pleasure in getting to know this mystery woman, who shared her interests and talked at some length about her regret for what she had done in the past. A portrait of a profoundly changed woman emerged from Kathleen's description of her encounter with April's mom.

Luke's energy and vision for restarting his construction business in Ontario were magnetic in drawing Eric out of his boredom. Even in his illness, Dan could dimly sense a future for himself working with his brothers. Sarah and Kathleen had much to share. Tom had designed Luke's website and started getting further web-design work. A sea change was moving through the family, although a severe winter had descended on the land.

11

Listening to Psychosis

Cerebral Non-Dominance

Daniel began to draw pictures obsessively. The separate forms of his self-concept, the entities that he conceived of as mounting repeated attacks on that complex identity, the images of holy beings who helped him and that he identified with all were recorded in primitive form on large and small pieces of paper, sometimes for hours at a time. He diligently coloured the circular mandala patterns I had printed from a website, three or four or more per day. When I read of the value of a circular space for composing the self-concept[1] I presented Daniel with letter paper on which I had upturned a bowl to trace a large circle. He immediately divided the circle into four sections and used different coloured markers in each compartment. He had been making these pictures for a couple of weeks before I noticed that the colours were a code so that the identities within the compartments leapt into meaning. His Native Canadian friend, Matt, who he also saw as crucified with Christ, was in the red section. He himself and "her" were marked by a silvery white cross. A friend, a woman of colour, was in another section marked by a black cross. Another friend, Dale, was yellow. One of the songs I had often sung to the children as part of their childhood bedtime ritual had the words:

Jesus loves the little children
All the children of the world
Red and yellow, black and white
They are precious in His sight
Jesus loves the little children of the world.

Daniel was singing on paper and his art, as he claimed of his music, was an act of faith as well as an effort to order his chaotic mental environment. The healing he sought for himself he sought, also, for all those he knew to have been crushed by dyslexic syndrome, by impoverishment, by social disadvantages of many kinds.

The Descent into Hell

During the first week on medication Daniel seemed better, then suddenly worse. He affirmed my suspicion that he had smoked pot the day before; but a night's sleep did not undo the damage. After breakfast, Richard came as usual to pray with me in the dining room; Dan wandered in and knelt at the prie-dieu Faith had given us (that no one uses) and started to pray. Richard joined in; then I. Dan said the prie-dieu was perfect for confession; and I thought it had accomplished its purpose, if only for these moments he had taken for repentance. I recalled the psychotic episode when his self-forgiveness had completed a healing process. What might be hoped for from his act of contrition now? He was calmer and more cognizant afterwards, but not healed. As Richard was leaving, he mentioned the mosaic for the sunroom that Dan had designed but not yet made.

Responding to his cue, I led Dan through the steps of assembling the tiles in order to work with them, and of altering his original design a little. That took about 10 minutes and was all he could handle for the time being. He was making progress on the task during intervals of normalcy, and floundering during his confused intervals. At one point he directed me when to speak to him so that he would be less confused. It was a tricky business for me to be sorting out the lucid intervals and speaking only during those; then, letting him flounder for two minutes; then, watching him collect himself and work for two minutes; then, watching him flounder again, after which point I could again make a suggestion or offer encouragement. After 10 or 15 minutes of this we had completed a useful part of the design process and stopped. Within two days the medication (yes, even the tiny amount he had taken) had obliterated the clarity of Dan's lucid intervals, making him more uniformly psychotic; the mosaic project was set aside.

The outstanding symptom of Dan's mania—quite noticeably different from his previous episodes—was his "pressured speech" or ceaseless talking (like many of the scenes played by Geoffrey Rush, in his portrayal of David Helfgott, in *Shine*[2]). Sometimes with great animation or anger, and during all of his waking hours, he conversed with what emerged as a pantheon of characters with both "real" and "concocted" names. His talking was accompanied by his constant pacing. In reference to the sounds he was making, these symptoms were similar in kind but more extreme than in prior episodes; he certainly had thought the same way and had behaved somewhat the same way before, but during earlier episodes of psychosis he had kept his mental world of "strange thoughts" largely to himself by muttering barely audibly or by not speaking unless spoken to.

Apart from the hoarseness he developed at first, he did not seem more distressed by his verbalizing than someone doing challenging mental/vocal work. However, the vocal stream of consciousness was a whole lot harder on those around him. From the moment he awakened—sometimes as early as 4:30 a.m.—right through meals and in the shower or on walks until he went to sleep around 10:00 at night he engaged in intense conversation between a central one and others of the fragmented parts of his self-concept or with the denizens of his hell or heaven. I wondered if the sound of his speaking voice could have a beneficial effect in the same way singing does for chronic fatigue syndrome. I had not heard such a monologue since my last car trip with Dad just before his death in 1997. Dad's hours of lecturing were rational, of course, but almost as disconnected from his audience as Daniel's "externalized interior conversation."

For much of the day Dan was on his feet and pacing. Have you ever spent time in a hospital waiting area longing for news of how a loved one has come through surgery? Knowing my urge to pace under those conditions suggested to me the level of stress Daniel must have been feeling.

Daniel seemed, also, to be clinging to Jesus as an organizing principle while he tried to pull together the disparate parts of his brain. Sometimes his thoughts ran together like the clothes in a washing machine, vaguely categorized but with flowing associations, like the sort of word-association games with which some psychiatrists have plied their patients to elicit subconscious memories. During the interval of his deepest confusion, for most of the day he "slew demons," talking to them and reasoning with them or rejecting them, and drawing pictures over and over again (often one on top of the others, in the manner of prehistoric cave drawings) to name and arrange and count and control the images haunting his brain. Sometimes, they "materialized" as hallucinations or "invaded" as delusions. For the most part they bore strong resemblances to the characters in Dante's *Inferno* and to the orcs, wargs, and balrog of Tolkein's fantasies. But Dan had also spent hours with the History Channel's horrors of war; with pornography; with the depraved values of the drug subculture: some of the characters in his tormented world harked back to the sins of the imagination and of the flesh. Now that he had confessed to the error of his ways, he was stuck nevertheless with the images that would not stay in his subconscious, out of sight and under his control. His subconscious thoughts had become entirely available to his consciousness and were unavoidable for the rest of us, too. His efforts to subdue the unleashed memories of horror and degradation consumed his attention. "Why did I do that?" and "That's not me; I don't do that any more" were the themes of his struggle to regain a "clean and sober" mentality and to reconstruct his spinning world. One of the first signs of his returning self-control was his spending time more privately in his room, stretched on the bed and drawing endlessly, as he contended with the horror in his mind, always talking or chanting or arguing with "others."

A Place to Recover

Recalling that listening and mandala-colouring activities had brought him out of his state of confusion during his previous psychotic episode, I tried from the outset of this episode to encourage Daniel to spend some time each day at such activities. I moved my computer to my sister's house so that I could continue to work and to write while supervising Daniel, sometimes staying the night—Richard would come from our house—so that Dan's incessant talking and early rising would not disturb the other people in our house.

The first couple of weeks I spent housecleaning; my sister was so pleased to see what had lain beneath the debris and dirt that her own sense of pride in home ownership was renewed. We had fun together, during the times she took away from supply teaching in Scarborough, reminiscing as she dealt with some of her accumulation of things from the past and doing what we could for a neglected house. For the first time, she was proximate to Dan during a period of constant verbalizing and pacing and saw the nephew she cared about as "a raving maniac." She appreciated better the risks we had taken to resume looking after him and she understood the importance of the house she had made available—not only for his efforts towards independence that had encountered a relapse, but for his recovery, as well. Freed of the pressures to conform to the expectations of the members of the family in our house, Daniel could find his own level—especially in the use of his voice and in the playing of his guitar, but also through recorded and radio music—for pursuing recovery.

For several weeks Daniel spent some time most days colouring a few mandala designs. At the point when his colouring became very detailed and well-organized (but rarely symmetrical), he reached a plateau of interest and turned to illustrate compulsively the world between his ears. He commented, "When I pretend, I become what I'm pretending"—his first glimmer of insight as to what one part of his brain was doing to another part, and he referred to "method acting." I knew from one young man who almost destroyed himself the dangerous aspect of pretending that can extend to young people who indulge in fantasy games; I also had known professional actors who became caught up in "The Method" of the Russian Konstantin Stanislavski and who had hurt themselves and others by losing track of what had been the more stable aspects of their personalities. I sensed that I was coming closer to a much more profound understanding of those kinds of events. Late one afternoon I gained a bit more insight into Dan's mental process when he observed, "I think I will draw a real picture, now that I've finished with my demon slays." I wondered how, when Dan obeyed the mental impulse to "be like Jesus," which is a form of pretending, his brain "skipped a beat" and made him think he and Jesus were identical, which was a prevalent theme of his verbalizing and is a typical symptom of schizophrenia/psychosis. I was closing in on the answer to that question and to many others.

How Much Is Too Much? How Much Is Enough?

That the medication of choice—on the principle of familiarity or experience—for Daniel interferes with his healing through listening had been demonstrated during several of his psychotic episodes. His personal assessment of the feeling—or non-feeling—the very small amount of the drug induced was, "It's like a mild hit with PCP"; in other words, it in some ways perpetuated the symptoms of psychosis. The fact that focused listening, at least sometimes, can remediate even this prescription-drug-induced damage also had been demonstrated by the way listening and colouring activities had put an end to his previous episode of psychosis—when he regained his non-dyslexic, non-psychotic status and no longer "flipped" back and forth between lucid and confused cognitive states. The amount of the drug overridden was minuscule, however.

To what degree did the medication prevent the Tomatis Effect—the feedback between the voice and the ear that strengthens laterality—from taking place? I knew from Wheatley's book[3] and from descriptions from the pharmaceutical company on the Internet[4] that Risperdal acted on the neuroreceptors in the brain.[5] Did it also affect the levels of dopamine produced in the components of the middle and inner ear? Did it affect the corpus callosum, the bridge between the two halves of the brain? During the previous episode of illness two years before, the Tomatis Effect took place in Dan *after* his medication had been reduced, first, by half (¼ of 0.25 mg every second day); then, to a third (¼ of 0.25 mg every third day). Now, we could see that even when he was moving back and forth between greater and lesser degrees of confusion, whenever Daniel put on headphones to listen to Mozart violin concerti or *Bach for Meditation*[6] these listening activities greatly calmed him. As his mood improved, his brain function improved during both cognitive phases—i.e., his confusion lessened and his more rational phase sharpened, although, overall, his cognition continued to be severely abnormal. Clearly, regardless of where "the problem" resided in his brain, the music improved his entire brain.

By early March of 2008 we had abundant evidence that Daniel's mood and cognitive function improved not only during the time he was wearing headphones and listening to Mozart or violin arrangements of Bach, but incrementally following these periods of middle-ear stimulation. Following a listening session he would produce much better-organized artistic renditions of his mental world. Several times a day, Dan also would turn on a small tape player with its Gregorian chant tape because he found the ambient music calming. We observed that on days when he spent an hour and a half with headphones listening to Bach or Mozart (or playing his guitar and singing) his mood was calmer overall than on days when he spent less or no time listening or playing and singing. The Listening Centre had determined upon an hour and a half as the appropriate length of time for the stimulation segment of the listening program. That amount of stimulation is grounded in aspects of neurological function, such as rate of fatigue and entrainment. Dan gradually was gaining ground.

Dan's singing and/or guitar-playing was spontaneous and sometimes quite rowdy. One night his singing in the shower sounded as though the entire "parcel" of Pogues were hollering in the suds and steam with him. On every occasion where he used his own voice to sing, hum, or whistle, or played his guitar, his cognitive state improved; but too often, Richard complained, it sounded as though he was channelling Shane MacGowan or Kurt Cobain. Sometimes he would sing ferociously for up to an hour in "rap" style to a monotonous version of "The Canoe Song" or "God Rest Ye, Merry Gentlemen" with rhymed verses he composed spontaneously, in rhythm, as he organized, named, and laid to rest the monstrous images pestering his sanity, all the while marching and stomping loudly. He told me these were hymns; I called them "The Long Rap." They were amazing acts of creativity, although sometimes ugly in content; he usually came away from them calmer, more lucid, and refreshed.

By mid-March, after 10 weeks of full psychosis and ceaseless (while awake) talk, Dan seemed to have adapted in one part of his brain to the deleterious effect of the prescription medication, but continued to verbalize very colourfully the imagery of his more severe psychotic intervals. On some days, his anxious efforts to rid himself of these images by drawing them and organizing them greatly overshadowed his intervals of relative normalcy. By the end of March he had become louder and sounded potentially threatening during some of these intervals of confusion (although all of his upset was expressed in terms of the symbolism of his fantasy/psychotic/confused world). We decided to raise his medication to ¼ of 0.25 mg every day, the level at which Risperdal had been reintroduced near the end of the psychotic episode of 2003 and the episode of 2005 (ended in January 2006). In 2003, remember, he had reacted negatively to 1/8 (1/2 of 0.25) mg of Risperdal and the doctor had thought he could not handle that amount and he was quickly weaned off it entirely.

After we increased his medication at the end of March he at first became more confused and more intensely psychotic, a reaction I had come to expect; within a week he showed greater calm, clarity, and normalcy in the less confused interval of his cycle, and became able to remain quiet for two or three periods of at least 20 minutes to an hour in the course of a day. His mood and affect improved generally. During his second week of increased medication and with continuing music stimulation, he began to take a greater interest in those around him, showing concern both for Richard and for my sister after each had fallen on the thick layer of ice on the deck and driveway. During the third week he sometimes began to sing the proper lyrics of songs and to use tunes he hadn't sung for years. His voice became more pleasantly modulated. He was able to conduct short conversations and to remember projects—such as the tile mosaic lighthouse—that had been abandoned months before. After three weeks he apparently had recovered from the negative effects of the increase. Concurrently, his psychotic symptoms remained essentially the same—confusion, hallucination, and pressured speech.

Daniel began to use printed words in his drawings, including his own name. He would make quite normal offers of help to me, such as saying he would make a pot of tea, although without following through. Eric noticed that he was taking better care of his golden retriever, Lox (no small task given the rambunctious nature of this breed). Occasionally, Dan's incessant chatter would stop. By mid- to late April, he had arrived at the level of coherence described at the beginning of this chapter.

As the lucid intervals of Dan's self-talk had become distinctly more complex and more rational (although that cognitive level would deteriorate with stress and fatigue), his voice became more textured with feeling and nuance. I knew I was observing the Tomatis Effect—Tomatis's observation that "the larynx can produce only what the ear can hear."[7] Daniel became calmer generally throughout the day, although not being able to find his cigarettes or some other stress would affect his mood and cognitive function—he would become quite upset and irrational and it would take him a while to calm down. His choice of clothing became less bizarre. He regained choice in regard to satiety, that is, he became able to turn down an offer of food because he felt satisfied. He stopped drinking (milk, juice, tea) constantly and paced less, and less intensely. He chose more often to sit and draw. His long walks with his dog left him calmer.

Should we have started him on the fractionally higher dosage of medication back in January? Or had Dan's determined struggle with the temptations of his addictions that he had symbolized as a pantheon of evil entities been of some benefit after part of his brain would again subside into the subconscious when he became well? He seemed to be on the mend with the least possible meds damage from which to recover.

Dyslexic Syndrome, Psychosis, and the Struggle for Dominance

Richard had finished reading the first eight chapters of a first draft of this book and I was considering his editorial suggestions, so Daniel's history was constantly before me even as I observed the moment-by-moment symptoms of his current illness. I began to watch him for similarities and differences. I noticed that he became *more* restless and agitated as his lucid phases improved. It seemed that the rational part of his brain increasingly was overwhelmed and upset by the less rational part and that the rational part worked ceaselessly to try to bring order out of the chaotic part. I could see more clearly that a desperate effort by the poorly functioning rational part of his brain had been poured out from the onset of his psychosis to "subdue" the vibrant subconscious. I began to wonder if this effort was a labour of Sisyphus, a task that could never be completed, because the subconscious reasserted itself according to some internal clock not subject to the will.

Another of the distinguishing features of Dan's more lucid interval was that he was able to verbalize with much more specific and grammatical English—which was part of the reason it could be identified as more lucid—along with the ability to attend (for two minutes), to think more rationally and with greater complexity, and to communicate more reasonably. This does not mean he was fully himself during these intervals, but obviously much more so than in his confused phase. I assumed from his previous episodes that he would remain anxious and on edge until the tensor tympani and stapedius muscles started working together again to produce the "click" that establishes "normality" in the language centre of his brain. I began to wonder what precisely constituted that "click" and precisely what constituted "normality."

I finished a complicated editing project. Major snowstorms early in the spring rendered us virtually housebound. The stress of listening to Daniel's dialogue with himself spurred me to explore on the Internet how else I might facilitate his healing. My central preoccupations over the years had been with Dan's dyslexic syndrome; the process of healing through which he had obtained non-dyslexic, normal brain function; and his struggle in his social context with his addictions. I had seen his state of normality restored again and again and again to end his psychotic episodes so that he could rejoin the battle with his addictions. He had remarked during a more lucid moment of this episode that he was "dyslexic again," recalling to me his comment towards the end of his previous episode of psychotic illness: "Of course I'm dyslexic." I had been so preoccupied with his addictions and with the *process* of his recovery that I had not answered basic questions about the *nature* of his mental illness and, for that matter, of his dyslexic condition. *Could his conditions of dyslexic syndrome and psychosis be parallel, concurrent, or even identical?*

I had forgotten exactly where the stapedius muscles were located. How were they linked to the auditory nerve? What other nerves were involved in the area around the ear? Where did the electrical impulses from sound waves disappear to when they could not get to the left side of the brain? For that matter, how did sound turn into electrical impulses or electrical impulses into chemical changes in the brain? I dove into my father's reference books on brain function and neurology and I reread Madaule; I spent days on the Internet tracking down answers to questions. I gained a more complete concept of the processes of normal hearing and of abnormal listening.

The Structure of the Ear

The ear is described in three sections: the outer, the middle, and the inner ear. The outer ear is the flexible piece of swirled cartilage on the side of the head that catches sound waves and channels them through a canal (meatus) filled with air to the specialized tissue of the eardrum (membrana tympani) where they are captured.

The middle ear is composed of several kinds of tissue. Made of osseous (bone) tissue are the three tiny bones, the hammer (malleus), anvil (incus), and stirrup (stapes). The hammer is attached to the eardrum by the eardrum muscle (tensor tympani). The anvil is attached to the hammer and the hammer to the stirrup. The stirrup is attached to the vestibule of the inner ear at the fenestra ovalis (oval window), the "window" being closed by the stirrup's foot. The stapedius muscle is attached to the neck of the stapes at one end and to a tiny promontory on the tympanic cavity at its other end. The stapedius muscle "is supplied by the tympanic branch of [the seventh nerve] the facial nerve."[8] The inner ear is composed of the vestibule, the semicircular canals, the cochlea, and the auditory nerve.

Captured by the outer cartilage of the ear, sound waves are focused through the meatus onto the membrana tympani (eardrum). Sound is trans-

Figure 11.1 The Outer, Middle, and Inner Ear

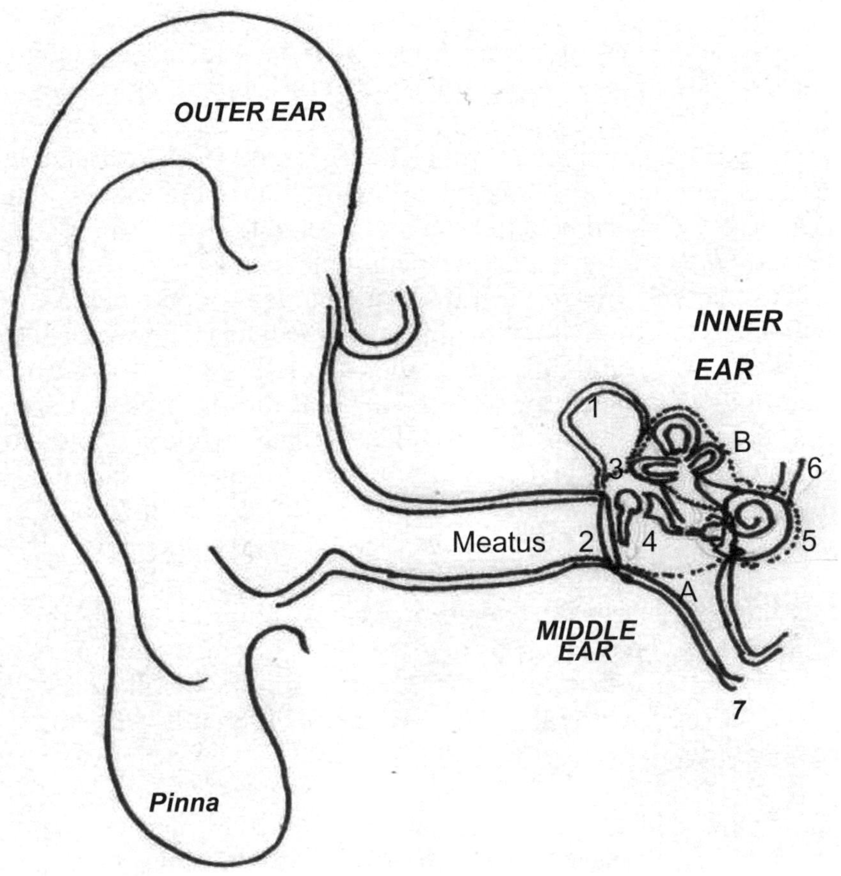

1 Mastoid cavity
2 Eardrum
3 Bones of hearing
4 Tympanum (tympanic cavity)
5 Cochlea
6 Acoustic nerve
7 Eustachian tube leading to pharyngeal cavity
A, B Dotted lines show bony casement housing (A) air-filled tympanum with bones of hearing and (B) fluid-filled housing of vestibular-cochlear system.

mitted not only through the bones of hearing within the tympanum, but also through the bony substance that encloses that cavity. The cochlea, pressed against the tympanum, makes its imprint on the wall of the tympanum so clearly that the surgeon performing a cochlear implant can locate its arches to drill through the tympanic shell to implant the electrodes. Microscopic surgery also is used to treat Ménière's disease, to remove infection, tumours, or a calcified stapes, to insert a stapedius prosthesis, and to repair holes in the eardrum. The schematized view of the middle ear in Figure 11.2 approximates what is seen by a surgeon through the eardrum. During surgery the eardrum is partially separated around its rim (annulus) and retracted; afterwards, it is replaced and held in position to reseal itself.

The tensor tympani, attached to the membrana tympani (ear drum) and to the malleus (hammer), flexes to draw the membrana tympani inward and thus heightens its tension. The stapedius muscle extends to draw the head of the stapes backward, and thus causes the base of the bone to rotate on a vertical axis drawn through its own centre: in doing this the back part of the base would be pressed inward towards the vestibule, while the fore part would be drawn from it. It is generally known that the stapes "probably compresses the contents of the vestibule."[9] Tomatis specifies the differences between the vestibule and the cochlea to express that concept more exactly:

> The vestibule measures movement of great amplitude, namely the movement of the body. The cochlea measures infinitesimal movements, those of sounds. It is designed to work with the vestibule, and is made of the same material, but it is organized differently to assume complementary functions.[10]

The range of movements this tiny processor in the ear can accommodate is vast—from the gyrations of an Olympic gymnast to the movement of molecules—and suggests the intricacies of its parts form a type of biological "microchip."

The Non-Linguistic Effects of Sound

The energy from sound waves, while it is being transmitted through the bones of the middle ear, also stimulates the tiny tensor tympani and stapedius muscles. The movement of the stapedius muscle causes changes in the vestibule through which sound energy also is passing from the base or foot of the stapes that is attached to the fenestra ovalis (oval window). The activity of both the stapedius muscle and of the stapes is transferred into the cochlea and the receptive branches of the auditory nerve.[11] These impulses connect not only to the language centre in the left brain; they also are involved in feedback loops in other parts of the brain: the cerebrum, the brain stem, the cerebellum, and so on.[12] In other words, the middle ear and inner ear mediate the conversion of sound energy to energy in various parts of the brain, which is then transmitted through the body.

> Tomatis estimates that, taken as a whole, the ear provides the nervous system with almost 90 per cent of its overall sensory energy. This includes energy coming from sounds—cochlear energy—and from the body movements—vestibular energy.[13]

Figure 11.2: The Articulations of the Tensor Tympani with the Maleus and Membrana Tympanus and of the Stapedius Muscle with the Stapes, which Alters the Status of the Vestibule

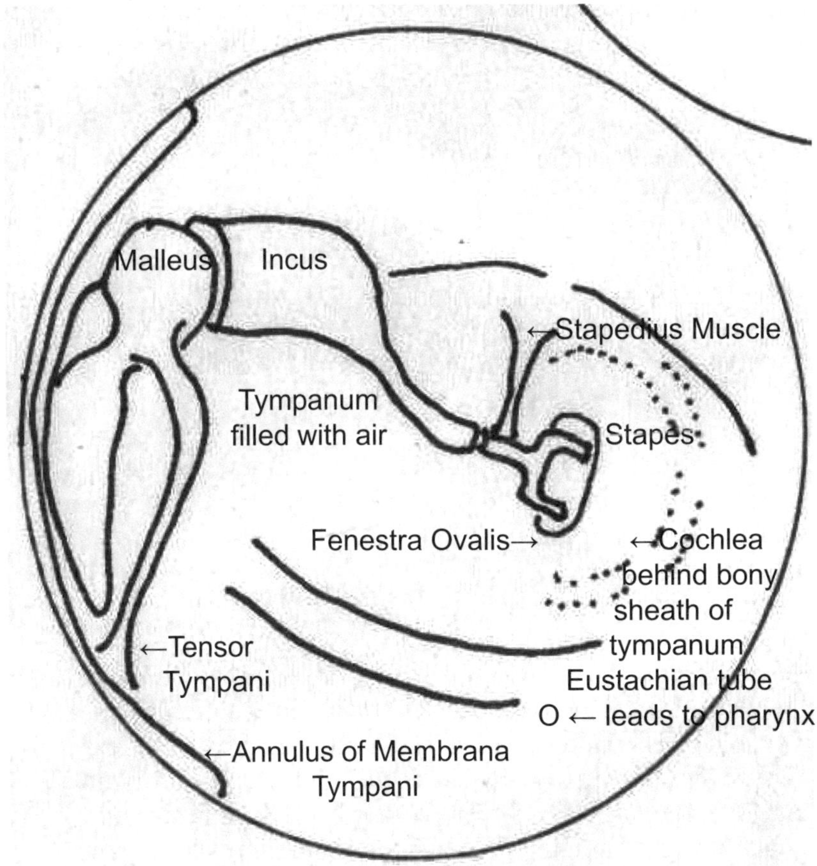

If these muscles fail to work synchronously, so that one or the other of them skews the biological mechanism, Tomatis was able to see—on the graph of his apparatus recording a listening test—"a dislocated curve":

> . . . a functional disharmony appears when the muscle of the stirrup, the extensor, takes over. The regulation in the internal chamber of the inner ear is no longer optimal, resulting in too great an absorption of the endolymph fluid and causing a muting of the high frequencies. Thus, the higher frequencies are not perceived and the hammer–stirrup block moves backward and outward, further decreasing acute perception of high frequencies while increasing the receipt of low frequencies.[14]

My personal experience, as someone who had suffered for 10 years with chronic fatigue syndrome, had been that with stimulation from music the regulation of the function of these minute muscles alters several aspects of brain function. The changes in my body included:

- the recovery of inactive nerves to facilitate muscle recovery during and after exercise;
- the raising of my body's immune responses to viral and other antigens (the ability to run a low fever, produce mucus, feel a beginning and an end to joint and body aches, to experience a transient fatigue);
- an interior (proprioceptive) awareness of energy and well-being;
- the ability to shiver;
- the ability to metabolize food more efficiently (without changing my diet I stopped gaining weight);
- better adaptation to changes in temperature;
- less brittle fingernails;
- the reappearance of my fingerprints.

I had experienced positive changes in all of these functions when my own stapedius muscle had been exercised back into action by the listening programs of The Listening Centre. The same revival and regulation of tensor tympani–stapedius muscle interaction had been brought about on other occasions through my sudden laughter and through my humming and singing. Madaule mentions similar changes in Tomatis's patients, for example, following his diagnosis and prescription of chanting for those French monks.[15] Dozens of similar sorts of beneficial effects from the production and hearing of music are documented by Don Campbell in *The Mozart Effect*. "Through the medulla, or brainstem, the auditory nerve connects with all the muscles of the body. Hence, muscle tone, equilibrium, and flexibility are also directly influenced by sound."[16] I think Campbell's statement can be refined, thus: because the nerves of the semicircular canals of the inner ear feed into the auditory nerve and because the auditory nerve connects with the brain stem, which in turn affects all the muscles of the body, the mechanisms of the ear transfer the effects of sound vibrations through the brain stem throughout the musculature of the body. I had come to that conclusion as a result of my healing from chronic fatigue syndrome and was not surprised by Tomatis's assertion that when "[t]he action of the extensor [stapedius] becomes greatly exaggerated . . . [t]his hypertension spreads throughout the body, impacting all of the extensor muscles . . . ."[17]

From Daniel's intermittently talking out of the side of his mouth, salivating to the point of spraying, and making a clicking sound on the side of his mouth as though the two sides of his tongue were operating at different speeds (the right side slower), I had learned that the malfunction of his right ear also affects the facial nerve to the right side of his mouth. Grimacing and winking and exaggerated sideways looks all are neurological effects of his ear dysfunction that I had noted in his previous psychotic episodes. Tomatis explains the effect of the condition of the ear on the face, mouth, tongue, jaw, and eye via feedback loops in the nerves feeding those parts of the head, although he usually was dealing with patients who had less severe symptoms.[18]

Although the neurological route from the ear to the brain is established in utero, it may be subjected to trauma of various kinds, for example, viral or other infections, loud sound, drugs, or psychological assault.[19] However, because a stage of the route sound takes through the body entails musculature—and muscles often can be exercised back into health—a mode of exercise might be determined that is suitable for reviving damaged stapedius muscles. Tomatis made the discovery that music rich in certain higher frequencies effectively exercises these muscles.[20] Enter Mozart, Bach, Vivaldi, and, for me, the chamber music renditions of R.E.M., a gift from Luke.

Relevant to our understanding of the function of the muscles of the middle ear is the fact that the inner ear can continue to function in some ways independent of the cilia in the cochlea. The hairs, or cilia, in the cochlea receive sound vibrations from the middle ear and transmit them to the auditory nerve for reception and interpretation in both hemispheres of the brain, but predominantly in the left brain's "language centre." The most common cause of deafness is damage to the cilia. My deaf great aunt had physical and emotional problems that were dismissed as "neurotic, "tics," and the result of "old fashioned" attitudes that may have been attributable to sound deprivation. However, tumours and their surgical removal are another cause. Professor of biopsychology John Pinel describes his own diagnosis, the surgical removal of his acoustic neuroma that destroyed his 8th cranial nerve, the subsequent severe fatigue, and his personal program for recovery that included African drumming.[21] The lack of interest in his post-operative symptoms on the part of his doctors did not prevent his designing a recovery program for himself. Fortunately, he was aware of the plasticity of the brain and had the information he needed to pursue recovery. It seems likely from Pinel's description that transmission of sound vibrations through his body from his regular drumming has to some degree replaced the sound energy formerly conveyed through his right ear.

In other cases of deafness, it is possible that sound energy passing through the muscles and bones of the middle ear continues to the vestibule through the stapes and stapedius muscle even if the cilia cannot further facilitate their transmission through the cochlea. That even the hearing process can in some cases bypass the cilia has been proven through deaf individuals who have their hearing restored via "cochlear implants."[22] Cochlear implants bypass the impaired hairs in the cochlea. Referred to as "cochlear implant" or "cochlear transplant" (although the organ itself is not being replaced), an electrode implanted into the cochlea receives sound vibrations from the vestibule and transmits them to the functional auditory nerve and thence to the brain. Even though the sound is transmitted through an electrode rather than cilia, the brain usually is able to decipher the sound and interpret it as effectively as sound travelling the normal biological route to the brain. Stefan Koelsch has ascertained that music also can be effectively deciphered via cochlear implant, although with reduced refinement.[23]

Since the perception of sound lies in the somewhat unique capacity of particular cilia to transmit particular frequencies, when those cilia have lost their capacity to function, some other mechanism remains responsive to particular frequencies. This, apparently, is the stapedius muscle that carries sound from the middle to the inner ear. Thus, it can be seen that the issues associated with dyslexic syndrome and chronic fatigue syndrome and other maladies are not related to the ability to hear per se, but to the ability of the middle ear components to respond to certain frequencies of sound and to convey those impulses through the auditory nerve in ways that affect other parts of the brain and the parts of the body influenced by them. The bony surrounding tissues are involved in that sound transfer as well.

Hemispheric Relations

The movement of sound through the brain is complex. I restrict my description here to the part that seems most relevant to laterality and hemispheric interaction. Let us consider both ears as they perform while hearing. The sound energy transmitted through the stapedius muscles attached to the vestibules makes *the neurological journey of hearing* (a) from right ear to left brain,[24] (b) from left ear by a circuitous route around the heart then across the hemispheric bridge to the left brain.[25] A simplified schematic drawing of these two routes is in Figure 11.3.

The corpus callosum is the main neurological bridge between the hemispheres of the brain. It is widely known that most people become "right-eared" and right-handed and that the development of the language centre in the left brain under that condition also defines normalcy.[26] While Tomatis does not explain how laterality is established in utero (he thought it began at birth), his understanding of the mechanisms of the middle and inner ear and of the audio-vocal loop has allowed me to discover how hemispheric dominance—which determines laterality—usually is reinforced in the left brain after birth. My understanding of the *nature* of Daniel's disability had progressed to my realizing that the dominance of his left brain, which usually is the "command centre," also, of a person's brain and defines another aspect of normality,[27] functioned that way for him only every two minutes.

Tomatis explains how right-ear dominance is reinforced after birth when the audio-vocal routes are established—the reinforcement through the ear of the sounds being made by the larynx[28]—a route that normally becomes self-regulating from birth. He also knew that the dominance of the left hemisphere of the brain is related to right laterality. But his emphasis remains on the *laterality* and importance of the ear, not on *dominance* in the left hemisphere and how that dominance further affects the function of the two halves of the brain.[29]

That Tomatis does not seem to have recognized the full implications of a strongly established left-*brain* dominance is revealed in his work with autistic ("infantile schizophrenic") children.[30] At that late stage of his career, Tomatis had become involved with autistic children as a willing acolyte of the psychiatrists and psychoanalysts who had charge of the children. These children, in Tomatis's appalled observation, were literally "caged" in high-sided cribs in a hospital. However, his vulnerability to their keepers' arcane philosophies ("I took my pilgrim's staff and went to visit colleagues in the field of neuropsychiatry, especially those with psychoanalytic training"[31]) seems to have sidetracked his thinking into the psychoanalytic theories of his colleagues and somewhat distracted him from the observable facts to which he usually turned his inquiring mind. Tomatis's unconventional childhood and delayed progress towards left-hemispheric dominance may explain his vulnerability to these theorists. For example, his mother was young when he was born and was rejecting; his father, an opera singer, took him "on the road" frequently; he did not attend school regularly until he was 11 years old when his father set him up in Paris in an apartment of his own. Despite his remarkable adaptation to circumstances, he wished for some experiences other children were having. A feeling of lack or of exceptionality is an avenue of vulnerability.

Issues related to dominance in the two halves of the brain are an aspect of Daniel's neurology that will become clearer to the reader as we watch his recovery. Hemispheric dominance is considered possible in only one hemisphere (or the other) of the brain.[32] (True ambidexterity is very rare; most ambidexterity is partial and occurs when left-handed individuals accommodate themselves to the inconveniently right-handed world they live in.) What was I to think of the condition of Dan's brain where neither hemisphere could assert and maintain dominance? While what Tomatis wrote about laterality (right- or left-handedness associated with ear preference) solved the question of how lateral dominance (ear, hand, etc.) is reinforced (he treated the condition of mixed laterality), he applied his knowledge about failures of hemispheric dominance primarily to persons suffering from dyslexic syndrome, vocal disorders, stuttering, and autism in children, not to drug-induced psychosis or to adult psychosis or other mental disorders. Especially in the most severe speech symptoms he dealt with professionally—those of autistic children—he did not have the most propitious subjects for observing the relationship between hemispheric dominance, laterality, and the production of syntactical speech. Unfortunately, he had just begun to succeed in helping his young patients when negative responses from the village folk where the children were to start school reverberated most unhappily on a colleague and put an end to his promising experiment treating the hospitalized autistic children.

Listening to Psychosis

As I took up again the collection of books and other publications I had read and shelved 11 years before, during Dan's first episode of psychosis, I realized that my reaction to this most recent episode of Dan's illness had been less a matter of my observing his symptoms than of enduring their assault on my senses, which was natural. Even after Dan had started to improve, friends who visited were horrified by the changes in him. Most of the other members of the family could not bear to spend much time in Dan's presence because of his "raving." Eric would come to my sister's house for supper, which usually had a calming influence on Dan—although occasionally just the opposite—and was a welcome relief for me, but he departed promptly after the meal. Even Richard found the couple of hours from dinner until bedtime—if he stayed—almost unbearable at times. I remembered Thelma Wheatley's commitment to her son Julian and took courage. I well understood why most people faced with mental illness choose to drug the patient to a point where the caregivers find relief, whatever the consequences may be to the brain—and ears—of the patient. As we began to see positive changes in Daniel's mental function through his listening, singing, and colouring and drawing activities, Richard and I supported one another in enduring his symptoms while trying to maintain the low level of Dan's medication so that it was not interfering with his recovery process.

No doubt using my own stapedius muscle to shut out some of Daniel's verbalizing, I read again of the synaptic junctions in the brain, of the all-or-nothing quality (the action potential) of firing in the neuron, of the need for certain chemicals to facilitate the passage of electrical impulses through the brain. Medical knowledge about the brain has often proceeded from the study of individuals severely damaged through trauma or by surgical procedures undertaken to ameliorate debilitating conditions, such as severe epileptic seizures. As I learned where particular kinds of learning are thought to be located in the brain, I learned also of persons who can survive with only one hemisphere. Recent findings suggest that the brain can "recreate in another part of the brain" a facility that has been lost in one sector. This ability is described in some detail by Dr. Jill Bolte Taylor in her account of her recovery from the cerebral stroke that shut down her left hemisphere.[33] A few so-called "dual-brain" persons have survived the severing of the corpus callosum between hemispheres with particular behavioural consequences. It became obvious to me that Dan's verbalizing defined particular modes of thought that pertain to particular parts of the brain. The tendencies for specialization in the brain fall into these categories:[34]

Cognitive Mode	**Right Hemisphere**	**Left Hemisphere**
Spatial Perception	Superior distance and 3-dimensional analysis	Poor
Thinking	Holistic, imaginative	Symbolic, analytical
Focus	Background, general	Foreground, specific
Awareness	Overall picture	Details
Better at	Open-ended tasks	Structured tasks

Once I had these differences firmly in mind, I saw Dan's behaviour, including his use of language, in a new light. Also, I understood much better the function of my own brain in which the left hemisphere dominates but has fairly ready access to the right hemisphere for someone who is left-dominant and right lateralized.

Dan improved to the point where Richard was more able to work on proofs while staying with him so I could get away from the house. My first outing was to the library to refer to particular books I'd read when Dan first experienced psychosis. The library, unfortunately, had culled some of these, but I stumbled across Don Campbell's *The Mozart Effect*, which I assumed had something to do with Tomatis's findings because a card I had received from The Listening Centre referred to him. Campbell indeed affirmed what I had learned about my own and my family's healing both through the Listening Centre interventions and my personal disciplines of humming and singing. He not only was a friend of Paul Madaule, but had met Tomatis and discussed with him the French specialist's theories. I found tantalizing those aspects of his explanations for the process of healing by music—I wanted to learn more. Furthermore, his dozens of examples in a wide range of applications of the efficacy of music to heal many problems in the body, including the brain, were encouraging for the process I was trying to facilitate for Daniel. An interesting emphasis in Campbell's reiteration of Tomatis's discoveries and method—because Campbell is, first, a musician—is the enlargement of the corpus callosum in musicians.[35] A musician's attention to the various components of music requires the action of parts of both hemispheres and a high degree of co-ordination between them. The effect of enlarging the corpus callosum is to make such a brain more flexible for other kinds of learning, too, a fact just now reiterated in radio commercials for the Ontario Conservatory of Music's program for toddlers.

Dominance Confounded

In Dan's music-making, particularly during "The Long Rap" that he sometimes strummed his guitar to, I could observe greater integration of the "rational" and "creative" aspects of his brain function. At last it dawned on me that the peculiarities of Daniel's psychotic condition—the "flipping" between levels of cognitive function—were the result of *the hemispheres of his brain trading dominance as methodically as the swing of a pendulum.*

My next question was, Why? Was that shifting absolutely unique to Dan? Or to persons with psychosis? Or to persons with dyslexic syndrome? Or, to some extent, to everyone? When Tomatis refers to a "flip-flop" of dominance he is speaking of the dominant ear,[36] that is, he is speaking of laterality, but not of the dominance of first one, then the other *hemisphere of the brain.*

Did this switching of hemispheres in Daniel start in utero? Likely Tomatis would have said that it did *not*, since he saw laterality as something dependent on feedback between the ear and the voice (larynx), which is not exercised until the baby utters its first cry. Sollier affirms that laterality is not firmly established until the child is six or even 11.[37] Or does laterality begin, in some sense, in utero with that enlarged language bump in the left brain? Does the exposure of a fetus to excessive sound tip a delicate balance already weighted towards left-hemispheric dominance to confound the normal developmental process towards right laterality that merely continues into a new phase when the baby comes into voice after birth? Sollier cites a study of underweight births of infants whose mothers located near an airport during their pregnancies; the fetus can be negatively impacted by unaccustomed intense sound.[38] The study compared these infants with those of other mothers who had spent their entire pregnancy near the airport and whose infants were not noticably impacted; thus, another critical concept to consider is "unusual" as well as "intense" sound. How did blasts of bone-conducted table-saw sound impact Daniel's prenatally established circuitry so that his brain became hypersensitive to certain frequencies? Is it possible—even though the eardrum is inactive when immersed in amniotic fluid—that the tiny muscles of the middle ear have begun to adjust to sound conducted through that watery environment? The eardrum may not be able to vibrate, but might not muscles move? A memory surfaces: of ducking underwater in the lake with my sister so that we could surprise one another with the sound as we struck rocks together. The sound is perceived as much louder under water where it is received through bone conduction than the same sort of percussion is perceived above water with only air conduction. Does the fetus try to protect itself with the mechanism that later provides protection from sound, namely, the reaction of the stapedius muscle of the middle ear? Or is some other mechanism affected by excessive sound, such as the hemispheric link, the corpus callosum, to inhibit the normal trading of sensory input and awareness between the hemispheres?

I had clear evidence that Daniel's hemispheric link was not operating properly. What had happened to the link between Daniel's hemispheres—the corpus callosum—that prevented inter-communication and synchronization of their functions?

> Ever since nerve conduction was understood it has been speculated that the great interhemispheric bridge, the corpus callosum (meaning 'thick-skinned body'), is essential for mental unity. When it and other smaller interhemispheric connections are cut, communication between the two sides must pass by the stem of the brain, which is nor-

> mally considered unconscious, or through external relations of the body with the world of stimuli and the effects of actions.[39]

Was Daniel's problem an effect of the loud sounds to which he had been exposed, of the street drugs he had taken or of the prescription drug Risperdal acting on the receptors in his brain? Or did one or more of the drugs affect some part of his ear—the cilia or the tensor tympani and/or the stapedius muscle—so that sound vibrations could not be transmitted normally and the sound stimulation never reached the left hemisphere or else arrived there partial or distorted? Potential side effects of the prescription drug included tardive dyskinesia, a result of the failure of the nervous system to maintain the body's muscles (which I perhaps had experienced in some degree during chronic fatigue syndrome), and a myriad of other possibilities, including tachycardia (which reminded me of Richard's heart palpitations), neuroleptic malignant syndrome, syncope (fainting), priapism (abnormal erectile function), antiemetic effect, disruption of body temperature, suicide, hypersensitivity, hypotension, renal impairment, and hepatic impairment, among others.[40]

Another possibility for explanations of Dan's behaviour came from other "loops" in the brain pointed out by Tomatis[41] that provide feedback within particular sectors that in turn affect the entire body. Did Dan's obsessive repetition of drawing activities and verbal ordering of his cosmos have something to do with these circular neurological paths, or was the disturbance simply in the cingulate system of the brain, as Dr. Amen's description of obsessive behaviours suggests?[42] Or were they "closed tracks" within the right hemisphere that had lost much of its ability to communicate with the other half of the brain? Or were directions from the "command centre" in the left brain not reaching the compulsive activity of the brain stem to quell its urges, which was the explanation given to one of my friends by her physicians for her obsessive-compulsive disorder? Why did Daniel, when ill, habitually perform a right-hand activity, such as stirring pancake batter, with his left arm and hand limp or his left hand in his pocket, although it could have been useful holding onto the edge of the bowl to prevent its spinning? That was a "dual-brain" phenomenon such as would be displayed by a person with no functional corpus callosum and inclined me to the "closed-track" concept.

> The separation of the two hemispheres creates, in some senses, two minds within the same body. The patient, sometimes called a split-brain patient, is able to get around well and can do most, if not all, of the things he did before, but some things he does in a subtly different

> way. For example, in performing tasks that require the cooperation of both hands, the patient may be severely impaired. . . .
> In split-brain patients, it has been observed that when the two hemispheres agree on evaluations of people and other matters, the patients outwardly appear calm and tractable, but when the two hemispheres disagree, they are outwardly hyperactive, agitated, and aggressive. It would appear that the two hemispheres are also capable of reading differences in each other. . . .
>
> Split-brain patients have difficulty with stereopsis (depth perception) when an object lies immediately in front of or behind the fixation point. This is because the images fall on the retina so that they are transmitted to different hemispheres, and determining whether to converge or diverge the eyes is impossible.[43]

This latter description sounded a lot like Daniel, and explained why, during psychotic episodes, he would be so distressed by oncoming cars when sitting in the front passenger seat of the car. Clearly, at least part of his problem lay in a separation—in some sense—that occurred at intervals between his hemispheres. Later, I would connect this effect with his nystagmus, also.

Madaule theorizes that the fetus moves in utero towards the mother's skeleton in preference for the sound of her voice and to "protect" him/herself from visceral sounds,[44] sounds that Tomatis, on the other hand, describes as pleasant.[45] As Petrofski points out, Tomatis's consideration of the hearing fetus is separated from his views on the development of hemispheric dominance and laterality; he relates laterality to the initiation and subsequent development of speech, that is, to "expressive listening," not to "receptive listening."[46] Sollier, also following Tomatis, adds that sound receptors in the skin function as a primitive ear. He theorizes that the fetus's turns in utero are in response to the developed vestibule, and notes that excessive "noise takes its toll, just as it does on those who are already born."[47]

Those already born have a defence in the middle ear that regulates for sound. The fetal middle ear is filled with fluid and is said, therefore, to be non-functional,[48] but the auditory nerve is developed to the point where it can transmit sound. In my experience of immersing my head under water, sound is conducted even more strongly through my ears (I feel pain) under water than in air. My eardrums may not vibrate, but my muscles work. If the fetus is learning to move its entire body in responses to voice stimuli,[49] might the fetus also attempt regulatory action with the middle-ear muscles when it receives such excessive sound, whether by bone conduction (which must also reach the middle ear) or through fluid? Does the perception in the fetal brain (both hemispheres) of excessive sound alter its inter-hemispheric function so that the bone-conduction thresholds are altered? "Daniel's responses initially had showed high sensitivity (low thresholds) on the bone conduction measure compared to his responses on the air conduction measure."[50] That was the condition of dyslexic syndrome that had been altered, as indicated by the post-listening intervention modified audiograms. Daniel's "high bone-conduction" sensitivity did not just mean "superior hearing ability" as I had understood from his "modified audiogram" at The Listening Centre, but must also have characterized his in utero learned (conditioned) response to damaging levels of high-frequency sound.

The conclusion of the Georgiana Institute (of Roxbury, Connecticutt), which also offers listening intervention, is that "the treatment apparently strengthens the muscles and improves the reaction involved in the prevention of sensory overload."[51] In other words, the perceived problem being treated by their AIT program is "hypersensitivity in the perception of some frequencies."[52] This analysis of the fundamental problem in dyslexic syndrome and autism and other problems undertaken both by The Listening Centre and by the Georgiana Institute and other facilities offering the Tomatis Method or some version of listening retraining is illuminating. For one, it explains the processing deficit in Daniel that existed from before birth, although the precise mechanism of that prenatal deficit is not described. Could such a "processing deficit" prenatally have produced a kind of "sensory overload" so that Dan's brain responded by losing its delicate edge towards left dominance? Might this hypersensitivity, when again and again exposed to excessive noise, have "blown the circuits" for whatever delicate tendency towards the left hemisphere had already developed? I was closer to understanding how Daniel's dyslexic syndrome had begun in utero. And so do autism and, reputedly, bipolarity and schizophrenia begin in utero. I pondered concepts of electrical charge and discharge. I wished for a chat with a bioengineer.

I had established to my own satisfaction that Tomatis was mistaken in thinking that a problem such as Daniel's was a product simply of faulty audio-vocal feedback; it had started in utero and without doubt as a result of unusual and excessive sound. His listening program had "reduced his sensitivity through the bone conduction . . . to allow better sensory regulation."[53] And how had that been accomplished? By feeding a toning sound via the *right* bone to the left hemisphere to focus his attention on the second avenue of perception: conduction of sound through air to the eardrum and on through the middle and inner ear to the acoustic nerve and to the left brain. How does that work? Bone conduction generally is diffuse, thus equally affects both hemispheres. Very strong bone-conducted sound would very strongly reinforce bi-hemisphericity and also predispose the hearer to reject or avoid such high-frequency sounds if possible. Air conduction begins shortly after birth and soon ceases to be diffuse as the audio-vocal loop is reinforced: it is specific and engages a discriminatory action in the left brain. But that experience alone perhaps cannot offset the prenatal conditioning from unusual, excessive sound that has programmed the brain to reject certain frequencies. It must be led gently under the influence of violins to again accept those frequencies into its processes.

The Hydraulics of the Central Nervous System

I recalled from my reading of Dr. John Upledger's *Craniosacral Therapy*[54] that changes in electrical potential could be measured on the skin of his patients. He had devised a method of testing his ability to observe (through palpation, i.e., sensing by his hands) changes in the closed (by the meninges) hydraulic system surrounding the central nervous system by measuring the mechano-electric changes in patterns on the patient's skin.[55] He did not provide enough information for me to make the leap to electrical changes affecting Dan's non-dominant brain hemispheres, but he did provide a theoretical explanation for the improvement in language function of dyslexic syndrome patients, which he claimed he could effect by altering fluid pressure in the brain through gentle manipulation of the temporal bones of the skull.[56] If he, too, could cure dyslexic syndrome, might it be because he could promote the influx of fluid proximate to the corpus callosum to "normalize" the condition for conductivity of electrical impulses across this neural network?[57]

After re-reading *Craniosacral Therapy* and contemplating the potential dangers to Dan of my attempting a craniosacral intervention I decided "it couldn't hurt."[58] In early May I spent 25 minutes one morning holding Daniel's head while he was supine on his bed. I am not experienced in such therapy, but I am an experienced potter. Wheel-thrown pottery requires the skill of exerting on the moving clay equal pressure with both hands simultaneously without allowing the clay to move the hands, except at the potter's intention, which is fundamentally the Upledger technique, but with much less pressure. The following techniques lasted from 5 to 7 minutes each:

- I sat in a chair beyond his head and exerted slight continuous pressure at the base of his head by supporting its weight;
- I exerted a slight pressure on the bones behind his ears;
- I used the "ear pull" technique as it had been used on me by a massage therapist who used Upledger techniques;

Figure 11.3 Schematic of the Neurological Routes of Sound Showing the Relationship between the Ears and the Hemispheres in Right-Eared / Left-Hemisphere Dominance and Left-Eared / Left-Hemisphere Dominance, with Strong Influence from Right Hemisphere

A. Right-Eared Dominance
Right Laterality
LEFT-HEMISPHERE DOMINANCE
RIGHT EAR PRINCIPAL PATH
Sound from Left Ear [<<] is nuanced in Right Hemisphere, following the path that is primary in B

B. Left-Eared Dominance
Left Laterality
LEFT-HEMISPHERE DOMINANCE, WITH STRONG INFLUENCE FROM R.H.; LEFT EAR PRINCIPAL PATH
Sound received in L.H. is principally via R.H. with Right Ear [ ))] diminished in its influence

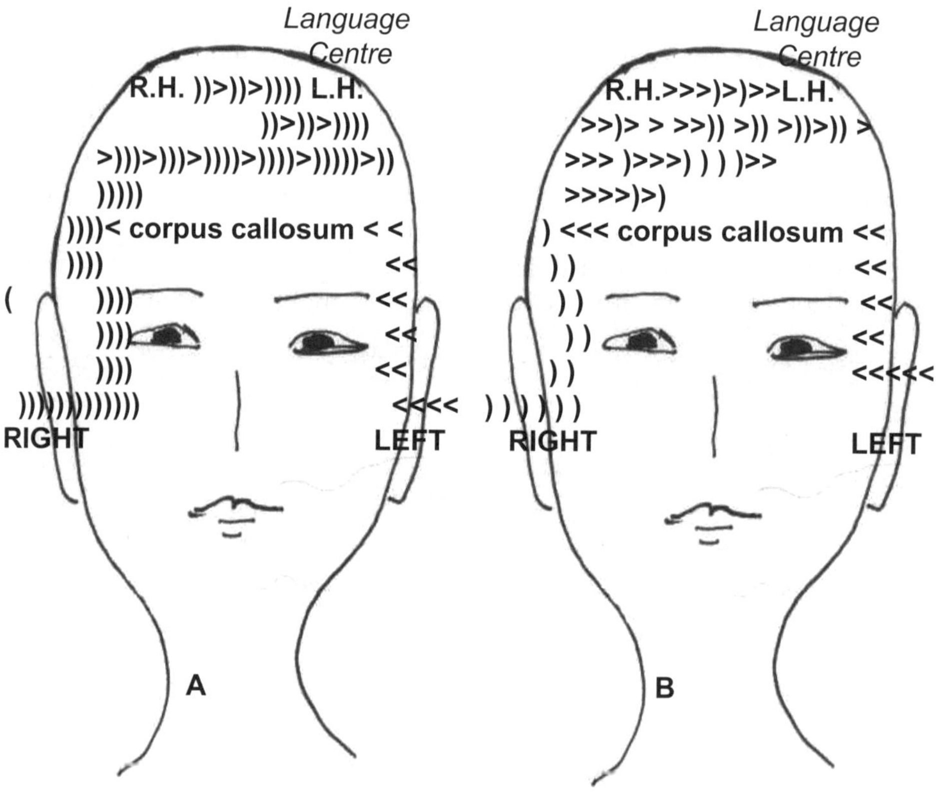

Notice that in right-ear dominance sound accessed by left ear is used by left brain as a subtext to its dominant activities, but the situation is not duplicated in left-ear dominance because the right ear is not fully functional.

- I ended the session by supporting his head again.

Remember that Dan was verbalizing constantly, virtually every moment of the day. During this intervention he became quiet and ceased talking almost entirely during the 25 minutes of the intervention. I noticed that his head had a tendency to turn towards the left. At one point he sighed. Later, he almost chuckled, which was remarkable because at this stage of his illness he had shown little or no ability to laugh. He was calmer for a while afterwards, which was the same condition of mood and of self-control that followed a listening session. In other words, my attempt at craniosacral therapy seemed to recapitulate the effect of Dan's listening to Mozart violin concerti. When I contemplated repeating the intervention a couple of weeks later, Daniel had entered a very touchy and self-conscious phase of his recovery and I felt anything that involved my touching him, even the supporting of his head with gentle pressure, would be inconsiderate and ill-advised.

The Psychology of a Non-Dominant Brain

Clearly, Daniel's "psychosis"—his confusion, his ever-present awareness of his "subconscious" of dream and imagination, and his peculiar artistic compositions—all derived from his right brain. He also associated this part of his brain with his dyslexic syndrome—"The Confusion" he had illustrated at The Listening Centre (Chapter 4)—and with his drug experiences ("my drug body"). His left hemisphere kept him at the task of organizing, making sense of, and attaching meaning to all that his metaphorizing right brain "came up with" every two minutes. As the left hemisphere also is associated with action, it kept him at his task persistently. The "holistic" world view contributed by the right brain influenced by the limbic system, however, did not meet with a warm reception from Daniel's left brain (he called it "The Attack") and was experienced as a sort of verbal assault by those in Daniel's environment (not usually directed towards them, but towards himself). The pictures and explanations of the symbolic and imaginary characters from his psychotic cosmos, while imaginative, could not, as far as I could discern in the early stages of this episode, help to connect him to reality. And the pull away from reality by the right brain was supported in some way by his limbic system: his right brain *wanted* to stay in the realm of the unreal, psychotic, and (to my view) irresponsible. In other words, he did not merely slip from one condition to the other peacefully. It sounded as though a battle was raging and each side wanted to win. The cycle that was reproduced every four minutes was fuelled by something in addition to the left brain's impulse towards rationality.

Dominance and Reason

I can remember how, as I child, I had thought that adults make their decisions "thoughtfully," that is, according to reason. As I matured I noticed that few people consistently rely on their reason in their decision-making and that more frequently their feelings provide a basis for decision-making. By the time I was midway through university, I had come to think that people try to make their decisions, at least the one's they deem important, according to their system of beliefs, which underlies both their rationality and their feelings. In Daniel's struggle over the vacillating dominance that doubtless occupied all his waking time, except perhaps when he was making music, his personal concept of God was prominent and perhaps pervasive. (I assume that for a person raised in a different religion the so-called "religious obsessions" would have featured a religious figure other than Jesus; for those raised without a recognized religion, whatever functions for belief in that person would dominate the effort to reassert rational thought.) Even his singing usually was about his interior world and, according to him, was "a kind of hymn"—i.e., was related to God even when a God-referent was not apparent to his audience. Later, when he was closer to healing, his right-brain utterances about religion sometimes became extremely negative about Jesus and ran most of the gamut of what the Church officially condemns as blasphemy, Satanic, and "possessed." He often moved from very-hard-to-tolerate (for me) images of "the ripper wars, abortion of babies" and venomous and vulgar rejections of Jesus to sweetly sung hymns and calm affirmations of his faith in God. If the non-Christians in the household found Dan's constant references to Jesus aggravating and "proof of the fanaticism of his upbringing," the two Christians closest to his twisted stream of consciousness needed patience, tolerance, and objectivity to endure the "blasphemous" rants. (It was during these rants that the similarity of these episodes with the symptoms of Tourette's syndrome became clearer to me.) What was abundantly clear was that Daniel's belief system persisted, albeit in distorted form, even when his cerebral hemispheres did not communicate with one another normally. I wondered about routes of communication in the corpus callosum and the "directionality" in the messages that did move between the hemispheres.

I looked for an explanation for *Dan's concept* of why his hemispheres did not work synchronously. I had read in *The Mozart Effect* that "providing the subject with a piece of paper on which a circle has already been drawn focuses him or her and elicits more revealing imagery."[59] I provided circles for Dan, along with the mandala patterns, and he more easily drew images of his own design inside these circles. His colouring exercises also reflected the lack of synchronization of his hemispheres in their asymmetries. He showed me an asymmetrical addition he had made (as he frequently did) to the mandala design:

> "Look what I did, Mum. It's like a feather; a Mohawk feather or som'p'n."
> "Why do you do that, Dan?"
> "It's a place that's hurt and beaten up that needs to be fixed."
> (23 April 2008)

His explanation of why he departed from symmetry in his mandala-colouring resonated with the four-part circle that introduces this chapter: his association of his Native Canadian friend Matt with the "hurt and beaten up," the downtrodden, the "crucified," and the outcast, and of his identification especially with Matt, but with each of them. Red and yellow, black and white—and a part of one of them was injured. Or perhaps all of them had been injured in similar ways and Daniel knew it.

12

Tomatis's Golden Key

The Neurology of Integration

> *"You will see wonders in the heavens*
> *and at My feet"*

Reason roams the corridors of raw imagination
rummages through rooms of memory
peers into cellars of dreams seeking
in her unruly twin for the roots
of preposterous suggestions that rise
unbidden to her ordered realm.

> *"You will see wonders in the heavens*
> *and at My feet"*

Whence the clarion hung like the full moon
eclipsed in its Japanese lantern gentle splendour?
What joins the distant marvel to our riven earth?
Between the uncontrollable halves of the universe
the immutable cosmos and interior mystery
ribbons of thought ripple and turn in every imaginable hue
Where do they collide, intersect, or weave
and to what purpose between the exploding suns
in the absolute zero of interstellar space
and in the dissolving flesh lit by splinters of the same Light?

> *"You will see wonders in the heavens*
> *and at My feet"*

The earth is your footstool, Lord.
Yours the knitting together of the crazy quilt of life
all beauty, pain, all mysteries, all journeys, joyous meetings
"Come to Me, go there, stand still
watch the river rose-golden flow
down and out from My holy hill
despair not of the flesh
I am in all things whispering
I am that I am
that loves."

(July 1996/ 8 July 2008)

Hemispheric Integration Disorder

The realization that I had solved the puzzle of Daniel's alternating cognitive conditions shone like a pearl revealed in its shell. The sense I had had when I first read in Paul Madaule's writing about Alfred Tomatis's discoveries, namely, that his learning was on the cutting edge of understanding about mental illness, now was confirmed in my own observation and analysis. That revelation, I knew, would apply to other behavioural phenomena I had observed, both in myself and in others. As I grasped that the alteration in cognitive conditions Dan was experiencing was the result of a switch in function due to an uncontrollable shift in the dominance of one hemisphere relative to the other, I could understand why Dan's constant verbalizing changed *in content* as well as *in coherency* within the four-minute cycle. I understood fully now that the right ear–left brain connection involved in focused listening was not simply a means of "strengthening the language centre in the brain" or of establishing laterality, but was the means of establishing the dominance of that left hemisphere. And that it is the dominance of that hemisphere that determines laterality. I had discovered that *the right middle-and-inner ear is the "microchip" for establishing the brain's left-hemisphere as the individual's "control panel."*

I could see that his "egocentricity" was a result of the loss of the essential mechanism of self-control and that the effort towards regaining that control was the necessary preoccupation of the rational half of the brain. I understood that megalomania had nothing to do with a desire for power but with a breakdown in the brain's capacity to assign the locus of self-control. And I could see that much acculturation reinforces the neurological tendency towards left dominance and why, in most people with healthy ears and a modicum of schooling, that process is fairly well established during the early years of life. (Tomatis himself received less than a modicum and showed signs of less well-established left dominance in childhood.) For someone like Daniel, the normally self-regulating process becomes so only when the ear is "retuned" to (and by means of) higher frequencies of sound. The correlative to established dominance is, of course, the loss of dominance, which might occur at any time in the life course from prenatal onward to produce the range of mental and other physical disorders. Those behavioural scales openned in my mind's eye like unfolding fans.

I could see how social factors militating against left-hemispheric dominance and fostering the urges of the right brain could undermine the level of stability of left-hemispheric dominance in a person, perhaps even to the point of panic attacks or more severe mental instability. Adding chemicals such as alcohol to the equation could ensure hemispheric instability—the sorts of behaviours our AA friends called "dry drunks." Offering a person social licence for certain kinds of emotion-based behaviours will tend not only to encourage those behaviours but also to diminish any potential for control of those behaviours by the left brain. To many, such an observation is a "no brainer," but the neurological implications of whether a society or culture intends to support one or the other trend in hemispheric dominance is of profound consequence—as is the importance of recognizing that some persons' ears do not *allow* them to develop hemispheric dominance in the normal way so that they *cannot* assert control over their behaviour even when they very much want to. This leads to the sometimes very elaborate rationalizations that make a virtue of necessity.

The shifts in mood predicted at The Listening Centre during Dan's "tuning," which would accompany that buildup towards dominance, were taking place every two minutes in Daniel. Some sort of "kindling" effect, or escalation of emotional (which is electrical) energy, seemed also to be attached to how that constant swinging of the mood pendulum was expressed if Dan found the circumstance of persistent alternation distressing. In fact, as the moment arrived for the uncontrollable switch away from his two-minute interval of left-hemisphere rationality and dominance, Dan would experience, and verbalize, the upsurge of the dominance of his right brain as "an attack" that he could not repel, except by entering into it as a soldier into battle. Was that sort of tension in his brain part of the reason for his dyslexic childhood fascination with Robin Hood and King Arthur that melded with the fantasies penned by C.S. Lewis and J.R.R. Tolkein and later with warfare and the History Channel? Did he look to those sources for ways to deal with the assault from his right hemisphere? Did all of life feel like a battle to him until he drugged himself? What differences in mood were implied if one or the other hemisphere were to assert dominance? Could one become stuck in a depressed mood by the body's automatic reversion to, or by capitulation to, or by seeking of right-hemispheric dominance (or, in right-hemispheric dominant individuals by acceptance of the left, non-dominant hemisphere)? Some of the alternatives implied in these questions also suggest a measure of willfulness being exerted; if that is a possibility, how so?

Hemispheric Dominance

As noted in the previous chapter, in normal people one hemisphere of the brain is said to dominate; but what, exactly, does that mean? For most people, according to Tomatis's theory and to evidence of the fetal language bump (the enlarged place in the left side of the developing brain of the fetus that becomes the language centre in the infant), that tendency towards left hemispheric dominance begins in utero. Dan probably had hemispheric dominance in utero before losing it. (My bookcase-building was in the late summer/early autumn, because I was obviously pregnant, and Dan was born in mid-December.) Following his healing and the establishing of left-hemispheric dominance at The Listening Centre, he again lost left-hemispheric dominance every time he took drugs. And that provided another reason, as well as addiction, for his return to using drugs, because they reproduced the dyslexic condition of consciousness and the world he knew best and had adapted to, albeit with severe deficits. Most definitely, that alternating dominance had characterized his dyslexic syndrome condition—the whole of his experience from some point before birth to the age of 16. The loss of his pre-condition—when he gained left-brain dominance—had introduced a new phase we can define as the "syndrome of healing" that began via his right ear through "listening retraining" at The Listening Centre. As the sound energy travelled new routes through his body he replaced every dyslexic syndrome symptom with its correlate in what we might call "the syndrome of the range of normal." Then, with the help of alcohol and drugs and medications and blows to the head, he had returned repeatedly to his condition of dyslexic syndrome.

Following Dan's listening intervention, as I have detailed, several of his early recoveries to the state of left-hemisphere dominance gave me the impression he had not lost his Listening Centre healing during his periods of psychotic illness; whereas, undoubtedly he had. However, by singing and playing the guitar, listening to music and creating art—all of which were among the few constants in his life—healing had been restored over and over again. During part of the last of his hospitalizations he actually had his guitar in the hospital with him and used it. Dan's deep desire to be like others prompted him repeatedly to reach out for what would establish that foundation for a broader fellowship even though he would become desperate for companionship and return to his friends in the subculture. And how would he have known how to reach out for healing if he had not experienced it at The Listening Centre?

As I reread Madaule's[1] account of his behaviour and feelings as a dyslexic syndrome individual, I saw, as Tomatis had, how issues of vacillating dominance plague all dyslexic syndrome individuals. Without assuming "psychological dependence" on me, we can easily surmise Daniel may have wanted a state of passivity, neediness, and limited responsibilities—because that partly defined the condition resulting from lack of dominance that had established his personal baseline for reality and the comforting feeling of old adaptations to his surroundings. Furthermore, the condition of the right brain when separated from the left brain produces a sense of contentedness and lethargy that is highly seductive.[2] Tomatis describes one version of this separation where the effect of "too much sound" on the right ear was transitory:

> Many years ago, while trying to prove that the right ear is the director I carried out certain experiments. The simplest one was to saturate the right ear with sound to see what would happen. Any singer who was subjected to this trial was immediately unable to control his voice. He thought that he was singing easily, as if he had been liberated. What had occurred, however, was that he became freed from his auto-control, just as if he were drunk. It was not a pretty thing to hear.[3]

The effect Tomatis calls "sound saturation" is similar in effect to "masking," in this instance in the right ear. Tomatis adds: "From some reports there is an exception: people who flip-flop dominance." He does not follow up on this recollection. Then he goes on to theorize about what happens when the effect of too much sound is not temporary:

> Functionally, it may not be exactly the same for the cilia cells of the cochleas as [the flashbulb effect] is for the eye, though we cannot be sure. Maybe intense sound temporarily shuts down the physiochemical excitations taking place between dendrites of the afferent nerves leading to the central nervous system. In any case it leaves a sort of negative remnant and for a time overwhelms the control. The same test applied to the left ear has no effect and may even facilitate the flow of sounds by getting it out of the way of the right ear.[4]

You will see later in this chapter that I came to a similar conclusion in observing Daniel in regard to his right ear and masking, although I could not avail myself of Tomatis's writing until the following winter.

I especially remembered the occasion of a full psychotic break when the recently frenzied Dan sat calmly on the deck waiting for the police and paramedics to take him to the hospital—that had been a remarkable turnaround from frenetic to calm, although the "normality" was a bit shaky, like an epileptic after a seizure, which had made me wonder if epilepsy is related to middle- and inner-ear malfunction.[5] Later, he was able to recall and talk about his trip to the hospital in the ambulance, which had been followed by his sojourn in the hospital "blissed out" on prescribed drugs.

Daniel certainly was not being offered heavy doses of pharmaceuticals; but it didn't take much Risperdal to throw the more rational part of his brain off-normal for a couple of weeks or longer. He wasn't blissed out: he was experiencing this psychotic break and it "wasn't fun," as he frequently remarked to the figments of his imagination. His overriding concern was to try to reclaim the dominance of his rational left brain. During the interval of right-brain ascendancy he "fought" desperately its demons and balrogs and vampires and orcs with whatever companions in this world of imaginary and symbolic beings he could find to stand with him, from Jesus to St. Francis of Assisi, to his Native Canadian friend Matt, to the goddess female he longs for as a mate.

What made it impossible for him to maintain this dominance? Two minutes of feeling some sort of success or respite inevitably was followed by conflict re-engaged. Particularly in the afternoons and evenings, he became too tired to maintain the fight for dominance and would permit his rational brain to rest while allowing the images of his irrational brain to dance to his brain's rhythms—a condition I increasingly regarded as the state he enjoyed when drugged, whether via street drugs or prescription pharmaceuticals. This situation harked back to the comment of Dr. G., who had observed that some persons *enjoy* psychosis. Well, if one could not *evade* it, one must somehow acclimatize to it. Alternatively, Dan ruminated with his rational mind on the horrors of this Hell, sometimes from the standpoint of his "identity" with the most evil dominators he knows from history, from Xerxes and Alexander the Great through wicked popes and Nero and Genghis Khan, to Hitler, Stalin, etc. In the earliest stages of his illness his identification with either good or evil figures sounded total: he *was* Jesus, God, Hitler, the original Adam, and so on.

One of the features of Daniel's illness—or one of the effects of his medication or some combination of both—was the loss of his short-term memory, which helped to explain why he fought the *same* battles over and over again. Herein, too, may lie the explanation for his perseveration in his childhood dyslexia that had so annoyed his siblings. During his recovery from his previous psychotic illness I had noticed the stage at which his short-term memory had recovered to operate "in slow motion." One of the signs that he was recovering from this episode of psychotic illness was that his total self-concept or identification with these symbolic figures developed fissures—he began to see himself, instead, as "standing with" them. The condition of his brain was improving in regard to his short-term memory, which allowed him to "notice and recall success" in his battles and to separate himself somewhat from imaginative figures. Daniel's ear–brain connections at this phase of his recovery are illustrated in Figures 12.1 and 12.2.

Tomatis places short-term memory near the temporal cortex, which may explain the synchronization of the recovery of Daniel's cochlea with his returning sense of temporality and the recovering of his short-term storage of memory. I tend to think the control centre for recall of memories may be located there but that memories are recalled from many places in the brain.

> Sounds are differentiated in the cochlear duct or organ of Corti. Each frequency is perceived in its own assigned location along the duct, where it distributes to the hair cells, which are specialized according to pitch. Then they are sent towards the cerebral cortex. The same localized differentiation is observed on the left temporal lobe, where the auditory area is located. This temporal area is very important for memory and concentration. It is divided into three zones.
>
> This is the area of nominative memory, especially important in singing. Information disseminated to the entire encephalon through the fascia originates in the preceding [temporal] area, flows in the direction of the pontine nuclei and thence to the cerebellum. . . . Finally, the entire information is redistributed to the whole body.[6]

Males have fewer neurological connections within the hemispheres than have females.[7] Some researchers have thought that a number of differences between males and females derive from this difference, including the tendency of the male mind to "compartmentalize" (i.e., to focus on a task while rigidly excluding other concerns), of the female mind to synthesize, of the male mind to focus to the point of obsession, and of the female mind to generalize to the point of exhaustion. About the same time as it occurred to me that Dan's recovery process might be inhibited by his options available for neurological input from either of his ears to his left brain, my research was interrupted.

Cilia Damage and Mania or Hyperactivity

The bondage of an unusually harsh winter had come to an end. Spring weather melted huge drifts of snow and a foot of ice. Dan's restless pacing was leading him out into the woods and along the road, sometimes with his dog Lox, who would escape from his collar and leash. As Dan would disappear into the woods in frantic search of his capricious golden retriever, the anxiety level was rising in all of us, so we decided to move Daniel back to our house where Dan's wandering was safer and Lox could roam in a very large, fenced yard. Just as that move was taking place, Luke and Kathleen and their three little boys needed to relocate and so Lainey's house conveniently became available for them.

The new setup for listening (attached to my computer) put Daniel opposite and facing me, rather than turned away from me as he had been at my sister's house. If I spoke to him while he was wearing headphones, he no longer "heard me from behind," but would see that I was speaking before he tuned in aurally. I noticed he was removing the left earpiece (the one wadded with tissue to reduce sound through his left ear and thus prioritize the right) to hear me more clearly and that he did that as soon as he had segued into his lucid phase. Hence, he could hear me and respond intelligibly at once, as long as what I had to say was brief and he had time to respond before he was "attacked by The Confusion." This sequence of events was repeated several times over a couple of days before I stopped to consider what he was doing. I had assumed Dan's increasing mental clarity was the result of gradual improvement in his right ear. He was showing me, instead, that his mental clarity resided more in his left-ear function. Might his hearing—not only his ability to attend—in his right ear be impaired?

I approached him to speak quietly into his right ear during his confusion phase. "Can you hear me?" Pause. "Can you hear me?" Pause. "Can you hear me?" Pause.

"I can hear you," Dan replied, "but only in between the interference."

I recalled from the website of The Spectrum Centre in New York[8] that Tomatis equated meaning in language with the ability of the listener to distinguish the high frequencies in sound. If Dan could not distinguish the high frequencies, he might be having trouble not only with flaccid stapedius muscle(s), but also in receiving meaningful sound from his right ear. The "frequency analyzing power of the ear," established by Georg von Békésy,[9] referred to the cilia in the cochlea. All at once Eric's comment from a month before came to mind—that his brother's ears could have been damaged from the sounds of the mosh pits in Montreal clubs and from the amplifiers at Punkfest as thoroughly as by the drugs he'd taken. Damage to some of the cilia would also make him "more susceptible to the effects of masking"[10]—the problem I felt he'd had as a dyslexic syndrome child in a noisy open-area classroom.

What happens when the cilia in the cochlea are damaged by sound? The effect on mouse neurons subjected to damaging sound has been studied by J.F. Willot and S.M. Lu and is summarized in this abstract:

> Responses of auditory neurons in the inferior colliculi of mice were studied longitudinally before and shortly after each animal was exposed to intense noise. Noise exposure caused expected losses in auditory sensitivity, but in 31 percent of the neurons studied, unexpected alterations of temporal patterns of action potentials were observed; certain suprathreshold stimuli that had evoked only transient "onset"

Figure 12.1: Daniel's Self-Concept in Alternating Dominance with Right Ear Damage: Right Hemisphere Interval; Sound Primarily Mediated in Right Brain Functions as he cannot order his consciousness

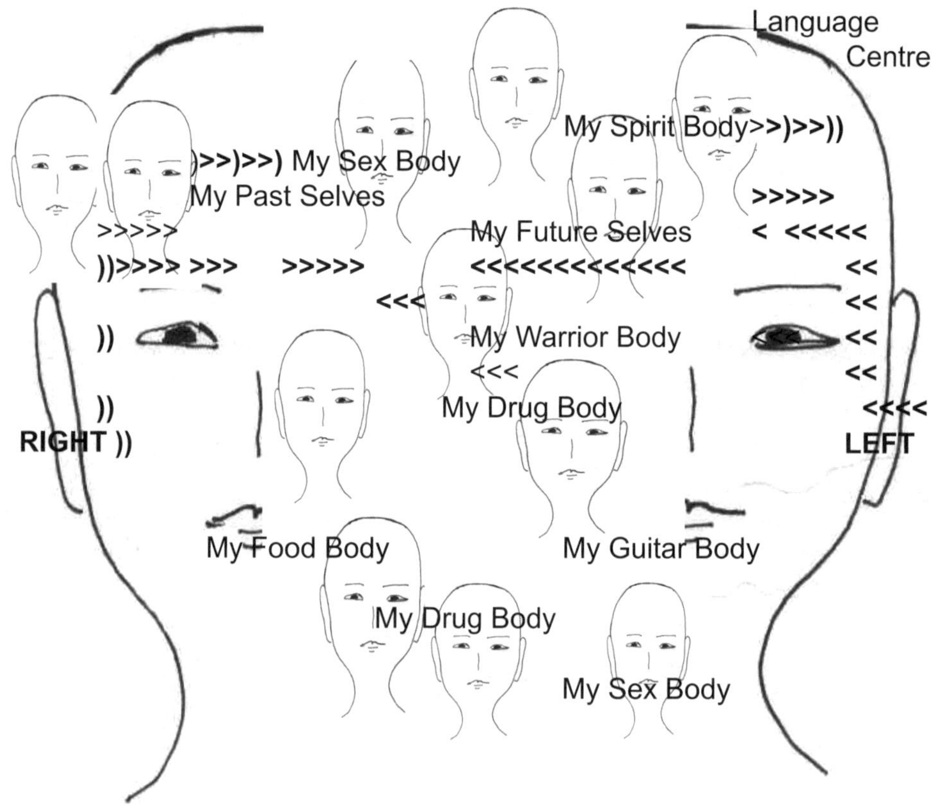

RIGHT EAR cannot register sound properly for transmission to the left brain, but sound in that ear continues to affect:

1. ear: balance; order
2. brain; see Figure 12.2 below
3. organs of the facial nerve:
   eye, tongue, salivary glands; mouth, chewing, aspects of voice production

LEFT EAR sound goes to right brain—

1. where it is interpreted imaginatively, open-endedly, with impaired visual components, and holistically—on its way to the left brain
2. Left ear hears better, which may be why it tries to dominate despite irrational or unlikely input from right hemisphere.

RIGHT HEMISPHERE DOMINANT

1. Interprets sound input imaginatively. 2. Cannot attend (pay attention), sometimes to the point of not hearing at all. 3. Self-involved.

4. Preoccupied with interior constructs.
5. Creates and invents.
6. Believes its perceptions are real.
7. Bizarre religious beliefs.
8. May be locus of rebellious feelings about the mainstream culture.
9. Makes bizarre choices of clothing, and in problem-solving.
10. Inability to control verbalizing of the stream of consciousness.

LEFT HEMISPHERE TRAILING

1. Not as perceptive spatially and tends to believe hallucinatory input from right brain.
2. Cannot control verbalization of stream of consciousness.
3. Assigns names and tries to rank and order the "perceptions" of the right brain, adding details.
4. Wants to preserve its learning in the structured tasks of drawing and writing.

Figure 12.2: The Relationship of the Ears and the Hemispheres in Daniel's Rational Intervals, Left-Hemisphere Dominant for Two Minutes as he can better order his consciousness and behaviour, including speech. The mediating brain structures depend on the right ear and temporal lobe to organize sound.

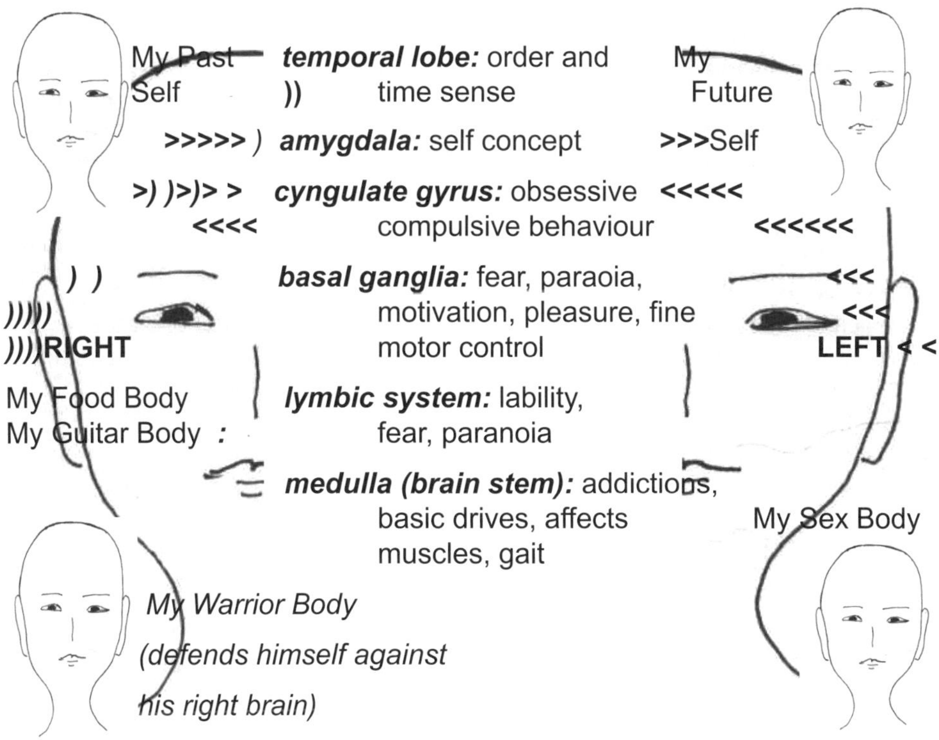

Figure 12.2 continued

RIGHT EAR sound signals cannot access left-brain control centre normally

1. effect on right eye: sends impaired visual signals to brain for interpretation
2. brain maintains fragmented self-concept
3. right ear maintains unregulated stimulation of temporal lobe, medulla, amygdala, basal ganglia, cyngulate and limbic systems, etc.

LEFT EAR sound goes clearly to right brain—where it is interpreted abnormally, yet holistically and open-endedly, on its way to the left brain

RIGHT HEMISPHERE TRAILING

1. Interprets sound imaginatively.
2. Presents entirety of subconscious to the left brain. Memories may be distorted.
3. Presents the entire subconscious as real in the literal sense—the "overall picture" is skewed towards the reality of the dream world.

LEFT HEMISPHERE DOMINANT

1. Can attend and communicate (rationally, but under persuasion by his right brain), with good syntax and vocabulary. Increasing ability to control verbalizing of stream of consciousness.
2. Horrified by changes in self. Tries to rationalize fragmented self and hallucinatory and irrational input from right brain. Speaks against right brain. Argues. Reasons. Is to some extent "persuaded" by right brain.
3. Assigns names to selves and to "attacking" concepts and images from right brain.
4. Tries to rank and order the "perceptions" of the right brain, adding details.
5. Wants to preserve learning by means of structured tasks, i.e., drawing, writing.
6. Carries on normal activities as possible, such as personal care and hygiene, feeding and attending to pet dog, good manners, getting food and drinks, caring for valued objects: art materials, clothing, doing laundry, sweeping bedroom. Performance varies greatly over the course of the illness.
7. Conventional beliefs.
8. Conventional choices of clothing, activity.
9. Co-operative in behaviours to improve his mental condition, i.e., taking prescribed medication; no other drugs or alcohol; listening to music with headphones; colour ing circular designs; doing puzzles; using a golf club for chip shots; chopping wood; sleeping at night; getting up in the morning; eating well; taking vitamin B and omega fatty acids capsules.

> responses or inhibition of spontaneous discharges prior to noise exposure came to elicit sustained excitation after exposure. Thus, noise-induced hearing loss can be associated with increases in neural responsivity and alterations of normal neural coding processes.[11]

The effect of sound damage on Daniel's right ear might be summarized in the title of the abstract: "Noise-induced hearing loss can alter neural coding and increase excitability in the central nervous system." Furthermore, it has been found that dopamine production in the cochlea is reduced in rats with damaged cochlea.[12] (I had wondered how Risperdal, which affects dopamine production in the brain, might also affect dopamine production in Daniel's inner ear.) What Dan called "The Torture" or "The Attack"—which I now realized was identical to "The Confusion" he had illustrated during his listening retraining—might be not only a switch in hemispheric dominance but, additionally, the continuous firing of electrical impulses from some damaged neurons that impacted his left brain without normal coding and thereby produced "excitability"—i.e., mania—in his central nervous system. Perhaps a continuous firing of neurons from the ear also was producing the condition that forced the shift in hemispheric dominance. And hadn't he had a cold the last time he'd gone off to Montreal? And wasn't he smoking heavily at the time (an addiction exacerbated by loss of hemispheric control[13]), both cigarettes and marijuana? Since the cilia of the cochlea may be damaged "by intense sound, lack of oxygen, metabolic disturbance, infection, or drugs"[14] and Daniel had been exposed simultaneously to at least two and maybe three or more of those factors, the possibility (on that evidence alone) of his right ear having become impaired for the processing of high frequencies seemed fairly great. Later, I would read Tomatis's own observations on the neurology of sound saturation on the right ear and the probability of that sort of saturation on the left ear "having no effect,"[15] although in Daniel's case of "flip-flop" dominance it had the effect of providing at least one clearer route of hearing/listening, namely, that from the left ear through the right brain to the left brain. That route is *not* the desirable one for mental stability.

I suspected that I had stumbled upon the etiology of hyperactivity in some dyslexic syndrome children. Most children today are exposed to far higher levels of sound than ever were experienced by children of my generation, less the children of earlier times. The same is true of visual stimuli, of course. In the early 1980s a Wang Chung video was pulled from rotation on some music video TV programs because its frenetic, quick-shifting images were found to have damaging effects on some young people's neurology, especially those with or having a propensity towards ADD. Today, that video is relatively slow-paced compared to many music videos and to some TV commercials, films, and video games. Perhaps diminished perception of certain frequencies in some children *as a result of sound saturation*, although not as easy to detect as outright deafness, explains the increase in the number of children being diagnosed with hyperactivity as a component of their dyslexic syndrome. The amount of time spent in front of a TV by young children also is being linked by some observers to a risk of dyslexic syndrome; television does not require the child to focus her or his attention with the right ear and provides a high level of visual stimulus to submerge or substitute for aural input. My friend Ruth's inability to hear certain frequencies of sound and how that disability affected her processing of information, especially of nuanced communication in social settings, is relevant to the symptoms of dyslexic syndrome, too. Although she remains one of the most voracious readers I have known, she also is left-handed, somewhat unco-ordinated, distractible (at tasks other than reading), and identifies with some of the self-concept issues attached to the syndrome. Daniel's fluctuating ability to appreciate how his behaviour affects others fits this sort of template as well.

The good news, contrary to former medical opinion on damaged cilia, is that cilia can sometimes recover.[16] Levinson prescribes various nutritional supplements to this end and Tomatis made dietary recommendations, too.[17] The Earobics[18] exercises that I obtained from The Listening Centre might also have a beneficial effect on damaged cilia. Keeping the volume low while Dan was listening with headphones seemed wise. And that same program on tape cassette, while stimulating the stapedius muscles, thus "retuning" the ear to the necessary frequencies, also emphasizes posture and breathing to improve the oxygen supply to the affected areas of the ear. Daniel had recovered before; he appeared to be recovering again. As he said to me in mid-April after an hour with the headphones, "That really helps, I feel really clear-headed now. If I could only stop talking to myself."[19]

Hemispheric Dominance: A Delicate Balance

Working late on an evening in mid-June, I was pleased to see that Daniel, who had been avoiding listening for over a week (partly because a house guest used my computer frequently), had come to sit in the rocking chair close to the computer hook-up for the headphones. He commented during the first couple of minutes of his listening session, "This really calms me." That appeared to be the case, until after about half an hour of listening he started talking and singing to himself, careless that the escalating noise he was making could disturb others who had retired for the night. He was less focused and refused to co-operate when I spoke to him; he was less able to control his voice and appeared stubbornly determined to persevere in his own singing.

A few incidents over the past months where Dan had chosen damaging sorts of music (raucous, lots of low-register sound, and played loudly) made me wonder just how eager—or capable of being eager—he was to recover. Interfering with his choices in such situations entailed unpleasant arguments, though I stood my ground in turning off the offensive sound. I had noticed that towards the end of some of his long listening sessions the calming effect had been lost. I had assumed this effect was the "energizing" result of stimulation to the stapedius muscles, but perhaps some other dynamic was in play. If parts of his ear other than the stapedius muscle, i.e., the cilia, were damaged and impacted adversely by his listening exercise, then he was getting two different effects from listening, depending on the duration and/or intensity of the stimulation. The right stapedius muscle could be responding initially by better integrating the hemispheres and perhaps is not hurt by persistent sound, but if damaged cilia become over-stimulated, "increases in neural responsivity and alterations of normal neural coding processes"[20] recur. Thus, Dan's left brain is unable to maintain the dominance encouraged by functional right-ear stapedius muscle, and he returns happily to the condition of intermittent—or perhaps even steady—right-hemisphere dominance, which for him is either the dyslexic or the semi-psychotic state. I now understood the reason for the "fragility" Dr. D. had warned of when Dan first was discharged from the Ontario hospital. As I began to think in terms of multiple factors impeding Dan's progress, I noticed his vigorous rocking in one of the rocking chairs in the sunroom and wondered if the rocking motion was a good idea for its potential effect on the vestibular part of Dan's inner ear. I replaced the rocker with his small bamboo armchair. I had come to think there were optimal conditions of duration, volume, and motion for Dan's listening exercises and that he might lack, or for some reason might want to override, an internal awareness of these optimums.

One evening in May, Daniel asked me a question barely distinguishable from his religious obsessions, but this time he waited for and listened to my response. We continued to exchange questions and answers, opinions and responses. The dialogue lasted past the anticipated four-minute cycle. We conversed for at least 10 minutes before Richard arrived with a weary plea that I come to bed. The next morning, Daniel still was capable of conversation, but he had lost that capacity to listen in a sustained manner as the day wore on. This breakthrough seemed to introduce the next stage of Dan's healing as he began to recognize many of the denizens of his right brain as symbols, himself as fully separate from his delusions, but drawn towards notions of good or evil, rationality or psychosis, by the fluctuating force in his body that put one or the other hemisphere in the condition of dominance. It occurred to me that a reduction in his medication at this point might be appropriate, but I veered towards the "safety" of conventional medical wisdom that people with mental problems should stay on medication for a good, long time once they had become stable. A couple of weeks later, near the end of June, Daniel experienced an entire day of quite normal behaviour—he maintained left-hemisphere dominance for more than 12 hours. The family held its collective breath in hopefulness. His stability did not survive the night.

A month later, however, Daniel had more self-control in social situations—we could see him improving in the way he spoke to Luke and Kathleen's children, for example, and he could maintain silence and polite replies during one course of a family meal. However, the pendulum of hemispheric dominance continued to swing as his more modulated, ongoing conversation with himself attested. He commented to me that he thought the effects of the "magic mushrooms" finally had worn off. When I asked him what it felt like to take them, he said, "It's like walking into a dream. Then you start learning things." In other words, taking hallucinogenic mushrooms is like walking into the right brain and burning your corpus callosum bridges behind you. That world of dreams had a powerful attraction for him; the evangelizing dealers in the drug subculture play on the sense of pride that can be invested in arcane "knowledge" caused by chemically altered perception. Then, the sharing of abnormal mental experience creates deep bonds among fellow-users, as shared exceptional experience does in all subcultures, including churches, ethnic groups, marginalized and persecuted groups, and so on. However, six months of separation from that milieu and away from mushrooms, Daniel thought the dream-like quality of his waking hours had disappeared. He may have been referring to his mental clarity during his rational interval. His comment hardly seemed relevant to his psychotic intervals, but perhaps he perceived change affecting that phase of his cognition as well. At the same time, however, his pressured speech—constant self-talk—continued more or less unabated.

Musical sound had a huge impact on the struggle accompanying shifting dominance, as was evident almost every time Dan put on the headphones to listen to Mozart or Bach. In January, sometimes he would be too agitated to listen for even a few minutes, but if able to listen, he would immediately become calmer. The manic impulse to verbalize would lessen, even subside. For a while following a listening session he would be able to maintain improved rationality, which sounded to me like greater dominance by his left brain. By April he had become able to control the flow of words completely while listening. During May, he gained the ability to somewhat control his spoken stream of consciousness even when he was not actively listening with headphones. In June and July, when he stopped the daily listening exercise—as he did for about seven weeks—his constant verbalizing returned. At the end of June, much of the time while he listened he was in constant, and often consternated, conversation with the familiar pantheon: Baylor, Jaylor, and Wardun; Cherulub, Jagular, Angularizine, and Medjulanium; Podamy, Zigulanium, Legular and Alphabetica,[21] and variants of these.

Three days after returning to about an hour per day of listening to Bach and Mozart with headphones, his self-control returned to about the level it had been when he had abandoned his daily focused listening. This rebound suggested to me a muscle, even so tiny a muscle as the stapedius, regaining its fitness with a return to regular exercise. How much more improved he might have been with more refined and focused sound stimulation, such as that produced by the Electronic Ear™, I could not tell. The overall improvement in his mood suggested to me that the neural pathways of his corpus callosum had been reactivated and reinforced. And, of course, he might still have cilia damage.

Schools of Thought on Dominance: The Trauma Theory

In Chapter 3 I discussed the theory prevalent in the mid-twentieth century that dyslexic syndrome was the result of "minimal brain damage." A consequence of this belief was that dyslexic syndrome was considered incurable. However, as many educators have pointed out, many dyslexics adapt successfully and some have made an amazing accommodation to their brain dysfunction. Others have not. In the fog of such an ambivalent prognosis for dyslexic syndrome, many felt that if the condition should not be altered, nevertheless it might be supported. One or another variant of such a view, of course, was the only alternative prior to Tomatis's invention of his method for listening retraining. It is well-established that children and adults who experience dyslexic syndrome may be taught to read, taught to be self-confident, taught to organize, and taught to tap into their mathematical ability. "Right-brained" as we commonly see that term in literature does not usually mean right-hemisphere dominant; people who refer to themselves and others that way are making a reference (whether they know it or not) to the processing of sound from their dominant left ear through their right brains on the way to their left hemisphere, which strongly "colours" the sound input with right-brain characteristics. These people are more likely to be slightly left-hemisphere dominant compensated dyslexics, with hemispheric instability, like the psychiatrists John Ratey and Edward Hallowell. Such people may excel at teaching dyslexic syndrome children and devising supportive strategies.[22] Thirty years ago, Griffiths actually refers to hemispheric dominance in her description of the characteristics of dyslexia in children when she writes that the learning problems of dyslexic syndrome children "may well be influenced by the relative development of dominance between the two hemispheres of the brain."[23] However, successful spontaneous adaptation to lack of dominance in the left hemisphere appears to be infrequent. Our educational institutions have only relatively recently become tolerant of left-handedness (I have spoken with many wounded adults of their anguished left-handed childhood), are unsympathetic to right-hemispheric dominance, and respond with confusion and negativity to non-dominance.

Behaviourism is a school of psychiatric thought founded on the same notion of trauma, extending the causes of mental disease from physical trauma to include a range of emotional pains inflicted on the sufferer. In Chapter 6 I discussed the behaviourist belief that we are neurological automatons to the stimuli we receive and respond to, and that fears and other mental attitudes are the result of conditioning: repetitive experiences that train the individual to respond in certain ways. In this school of thought the sum of one's "training" would explain the dominance of one hemisphere over the other.

How is it that some mentally ill people in the past have been able to recover? Two aspects of the historical treatment of the mentally ill bear on what I have learned while observing Dan. Just as the highly charismatic teacher may be able to penetrate the dyslexic syndrome listening disorder, the highly charismatic psychologist, psychiatrist, or psychoanalyst may be able to penetrate the psychotic listening disorder by strengthening the listening ability of the patient until the left brain concepts are sufficiently reinforced, including the belief that cure is possible. The patient might to some extent be taught also to "subdue" or turn attention consistently from the troublesome hemisphere. Some of these cases also may have been fortuitously affected by ambient sound. Music was an integral part of the treatment of the hospitalized mentally ill. Some readers will recall the musical event dramatized just before her character's healing by Olivia de Haviland in *The Snake Pit*. The Canadian novelist Maureen Jennings writes in *Journeyman to Grief*, "They claim that music heals you, so in the evening there is singing together or a musical entertainment that consists of caterwauling violins most of the time" (p. 126). My friend Ruth queried the author: "In your discussion of the 'rest homes' where the wealthy were treated for alcoholism I was particularly struck by this sentence." To which Ms. Jennings thoughtfully replied: "In one of the reports the director [of the Toronto Hospital for the Insane at 999 Queen Street West] mentions that church choirs visited regularly and that was very good for the inmates. He didn't mention instruments there, however, just singing." She continues, "*Slaying the Dragon* . . . is a fascinating history of treatment programmes for addiction.

. . . 'Music was used regularly for its perceived therapeutic value . . . pianos, organs, violins and harps in every asylum.'"[24]

Tomatis himself felt he was lacking in training in "psycho-pathological disturbances" and opened himself to the philosophies and theorizing of these psychoanalytic professionals, as noted in the previous chapter. These colleagues did not respond until a noted Parisian therapist brought him an autistic boy for consultation: "At this moment a woman I did not know entered my consulting room, a woman of admirable style, remarkable moreover for the richness of her vocabulary and her power of expression. It was Francoise Dolto."[25] Tomatis's admitted vulnerability to such an impressive woman, in my view because of his minimally developed knowledge of the thinking of such practitioners, derailed his confidence and inquiring mind in relation to his own science.

Theories from Evolution

What causes dominance? In the speculative realms of academic jigsaw puzzle-building within the Darwinian precepts of natural selection and the survival of the most highly adapted, some theorists propose the dual-brain theory.

> The dual brain theory claims that the two hemispheres of the brain may sense and react to the environment independently from one another and that as a result of traumatic experience, one half may dominate the other in order to reduce the traumatized hemisphere's exposure to harm.[26]

Other researchers, notably psychiatrist Tim J. Crow of Oxford University and his British colleagues, theorize that the separation of functions between the hemispheres of the brain is a result of the genetic evolution of the human species:

> . . . X-Y transpositions (e.g., in mammals) are relevant in that any such change has the potential to generate novel dimorphisms that will be immediately subject to mate selection in males by females; because the same characteristics are coded for and separately modifiable on the X, they are also subject to selection by males. Such a change on the sex chromosomes could introduce novelty into the mate recognition system that would be open to rapid and differential modification in the two sexes.
>
> . . . According to this concept language is the specific mate recognition system for Homo sapiens.[27]

However speculative Crow's theory,[28] we find him and his colleagues building upon it a theory of language-processing in the brain that has implications for their school of psychiatry and, as they leave no doubt, their understanding of their schizophrenic patients:

> We outline a bi-hemispheric theory of the neural basis of language that emphasizes the role of the sapiens-specific cerebral torque in determining the four-chambered nature of the human brain in relation to the origins of language and the symptoms of schizophrenia.[29]

The bi-hemispheric nature of the production of language that is generally known need not follow the diagram developed by these researchers. To speculate about the origins of language in the species is one thing; to impose that four-chambered paradigm (referred to as "cerebral torque") upon the symptoms of schizophrenia is quite another. I comment further on this model in the next chapter.

The Tomatis Theory[30]

Much more likely than the behaviourists' notion of "trauma" or the neurologists' theory of "cerebral torque," I think, is Tomatis's theory that left-hemisphere preference is developed as a matter of efficiency (another Darwinian principle) in processing sound input from the ears—the right-ear-to-left-brain neurological route being the shorter one. From this innate preference for efficiency derives the tendency, also, to left-hemispheric dominance and right laterality (the dominance of the right ear, right hand, right leg, etc.). But when that system of development experiences trauma, as when my child in utero was exposed several times over a period of several days to very intense sound, the normal process of listening and of hemispheric dominance, which influences laterality, is impacted and even interrupted.[31] Another principle of science ("Occam's razor") is that the simplest explanation—the "elegant" explanation—is most often the correct one; Tomatis's theory of the connection between the right ear and the left brain and all that derives from that connection is elegant in its simplicity.

Language Learning and Hemispheric Dominance

The dominance of one hemisphere "over" the other is considered to be attached to the locus of language analysis and production—most people are right-handed and left-hemispheric for language and half of left-handed people prefer the left hemisphere, too. Of course, right/left-handedness is not the only possibility; some people are rather ambidextrous and do some things righty and some things lefty. The implication seems to be that reasoning and language somehow inevitably "top" creativity and imagination and holistic or intuitive thinking and the other tendencies of the right brain. However, the command centre does not always reside in the left brain, which to some extent confounds the theorizing from language and rationality. A significant proportion of people are right-hemispheric for language, also. And even among left-handed persons, half of them are right-hemispheric for language. As Dr. I. Derakhshan explains:

> Laterality of motor and sensory control has been the subject of a recent intense study and review. . . . It turns out that the hemisphere of speech is the hemisphere of action in general and that the command hemisphere is located either in the right or the left hemisphere (never in both). Around eighty percent of people are left hemispheric for speech and the remainder right hemispheric. Ninety percent of right handers are left hemispheric for speech but only 50 percent of left handers are right hemispheric for speech (the remainder are left hemispheric). The reaction time of the neurally dominant side of the body (the side opposite to the major hemisphere or the command center, as just defined) is shorter than that of the opposite side by an interval equal to the interhemispheric transfer time (IHTT). Thus, one in five persons has a handedness that is the opposite for which they are wired (per laterality of command center or brainedness, as determined by reaction time study mentioned above).[32]

The teacher of Grade 1 children who is concerned with their laterality in reference to teaching reading and printing may not be aware of the neurological complexity behind "learning styles" among the pupils. Children's choices and abilities to understand reside in their neurology—their "wiring"—as Daniel's Grade 8 principal had observed. Dan's delayed ability to sort the sounds of language and reproduce them as intelligible ideas; his tendency to ambidexterity; his ability to read as easily in mirror image and upside down as in the conventional model; his strong imagination and creative tendencies; his extremely acute hearing; his exceptional spatial reasoning ability; his need for "holistic" understanding and youthful commitment to a world religion (as a node of psychological stability) all suggest strength in the right hemisphere together with a continuing tug-of-war for dominance between the hemispheres from the time he was young or before he was born.

The tendency of children who are having trouble with language to specialize in art and music is not simply a matter of their opting for something "easier" and "less challenging" or of inferior intellectual content. They are making their choices in the direction of the hemisphere that is claiming dominance, whether they "like" the results of that impetus or not. In fact, such choices may be fraught with social issues for such young people in some families (as my family became very anxious when I began a career in theatrical production and design, which they associated with art, mental instability, inferior earning ability, and unsavoury companions).

My interest in art has made me acutely aware of the alterations in Dan's art as he moves from non-dyslexic to dyslexic modes and from rationality into psychosis and out again. The observation that "the linguistic abilities of the right hemisphere . . . resemble those of a nursery school child who understands language best when it is fitted into the world of objects, interpersonal acts, and events, all of which sustain the meaning of what is said"[33] applies also to the art produced by Dan during his psychotic phases. In the artwork he created as he gradually recovered one could see without difficulty the developmental stages of the child from preschool, on through all the grades, to left-brained "maturity." That the maturational stages of the child usually reflect increasing left-hemispheric dominance is the message one reads in children's artwork throughout their schooling, although that process is not usually recognized in the neurological terms of hemispheric dominance by those teachers and parents who tape it to the walls and onto the refrigerator. In Daniel's colouring of mandala designs, his frequent choice of asymmetrical patterns of colours persisted as a demonstration of his awareness of inconsistency in the dominance of rationality in his brain, even while his brain was being assisted in dominance and rationality by musical stimulation.

The established associations in our culture of artists and musicians with the functions of the right brain, which may be subjected either more or less in individuals to the control of the rational left brain, may explain why so many artists, musicians, and people thought of as creative "live on the edge." They may have neurology with that tendency, and social expectations in our day have become sympathetic to the "liberated," i.e., less self-controlled, lifestyle. Left-hemisphere dominant people also become artists and musicians, certainly; but the differences in the emphases in the art they produce may appeal to different audiences. Some artists may be studied in terms of their hemispheric function. A striking example is Vincent Van Gogh, whose early life displayed some of the symptoms of dyslexic syndrome and who exhibited in later life many of the phases on the spectrum of mental illness (see Chapter 13). His self-mutilation (he cut off part of his left ear) may have been not merely symptomatic of generic mental illness, but a somewhat conscious effort (perhaps to suppress "voices") of a hemispherically unstable person to alter via a change in his left ear the condition in his brain and to regain his mental stability.[34] A final desperate attempt at self-help; a last cry for help.

The Origins of Hemispheric Dominance

The passage from Derakhshan above suggests that handedness proceeds a priori from the lateralization of the brain. How does laterality come about? To answer that question with reference to hemispheric dominance raises the question as to *when* dominance comes about? The Tomatis theory as to how dominance and handedness develop is that dominance comes through the efficient preference of the child for listening with the right ear. That this idea has not yet permeated the medical mainstream is indicated by neurologist Michael D. Mann in the following passage, where he comments on the apparent development of hemispheric dominance (asymmetry) as it is expressed in the use of "asymmetrical language" as the child matures:

> It is not known how this asymmetry in the cerebral hemisphere comes about. Some investigators believe that the dominance of one hemisphere develops as a result of experience in early life. One theory links the development of language skills to reinforcement of handedness by parents, but the existence of left dominant, left-handed people seems to contradict this theory. Other investigators maintain that the asymmetry is part of the heredity of the individual and that it merely shows itself gradually as behavior develops. Clearly, there are some asymmetries in handedness and in neural responses to speech sounds in infants. These would appear to be innate. Yet, some verbal asymmetries have not been found in boys until 5 to 7 years of age and in girls until 9 to 11 years of age. Whether this late appearance is innate or learned, the difference in time of onset in the sexes may help to explain the greater resistance of girls to permanent damage to language function caused by early brain trauma. It may be that either hemisphere can become dominant, but only one does under normal circumstances.[35]

In my experience, the relative *degree* of dominance varies considerably among individuals. Dominance is developing during infancy and childhood and in the normal child is considered to be well-established by the age of about six.[36] However, it seems to me that reinforcing hemispheric dominance is a lifelong process and that left dominance never becomes strongly developed in some individuals. Consider this description of his brain function from a retired English professor:

> I've been tested several times for my dominant hemisphere and two out of three times I was classed as a "hemisphere hopper." Neither is dominant. They take turns! I'm supposedly fertile cranial material for schizophrenia or bi-polarity. . . . I've tried to be aware of my electrolytes; take my daily potassium, calcium, magnesium, etc. Maybe on those rare days when I approach satori, I'm actually sliding (electrolyting?) into right hemisphere dominance!
>
> When I had the surgery done on my left eye first, it became my master eye until I had my right eye done. Thus my master "seeing" switched (for several weeks) to my right hemisphere versus my left. I actually started having different dreams.[37]

The Origins of Hemispheric Non-Dominance

Next, we may ask how that condition of non-dominance—that is, of alternating hemispheric dominance—comes about and what its characteristics are. Answers to those questions may lie in what is known of Daniel's conditions of dyslexic syndrome and of psychosis and how they came about. His symptoms during each of his episodes of psychosis produced dramatic illustrations of the characteristics of the bi-dominant brain. I have no doubt that part of the explanation for his symmetry lay in his exposure to damaging sound, and that drugs (both street- and prescription-sourced) and "psychological" factors also played a part in his not always wanting to alter that symmetry towards left dominance.

Beneath the symptoms in Daniel's speech there may be a relationship between the hydraulic and electrical systems within the brain that pulse with a constancy of rhythm that continues, even once laterality has been established, to promote inter-hemispheric exchange. I think this mechanism operates in all of us. It is the tendency to switch back and forth that is gradually trained out of most children. In essence, we have two brains and this mechanism of shifting provides for the use of either one as an alternate should damage occur to one of them. Mann observes this possibility for brain-damaged little girls and it applies to all situations in which laterality is assaulted, whether via the ears, through drugs, disease, accident, or some other misfortune. I think the underlying pulsing, because it continues throughout life, provides us with continuing ease of access to our less dominant hemisphere. Under the influence of that pulsing we may pause to daydream, explore creative solutions to problems, or rest from left-brained focus while we explore the memory of a pleasant scene or experience. *It may be the same mechanism that promotes hemispheric integration even under the condition of left-hemipshere dominance.*

The Tomatis theory explains many phenomena and goes a long way towards answering the questions about dominance raised by Mann. My question as to whether our dyslexic syndrome child's "innate" problem—as it would appear to have originated in utero—was consonant with his drug-induced semi-psychotic state was asserted by Daniel when he claimed during this period of his mental illness, "I'm dyslexic again." In fact, many of the symptoms of his former dyslexic syndrome are evident once more:

- inability (every two minutes) to attend (ADD);
- slouched when sitting; stooped when walking; balance problems; dizziness;
- complete inability to read during the phase of confusion (dyslexia);
- disordered speech; mis-ordered syntax (milder as dyslexic; severe as psychotic);
- inability to organize belongings in space or activities in time;
- poor fine motor control—indicated by a circle of crumbs and spills around his plate at the end of a meal; less control in using a pencil or coloured marker (trouble keeping colour within the lines, lots of broken pencil points and quickly worn-out markers);
- primitive artistic renderings of forms (in terms of normal age-expectations in the rendering of details) or representations of the human body in artworks (right-hemisphere dominant in artwork);
- perseveration in speech and in activities (milder in childhood; full-blown manic compulsion and obsession in psychotic illness);
- emotional lability;
- poor self-image;
- gesturing with the left hand and arm and twisting his body to the left while lecturing the denizens of his psychotic world;
- speaking out of the right side of his mouth, lisping in childhood, more severe in psychosis: sometimes with a clicking sound, salivating profusely, and showing some degree of nystagmus (involuntary movement of the eyes caused by disorder in the labyrinth of the ear);[38]
- inability to converge or diverge the eyes on approaching objects with consequent fear of collision and avoidance behaviour.

As a dyslexic syndrome child Dan had tended to lisp, which has been evident in his psychotic condition, but much more severely. These last symptoms—which arose in the fourth and fifth months of his most recent episode of illness and recurred when he declined the focused listening and colouring activities for about seven weeks and fairly often even in the latter stages of his recovery—are related to the facial nerve that connects with the middle ear and which does not carry sufficient neural impulses when the ear is dysfunctional. All of the symptoms listed above ended abruptly when Dan regained left-hemispheric dominance, although issues such as self-image and self-confidence remained in a lesser degree.

I saw that I must revise my original impression that Daniel's Listening Centre intervention and healing from dyslexic syndrome had survived his early psychotic episodes. Instead, his return to the left-brain-dominant condition was what constituted the end of a psychotic episode just as the left-brain dominance induced by the listening intervention at The Listening Centre had terminated his dyslexic syndrome. In fact, his Listening Centre experience had taught him how to reach out for healing sound in order to restore the condition of "normal" (although not so very "normal" from Dan's perspective!) left-hemispheric dominance.

It should be noted that The Listening Centre did not consider that Daniel's hemispheric dominance was being impacted. In line with Tomatis's attention to the audio-vocal relay that reinforces laterality, the methodology of The Listening Centre does not consider that laterality is being reinforced until the audio-vocal sessions—that Daniel never received—have been undertaken.[39] Similarly, Tomatis did not consider that laterality had anything to do with the pre-born state; he supposed that it developed after birth within the audio-vocal feedback system. However, the existence early in gestation of the left hemisphere's "language bump" suggests that the fetus is equipped with a predilection to the left hemisphere—and therefore to the right ear—by the time the fetus is hearing (by bone conduction)[40] in utero.

Derakhshan, as quoted above, equates the "command hemisphere" with the hemisphere of speech and language without reference to people whose hemispheric dominance shifts, likely because the psychology laboratory settings in which testing is conducted to determine such information are geared to normal people and their ability to speak, both to instruct and to respond to instructions. The test subjects are deciding whether or not to act (the command function) under controlled conditions. The sample of subjects likely is "normal" and probably does not include dyslexic syndrome children or psychotic teenagers or schizophrenics. According to Tomatis's theory and to the generalizations Derakhshan freely makes, we must assume these test subjects already are strongly right-eared or left-eared, which expresses both their handedness and their dominant hemisphere.

Children are no longer forced by their parents and teachers to use a hand for which they are not predisposed and for the very good reason that brains balk under such coercion. They would balk, by extension of the Tomatis theory, because those children lack the aural input to activate the half of their brain that easily accommodates right-handedness. Many of us remember children coerced towards right-handedness in the classroom and their unhappiness has given us an inkling of the feelings of the person who has had an established dominance that was interfered with. What happens when a child who has become left-eared and left-and-right-brained encounters a classroom situation where he does not want to appear different from his classmates and attempts successfully to become right-handed? How does that activity impact the locus of control in the brain? Is that the point at which the child has to "hold the picture of it in my mind and then draw it"—the characteristic in Daniel that alerted Paul Madaule to his "suitability for listening retraining"? Madaule[41] and Tomatis[42] seem to think so. Daniel's symptoms support that explanation.

An interesting finding about laterality is raised by Dr. Jerome Goldstein in his recent presentation to the San Francisco Neurological Society:

> Qazi Rahman of Queen Mary University of London felt that handedness might have something to do with being gay. [Of] 23,000 men and women in North America and Europe studied in the year 2000 [it was] found [that] being non-right handed . . . [carried] an increased chance of being gay in 34% [of] men and 90% [of] women. He concluded that his findings might have indicated a biological "error" where he more appropriately should have concluded that his findings were a biological "fact".[43]

While the wording of this report is ambivalent in terms of predicting or describing sexual orientation, it would appear that the condition of the right ear and issues of hemispheric dominance that determine laterality play a role in sexual choices and behaviours. It is not difficult to see how a lack of strong dominance by the left hemisphere would affect the ability of that hemisphere to act upon the impulses arriving in the right brain. The other relevant issues of belief system, traumatic experiences held in the right hemisphere, structures of rationality, and environments favouring one or another predilection are too complex to address here.

Altering Hemispheric Dominance

Taking steps through listening interventions to alter the ear preference and brain dominance of a small child still developing hemispheric preference is accomplished without apparent ill effect, as thousands of happy clients of listening centres throughout the world can attest and as I had observed in my friend's little boy, whose behaviour and mood altered in the socially desired direction and apparently to his own comfort and well-being. Altering the hemispheric dominance of a teenager may create a mental shift of major proportions, although Madaule himself weathered the experience with great success. But what happens to the "command hemisphere" in a teen who also has become an addict, a condition in which the impetus of the brain stem has formed particular neural pathways linked to unhealthy behaviours? Is it possible under those conditions that neither hemisphere, whether it contains the locus of speech or not, confers cerebral dominance, but that the impulses towards action pulsing from the "basic drives" of the brain stem are regulated in the frontal lobes of both hemispheres, not necessarily in a dominant hemisphere (although they will be rationalized by the language centre)? That appears to be how Daniel's brain functions in dyslexic syndrome, under the influence of drugs and alcohol and medication, and in psychosis. No wonder Daniel said when he gained left-hemispheric dominance (although The Listening Centre did not regard his improvement in those terms) following his treatment at The Listening Centre, "It's like brain surgery." Note, however, that Daniel was accustomed to left-brain dominance in the sense that he experienced that condition every two minutes and had all his life. But he was equally accustomed to right-brain dominance and through drugs had accentuated that condition and made it more persistent.

That Daniel has a verbal command centre located with his language centre in the left brain, "which builds the reality structure through analytical processing, [and] responds to stimuli by providing very specific knowledge and past experience,"[44] has been demonstrated by his constant effort to subdue ("slay," "kill," "deal with," "put down," etc.) the preposterous images rising from his right brain. During the early phase of his recent episode of illness when his grip on reality was most insecure, his intent on obliterating these "attacks" of his "subconscious run riot" was quite alarming. We kept him in sight at all times and kept visitors and other stressors away.

> . . . the left hemisphere integrates our perceptions into a localized matrix area. This makes it more conservative; it works well in a stable and familiar environment. In contrast, the right hemisphere can anticipate and adapt to change, and can maintain a stable matrix even in fast changing and unpredictable environments.[45]

This information altered my view of the video games to which Daniel had devoted so much time. Rather than opportunities for high-pressured rational decision-making, they were exercises in right-brain adaptation to "fast changing and unpredictable environments." Fast-paced television and film doubtless had the same effect. As such, they tended to *discourage* the dominance of the left hemisphere. And this may explain the connection some observers have made between television viewing by small children and dyslexic syndrome. Daniel's speech and conceptualizations were continuous throughout the four-minute cycle but they altered substantially according to their points of origin. Since the ability of his right brain to maintain a "stable matrix" requires communication with the left hemisphere (via the corpus callosum), which was diminished in Dan under the residual effect of magic mushrooms and marijuana, its matrix offered its dream world to a left hemisphere that could not adequately access or correct it. At times (in the third month of his recent illness), it looked as though the right brain was having some success in persuading the left brain of the "reality" of that dream world, causing the rational left hemisphere to forgo more and more of its grip on reality. His lack of care for his overall condition—a trait that had been expressed in some ways in his dyslexic condition and perhaps explained his inappropriate choice of music—was like that of a victim of stroke to the right hemisphere, who has lost his/her capacity for global patterning:

> Victims of right hemisphere stroke, therefore, have poor awareness of global pattern and possibilities. Their failure to follow treatment derives not from a lack of understanding, but from indifference to the adverse effects of their condition.[46]

Other notable features of Daniel's early weeks of illness were the monotony of his voice, the limited range of his musicality, and the loss of his sense of humour, all typical symptoms of schizophrenia, but also typical symptoms of damage to the middle and inner ear:

> Damage to the right hemisphere can also cause subtle language deficiencies. Although it is not involved in decoding the structure of language, by connecting experience with the overall cognitive structure, the right hemisphere can grasp extralinguistic context, such as emotion, humor and metaphor. Consequently, patients with right hemispheric damage may fail to understand a joke, or may develop a robot-like, unemotional tone in their speech.[47]

It would be interesting to know if these patients with "right hemispheric damage" had sustained a generalized damage to the right side of the head and thereby might also have sustained right middle- and inner-ear damage in the events that traumatized them, affecting their receipt of aural input to the left brain and disturbing the integration of their hemispheres.

Daniel's endless conversations with himself were not simple exchanges like a conversation between two normal people. They were more like a one-sided conversation where you knew about the "shadowy others" by how and what he said to and about them. Where did his multiple "opponents"—nine or more self-concepts who challenged him during the first three months of his recent episode—reside?

> In humans, stimulation of the amygdala leads to sensations of fear, disturbances of mood, a sense of unreality, and distortions of bodily perceptions.[48]

Evidently, the amygdala, which is part of the limbic system, the seat of the emotions, was malfunctioning in Dan's illness, too. Why? Like the turning of the maple leaves from green to red in autumn as they lose chlorophyll, could this effect in the amygdala be the result of the *absence* of a balancing neurological impulse from the ears? As some sound energy from the ear normally passes through the medulla and limbic systems, could the residual activity there, in the absence of sound energy crossing the corpus callosum, set up an abnormal amount of feedback from those regions of the brain? Or was this part of the brain receiving the electrical "spray" from damaged neurons and becoming over-stimulated? Of course, the brain stem was tangled with the addictive practices that precipitated this condition, and was influencing the limbic system as well. These are questions I cannot answer, but when normal left-hemisphere dominance was restored in Daniel's brain, all of those symptoms stopped immediately.

What Music Can Do

When I contemplated, even with my rudimentary lay knowledge, what had gone wrong in Daniel's brain, I was astonished at how he recovered. Tomatis recounts stories of amazing recovery, too.[49] One of the delightful signs of Daniel's healing in the past had been the renewal of his ability to see the humour in a situation, the light in his eyes when he caught a joke, the giggles that would assail him as some of the images from his right brain became the objects of humour rather than of horror.

Professor Lars Heslet, in his article explaining the applicability of the medical musical program developed in Denmark and being used in hospitals around the world, describes the role of music in integrating the hemispheres of the brain:

> A number of studies of the electrical oscillations of the brain under the influence of music have now been done which appear to confirm the "Mozart effect." From a neurophysiological perspective, one could say that the brain's electrical oscillations have an acoustic structure which swings like a pendulum around a point of equilibrium, from chaos to order, from disharmony to harmony. Such structures usually influence us unconsciously.[50]

I was not able to study the electrical oscillations of Daniel's brain, but under conditions both with and without the influence of music I had observed the swinging of his cognitive state from irrationality to greater rationality—with almost perfect regularity: the four-minute cycle. In Daniel's recent episode of illness, the unconscious had become not only conscious but constantly expressed verbally, the antiphon to his efforts to subdue it. His brain's swing from chaos to greater order was out in the open for anyone to observe. And I had observed the same oscillation to take place about every two minutes under the conditions of entirely separate psychotic episodes, and twice had recorded them while noting his functional ability both in counting numbers and in reading during earlier, widely separated times in his life. I knew, too, that music, especially when combined with circular-form art activities (e.g., colouring mandalas or drawing within a circle), influenced that swing to produce dominance of the ordering mind (the left hemisphere) and the recession of the imaginative, anticipatory, and adaptive mind (the right hemisphere) including its "psychosis"—"The Confusion" and chaos.

Heslet continues:

> It has been shown that music causes better synchronisation, resonance and coordination between the two halves of the brain (left and right), by stimulating the corpus callosum, the electrical and nervous connection between the two halves of the brain. It has also been shown that the distribution of oscillations between the two brain hemispheres can be influenced using music, especially in patients with depression, where the electrical activity moves to the opposite brain hemisphere as a result of music.[51]

Here is further evidence for the link between mood swings (the recognized symptoms in bipolarity) and alteration in hemispheric activity. Evidently, the distribution of oscillations via the corpus callosum in Daniel's brain was temporarily reinforced when sound stimulation was applied to his right ear—the short-lived improvement in mood and cognitive function evident in each listening session. As Heslet reminds us, "the effect of music [on brain wave patterns] is of short duration."[52] However, a *cumulative residual effect* in Daniel was evident concurrently. This improvement could be a result (partly) of a masking effect by the music of the "static" created by the firing of damaged neurons in the cilia, and it could be a result (partly) of an improvement in hemispheric co-ordination due to stimulation of the corpus callosum, because even his left ear (while stimulated) carried that stimulation to the left brain via the corpus callosum. Primarily, I believe, his right middle–inner ear muscles were developing sufficient tone to send better-regulated neurological impulses to the left brain, although they were not sufficiently fit to continue that action without musical input.

When the cumulative effect of the appropriate frequencies of sound stimulation has corrected the damage to the inner ear so that the stapedius muscle functions spontaneously and steadily transmits comprehensible sound messages to his left brain, the right brain "subsides" under that transmission of energy. It no longer vies for dominance with the left hemisphere, unless unusual stimulation from the limbic system—the emotions—occurs. Consequently, the levels of consciousness are reordered and the equation of the right brain as the subconscious and with sleep or deep rest is reasserted; the creative impulses are available, but under control so that they may be called on more or less at will, rather than obtruding compulsively on his consciousness.

The regulation of the aspects of brain function just mentioned occurs close to the limbic system and implies that the limbic system thereby also is regulated. As I noted in March as a sign of improvement in Dan, he stopped eating compulsively and expressed reasonable satiety. At the outset of this episode of illness he was underweight and had been eating very erratically as he careened towards illness and full psychosis, so a certain amount of nutritional catching up was healthy. Nevertheless, I had regulated the amount of food available to him, because he tended to consume whatever was in sight and drank (milk, water, juice, tea) more or less constantly. As his brain function improved, his self-control in regard to what and how much he ate and drank improved, too.

> The acoustic nerve runs to the middle brain [which connects the pons and the cerebellum to the cerebrum] and is located very close to the

> centre which regulates hunger, satiety, metabolism, and level of consciousness, as well as the centre which controls overall hormone regulation. One can therefore easily imagine that stimulation of the middle brain, which is directly connected to the ear, could affect our degree of alertness and our hormonal status.
>
> The middle brain is our emotional brain. It plays a central role in the regulation and mediation of the effect of music on our basic mood. One could imagine that the activation of codes located in the middle brain's emotion region, the "limbic system", triggers positive feelings. [This would provide a] basis for the belief that this part of the brain is receptive to harmony and consonance, and that harmonic musical stimulation can provide a foundation for the integration of sensory perceptions. The most consistent clinical effect of music therapy on conditions such as schizophrenia and depression is improved communication ability and social interaction.[53]

While I think a more fundamental mechanism is operative in the influence of music on mood than Heslet here suggests, nothing I have encountered implies that both avenues of influence might not apply. My own listening, singing, humming, and exposure to Gregorian chant were a significant part of my survival strategy for maintaining my energy and my optimism. I am sure the Gregorian chant cassette tapes that Daniel soon learned to turn on by himself made an important contribution to controlling his mood and soothing mine.

Questions about Daniel's Medication

Excitability in someone who does not have control of his or her brain is alarming, and that is why we had increased Daniel's medication by a small amount to control his lability (emotionality) even though it may have impaired his ear function and probably slowed that aspect of his recovery. Only when Daniel had made significant gains did it occur to me to look again at Dr. Huggins's protocol for Julian Orchard. Thelma Wheatley writes of her discussion with the doctor before she agreed to his prescribing Risperdal for her son:

> But how do you know which neurotransmitters Julian produces to excess? After all, even Dr. Huggins could not see inside Julian's brain.
>
> By his symptoms and behaviours. The behaviours reveal what is likely going on in the brain. For example, hallucinations, paranoia, delusions such as believing you are Jesus Christ and can walk on

> water, indicate 'positive' symptoms of schizophrenia and a likely excess of dopamine [causing] dysfunction. On the other hand, obsessive-compulsive disorders, panic, agitation, anxiety, etc., indicate a noradrenaline and serotonin [presence causing] dysfunction. It was not difficult to distinguish which Julian exhibited at any given time once you recognized the abnormal signs and symptoms.
>
> . . . [T]he advantage of Risperdal was that it possessed the capability of down-regulating, or blocking, noradrenaline, serotonin and dopamine-driven dysfunction at a low dosage range (0.25-1mg/day)—but only where the noradrenaline and serotonin-driven dyscontrol exceeded the dopamine-driven component.[54]

The problem with Dan's oscillating brain was that the observer would be hard-put sometimes to know when the dopamine-driven component of his behaviour was lower. Since he needed the medication initially to reduce the effects of too much noradrenaline and serotonin, when he began to take the medication (and again when his medication was increased), it presumably tipped the balance towards an excess of dopamine—his schizophrenic-like delusions and hallucinations became stronger. Once he had accommodated to the medication after a week or two, the balance seemed to be again in the direction of slightly greater noradrenaline and serotonin. But only slightly. Stress reducers like music, food and drink, or a long shower altered the balance in a positive direction. Stress and fatigue had an opposite—and negative—effect. As he gradually improved, it looked to me as if the medication increasingly was part of the problem rather than a benefit. Dan had recovered in the past, especially when on very low dosages, either with the help of its benefits or in spite of its side effects—or as a result of both those pushes and pulls. Now we were gambling yet again that he would recover while under its influence.

Variety in Music for Focused Listening

In early April I saw that Daniel was getting less out of his listening sessions. He no longer reacted with calm or with the energizing effect. I recalled from the movie *Lorenzo's Oil* the block related to age-appropriate reading material that Michaele Odone had encountered in her reading to Lorenzo. Could Daniel be bored or no longer effectively stimulated by the music he had been exposed to almost daily for months? I switched his listening program to a CD of *Antonio Vivaldi Violin Concertos*[55] and Daniel's pleasure in sitting down to listen returned immediately and the Tomatis Effect became evident once more. My analysis was affirmed when I came to Campbell's observation on "entrainment," i.e., the openness or receptivity of the listening ear to the offerings of sound:

> Just as one can develop a tolerance for certain medications, one can become habituated to music. Thus, listening to the same auditory stimulation repeatedly—even the most magnificent compositions—can interfere with entrainment.[56]

In this way, I learned that variety in Dan's listening program was important for a number of reasons, and after that I introduced different CDs of Mozart, Bach, and Vivaldi, although for him the Vivaldi tended to be over-stimulating and he frequently chose Bach over the other composers.

Wanting to "Have It Both Ways"

The notion is commonplace: we want strength of mind and purpose, and we want pleasure. Most of us will optimize that trade-off if possibly we can. And this episode with Daniel demonstrates the way we try to achieve that compromise of "having one's cake and eating it, too."

Daniel experienced two days about two weeks apart (in May) where his mental function appeared to be quite normal, but each time he relapsed to a prior level of function. Richard felt an element of will had entered into the picture and that Daniel's "psychology" was operative in this process. Madaule recounts Tomatis's observations of ear function fluctuating according to the overall emotional state of the individual.[57] Just as we can close our eyes by dropping our eyelids to block out visual stimuli we don't want to see, we can close down our ears by allowing our stapedius muscles to block out or minimize sound. This action also has the effect of focusing our attention elsewhere, that is, into the non-dominant hemisphere. "I'm not listening; I will do whatever I want to," is thus enacted. By this means, all of us retreat into the "daydreams" and the world of the imagination in our right hemisphere as a form of escape. Some of us can do this more effectively than others. And persons in Daniel's condition will have to work patiently and persistently to escape from the overpowering desire to stay there. By mid-June, Daniel was very resistant to daily listening sessions, although he occasionally hummed or sang on his own. A guest's presence in the house and her frequent need to be at my computer (Daniel's listening station) probably had a bearing on his choosing not to listen. In the same interval he decided that his colouring of mandala patterns was a childish activity. Concomitantly, Daniel's behaviour deteriorated. He became more labile, more openly rude and rebellious; he reverted to talking to himself for longer and longer periods of time, and the conversation included the familiar "demons" that we had not heard much about for several weeks.

I thoughtlessly pointed out some of these changes to Daniel when others were present. His reaction was rude and resistant. The listening station in the sunroom was less private than it had been, with his aunt's presence overlapping the visit of our other house guest; but he sat down to about 10 minutes of listening. He immediately became calm and more rational, and although the session was brief, it had some effect on his mood, for he later apologized for the way he had spoken to me. And I realized I had been tactless to challenge him in front of other people. His lack of self-control did not mean he entirely lacked an awareness of others and how they perceived him.

The next morning he put on a "heavy metal" CD that was playing when I came in from the kitchen. When I told him that he was listening to dangerous noise, he became angry and argumentative. I did not back down, although the ensuing conversation was very loud on his part and disturbed our guest, who went to shower, and my sister, who went to clean the cat litter box and sweep down the stairs. The counselling session with Daniel that followed made some important connections with particular psychiatric points of view. One of the aspects of Tomatis's teaching I have found problematical is his "not fully born yet" perspective of "sonic birth" on some of his patients.[58] My discussion with Daniel proceeded along these lines:

1. I objected to the music and turned it off.
2. Dan was angry and told me I had no business controlling him.
3. I replied that I did not want to be exposed to damaging noise and did not want him to be, either.
4. He said he didn't care. (We had been hearing that phrase frequently the last month.)
5. I countered that if he didn't care I might as well send him off to the Ontario hospital. My sister intruded and said that he'd better be grateful I cared because no one else would ever put up with his mental illness.
6. He started to say I might as well send him; but suddenly took that back emphatically with, "All right then, I'll get better." I was struck, of course, by the volitional capacity implied in his statement. (Do I sense a faint smile on the faces of the behaviourists?)
7. A few minutes later Daniel came into the sunroom where I was working, plopped down on the futon still on the floor from my sister's overnight on it, and said he didn't want to go back to the Ontario hospital but that he wouldn't mind "resting for a while" in the psychiatric ward in the gen-

eral hospital. All those years ago, and yet the siren invitation of the sweet nurse still beckoned!

8. I reminded him that life in the hospital is not having a life. I asked him why he did not care about the health of his ears and the health of his brain?
9. He said that he had, "through God and Jesus," risen above such forms of damage. He elaborated on this cosmology according to the quite consistent statements he had been making for months about himself, God, and Jesus that also do not conform to orthodoxy because they involve statements about who he was and what his knowledge includes that others cannot comprehend (especially Richard and me) prior to his own life on this planet. His equality with Jesus didn't come into the argument this time. He continued to say that he was not really born of me. That his father was not his father. That he was "bastarded into this world by God" (a very familiar phrase meaning he is "the bastard son of God" as opposed to the legitimate son of God, Jesus) and that I was the mental infant and that his father is just a dumb 12-year-old. This diatribe had been repeated fairly often and I regarded it as the right-brain version of "reality" that he had of late appeared not only to be stuck with but to be choosing. It bears strong affinities to a number of sects, subculture heroes, alternative and New Age beliefs, and the vituperative anti-Christian stance familiar to all of us. Richard had remarked a short time before that he thought the observation of Dr. G., i.e., that "some people enjoy psychosis," was operative in Daniel to at least some extent. Daniel's reasons for wanting to be normal diminished with the receding of his friendships, his alienation from his siblings, and with the onerous tasks he saw at hand and that lay ahead.
10. With my rereading the previous night of Don Campbell's *The Mozart Effect* fresh in my mind, I asked Daniel why he wanted to keep on damaging his ears? Why did he choose, even after all he had suffered, the wrong kinds of music? Why did he tell me that he had his earplugs with him when he went to Montreal, but had not used them? Why did he do some of his Mozart listening with the volume turned up too loud? Did he not want to "grow up" and become rational? Did he not want to become responsible and independent? I read him the paragraph from Chapter 4 on how my hopes for him had risen at the prospect of listening retraining. Did he not want to be able to have a wife and family and take care of himself instead of having to be cared for by a mommy?
11. A cascade of his right-brain images played itself through my memory as he began to speak bitterly and scathingly of the way I had "invaded his

brain" by suggesting that the woman he might one day marry might possibly have been married previously and could have children already. Those were remarks I had made in a conversation we'd had several months before his psychotic break. He mentioned a variety of imaginative ways these realities could be explained away—the children being the "spiritual results" of masturbation, the prior husband a "spiritual form" of his reincarnated self, and so on. I perceived these explanations as the "irrational" (but logical) responses of someone who cannot bear to face reality.

12. I explained to him that every time he damages his ears he sends himself right back to the prenatal condition where he was inner-ear damaged and had not developed the propensity for hemispheric laterality that most children are born with. I pointed out that the healing he has attained over and over again he has also rejected over and over again to return to that infantile state.
13. He argued about that for a while, citing various things "he knows" from drugs that I don't and can't ever learn. His defence dwindled into an admission that "life is hard."
14. He brought tears to my eyes. "Life *is* hard, Dan. I'm so sorry I gave you such a rough start into it. But you *can* find healing and become a responsible person."
15. He acknowledged that the kind of music he likes is damaging, that the state to which it returns him is pre-healing (i.e., at The Listening Centre) and dyslexic, and that he prefers it.
16. At this point I came into a measure of agreement not only with the psychiatrist who had commented in front of Daniel that some people enjoy psychosis, but also with Tomatis on his view of the autistic child's not being able "to complete his sonic birth process"—albeit under a comprehension different from both these doctors of the psychotic individual's overall condition. Dan had been born with hemispheric non-dominance, which had exposed him during 50 per cent of his waking life, until he was 16, to a state of right-brained consciousness. As an unhappy teen, Dan had sought the psychotic state through street drugs, the state of remaining in his right brain while disconnected from his left brain; but he wanted that state to be transient and under his control. Dan enjoyed being drugged on prescription pharmaceuticals and generally didn't mind most aspects of the "stable" phases of his psychotic illness.

If some readers consider this attitude a "developmental delay" or "emotionally immature," I would reply that I think the left brain reflects a level of control over limbic (emotional) impulses due to its neurological exposure to positive learning experiences (including advantageous sound exposure and a resilient belief system) and that in any person a quite specific level (whether little or great) of functional control in regard to *particular* impulses can persist to any age.

How the society surrounding the individual "rates" these emotional stressors and levels of rational control will vary tremendously. If you want to understand a person you will have to try to walk in his or her moccasins, not try to force your own boots or high heels onto the other person. "Maturation" and "maturity" mean very different things to different people. Likewise, progress along some one or another theoretical track of maturation (age-possible behaviour, a psychologist's yardstick) and maturity (age-appropriate behaviour, which could be anyone's yardstick) should also be considered in relative terms: where did this individual start and what were the environmental forces acting on that starting point and at every point along the way? What experiences inspired change and what experiences crushed the impulse to change?

In some ways, Dan has shown great emotional maturity, both in his pre-Listening Centre stages of development and since the listening intervention, and even to a perhaps surprising extent when severely impaired by hemispheric dysfunction. Those behaviours some might call "mature" were motivated and exercised through his belief system and operated to what degree they could during his left-brain part of his cycle, despite his inability to accept and maintain left-hemisphere dominance. Accepting left-hemisphere dominance also meant taking on a conscious burden of deficits and working with humility and persistence to become responsible in a world of left-brained people who learned those skills as children. Maintaining dominance meant dealing with his addictions and taking care of his ears.

I talked to Daniel about Tomatis's theory and about his not having been "fully born" in the sense that he had been born disabled and was still learning how to accept that reality and come to terms with the humiliating process of "going back and learning some of the basics" that other people who are born without that disability learn in childhood. I reminded him that although his actual birth was the easiest of my birthing experiences, it certainly had not been for him.

17. He asked what "prenatal" meant.
18. I explained that it means "before birth."
19. He said that his specialized knowledge goes back even before that and it is in that realm he "knows" his wife. And besides, he said, "I'm never

going to marry. I don't care."

20. "Don't you, Dan?" And I countered that just because he doesn't see how he can get to the point of responsibility where he could have a wife and family doesn't mean he can't get there. I waited a long time to marry and he might have to wait a while, but that doesn't mean it can't happen.
21. He still thinks she is really pure and holy and perfect and made by God and from another galaxy (as indeed such a vision of perfection would have to be). He returned to the music argument. He said he knows some of what he wants to listen to is damaging to the ears, which he could see himself throwing out, but some of what he likes "you would call damaging to the soul," such as Bob Marley (because of his associations with marijuana). And he is less sure he wants to put those behind him. And he thinks he is as sinful as Hitler and Stalin and . . . Alexander the Great and Spider.
22. I suggested that it was up to him to get rid of the things in his life that are holding him back from healing.
23. He said that he would put the offensive CDs in a box.
24. "But that's just another way of not letting go of them, isn't it? Just like pushing things you don't want to think about into your subconscious. But they are still there and you can still revert to them rather than going forward into a healthier life without them. Right?"
25. Dan's sideways, shamefaced look said clearly that he darned well knew that, too.

Obviously, our discussion lasted for the duration of several of Dan's cognitive cycles without any interference in continuity; he seemed to have been able to maintain left-hemispheric dominance for a short while if he wanted and chose to. An hour after our talk, Dan was arguing aloud with himself in his room. He affirmed that he deserves Hell; that he has been there. Being proud of his journey through Hell seems to provide a reason for staying there. But he was looking through the CDs in his carrying case, culling some. Yet another hour passed and I noticed a couple of CDs in the kitchen garbage—barely a start; and he was outside with the splitting maul, pounding away at a large elm log in an effort to create burnable firewood though mostly, it would seem, because the "violent" physical activity was a help in wrestling with the other voices. The co-ordination needed for wood-splitting is an excellent integrative activity. He was loudly "sending away" some thoughts, and the argument with himself emphasized the dominant, left-brained side of the debate; I could imagine the other voices, but they, hopefully, were in retreat.

What might come into his life to motivate him to become all that he can be in our predominantly left-brained world? The concept of Jesus gave him a centre when his hemispheric dominance collapsed, as the belief system in each of us seems to underlie the processes of reason in the left hemisphere that is fed by both hemispheres. His recovery to this point had reached a sufficient state of balance to create a crossroads. Dan thought he could survive with either or neither hemisphere dominant. Maybe. Or perhaps he could not yet remember the attractions of left-brain dominance. Or maybe his experiences of left-brain dominance had never given him enough satisfaction to reinforce a desire for that condition so that even now he was willing to accept the slide all the way back to the psychotic state from which he was climbing. Which way would he choose?

I decided to take a step back and let him deteriorate. Richard's view was that I was at the end of my tether, stressed, and perhaps also that I was not giving him (Richard) and other family members the attention they deserve. Furthermore, "you can lead a horse to water, but you cannot make it drink." If Daniel had decided not to listen, he could probably shut down his middle ear even if I pleaded or coerced him into listening. After almost a month of very little listening—perhaps three sessions of about 15 minutes—Dan wandered into the sunroom and sat down to an hour and a half of Bach and Mozart. The change in his mood was refreshing, to say the least. Remember, too, that even focused headphone listening is a bland substitute for the Electronic Ear™ and the LiFT® of The Listening Centre. The following morning he came back to the headphones and asked for more Mozart. I'm not sure why my computer responded with the chamber music rendition of R.E.M. tunes, but Daniel didn't bother to mention the mistake. Later in the day he showed signs of over-stimulation. A delicate balance had been tipped. Had he tipped it deliberately?

13

Language in Psychosis

Speech and Integration Disorder

The remnants of a September hurricane that did serious damage in Texas swept through our area with strong winds and a heavy rainfall. I lay awake in the night following a discussion with my husband about Dan's few days of unsettled, unco-operative behaviour marked by sudden outbursts of anger nonsensically expressed; again, his medication was at issue and we disagreed about altering the dosage. Richard had become anxious; I assured him we should stay the course. At last, Richard slept. The rain fell more gently in a patter on the tin roof over the bay outside our bedroom window, reminding me of other hurricanes and storms we had survived. Every few minutes the wind would rise in a gust that seemed about to raise the house from its foundations. In his less anxious moments, Richard knew, too, that Dan's unsettled behaviour was a sign of his coming closer to integration and, like the gusts of wind, marked the end of a great storm, not impending disaster. The following day, Dan saw me pitting plums at the harvest table. "I want to do that! Can I help?" and so he prepared a kettle of plum jam and sealed it into jars. He next made a large plum flan, and continued to help with various small tasks about the house.

My printer was down and the mandala patterns all had been used, so Dan was making his own coloured drawings when he sat down to listen. Over the next three days he produced a series of drawings that reflected maturational stages all the way from a recapitulation of preschool to the junior grades. Several of his images were careful, legible renderings of his efforts at "psychological" control: "This is a concrete box made out of very fragile kinds of block and I'm putting Luke in it; it's an upside down birthday cake." I was fascinated to see that he had again made a lattice, then a series of lattices,[1] very much like the simpler one he had made just before recovering from his previous psychotic episode; the series ended in the much elaborated, three-dimensional "concrete box." His voice became lower, calm, sometimes gentle, and contemplative: "Why do I keep drawing circles with things in them?" "That's an interesting face I've drawn. That one is frightening." His manners became considerate, "Are you OK, Mum? Look, let me help you with that; you don't need to be wearing yourself out with my laundry." His words fall like the gentle rain on the roof, loosening the knot in the centre of my chest that persists from constantly remaining alert, braced, caring, discerning, observing. He has gone out into the cloudy day—"my favourite kind of day," he enthuses when he returns with a huge pot filled with tomatoes from his garden.

Re-Integration of the Hemispheres and the Belief Structure

Tomatis's teaching regarding the connection between the right ear and the left brain in establishing laterality leads us to ask, "What happens when that connection fails?" Tomatis was writing primarily to singers when he answered, "the systematic repetitions found in the babbling of babies and in stuttering . . . and what I call a left voice or a right voice."[2] These results, in my observation, apply to a minor impairment of the connection, and include such symptoms as poor posture[3] (the "pothead slouch"; Dan's inability in dyslexic syndrome to stand erect), a general malaise,[4] such as chronic fatigue syndrome, and so on. The deliberate loosening of the left brain's dominance also produces the "mystic" consciousness of prayer, the semi-controlled "ecstatic" speech of glossalalia, and other phenomena familiar to religious persons. However, Tomatis acknowledges that very serious conditions result from the loss of sound stimulation to the brain: "The consequences for the nervous system of being deprived of such healthy stimulation [found in singing] are dramatic," and he notes that a great deal of singing alters the singer's "state of consciousness" and "mind-body integration."[5] When the energy-transmitting function of the middle and inner ear is not just poor but close to failing, the left hemisphere loses its ability to maintain its dominance. I suspect this also affects the normal routes of neural impulses, a general "directional pull" between the two hemispheres and allows the right hemisphere to more strongly assert itself during equal intervals of dominance.

Clearly, the left brain rationalizes, including a person's system of beliefs.[6] The failure (inability) of the left hemisphere to maintain dominance renders its belief system less robust and open to the influences of whatever arises into consciousness from the right brain.[7] Therefore, we can predict that individuals experiencing constant hemispheric imbalance or the loss of left dominance will be likely to develop less control through their systems of belief. This is not to say their beliefs are less sincere or well-intended, but simply that their middle- and inner-ear problem prevents them from enacting their beliefs as effectively or as easily as do normal others. Of course, this "weak-ness of will" (usually judged as "weakness of character") leads to profound feelings of guilt and lack of self-worth in the person so judged, and segues rapidly into despair. It becomes evident, too, why adherence to a particular belief (or system of belief) may tip a fragile balance either way, depending on the tenets of the belief sysem towards rationality or towards right-brain tendencies.

This same syndrome operates in areas of belief that have nothing to do with religion. For example, a profoundly deaf young man in one of my pottery classes demonstrated the same behaviours in his efforts to learn wheel-thrown pottery. He wanted very much to learn and believed he could, for he kept on trying; but his unco-ordinated efforts frequently left him in tears (a reaction for which his attendant blamed his mother's caregiving—a prejudice now familiar to me). Although he made good progress, his unreasonably high expectations for himself left him disparaging of his own performance. He was unable to get enough reinforcement from praise and from regarding his own successful work to want to continue. My encounters with deaf persons have been few, but suggest to me that the discoveries of Tomatis and those I have made regarding the needs of the brain and the entire body for sound stimulation might be relevant for the deaf. Perhaps the vibration from high-frequency sound, although not heard, might nevertheless be conveyed partially via headphones to the left hemisphere of the brain. Here is another avenue for research.

The hemispherically dys-integrated believer comparing his or her "success" as a believer with others operating in the same framework of belief is at a profound disadvantage and is likely to become discouraged. Conversely, the healing of the middle–inner ear that re-establishes the connection between the right ear and the left brain and restores its dominance may also render the individual very vulnerable in the left-brain area of beliefs that has been weakened by having been deprived for so long of strong reinforcement (if, indeed, his or her beliefs ever had experienced strong reinforcement).

A "crisis of belief" may be trivial and fleeting or may be extensive and prevalent. An example of the trivial (although its potential consequences could be far-reaching and long-lasting) is a misapprehension of a momentary perception, such as Jean's (with unsteady hemispheric integration thanks to her Alzheimer's disease; she experienced psychosis) when she suffered a heart attack that would prove fatal. She cheerfully explained to the ambulance paramedics that her heart attack was "the result of the rat poison my daughter-in-law put in the baked beans I ate for lunch." She knew Warfarin, used in rat poison, also is a medication used to thin the blood to prevent heart attack and she also had known it could be used to commit suicide. Her comment occasioned an immediate blood test when the paramedics and we arrived with her at the hospital, which exonerated me from her "joking" (if it was that: she formerly had an ironic sense of humour) accusation.

When one's perceptions are being misapprehended by one's reason in an ongoing, albeit intermittent, manner, such as Daniel's perception that the objects in the room were "wavy" and moving around, the cumulative effect on the left brain is to undermine its confidence in the structures of knowledge that established its norms of belief prior to the breakdown. When this interplay of hemispheres is mild and under the control of the left hemisphere, the result for the speaker and his or her hearers is considered joking or humorous. When the interplay also involves a hemispheric disconnect, people usually stop laughing. Andy Kaufman's career in comedy that careened out of orbit provides examples here. The occupational hazards for actors with hemispheric instability become clearer from a study of his life. Similar hazards apply to the efforts to minimize hemispheric dominance practised by Zen and Christian "mystics," although their highly disciplined environments tend in some other ways to support left-dominant control.

One might speculate that the re-establishing of a damaged connection between right ear and left brain, as occurred not only in Daniel and me but in several others I have known, might render the person who has been healed nevertheless very vulnerable neurologically in the profoundly fundamental area of reasoning, i.e., the reasoning of the left brain specific to the belief system. The mechanism of mental deterioration in "brainwashing" and other forms of mind control becomes evident when the right ear–left hemisphere connection is understood. Of course, strength of prior reinforcement in that regard will make the transition to restored dominance or to new dominance less problematical. The restoration of dominance returns that person to a condition of *prior* strength, or some state less than that. The situation of new dominance may give the individual a sense of power and self-confidence, but that is untested and not yet ameliorated and consolidated by experience.

As my understanding of Dan's neurology increased, I became better able to observe how co-ordination between the hemispheres of his brain was improving and where his belief system functioned in that better-integrated brain. Thus, a gradual improvement in his self-concept and mental order was observable through comparisons of his artwork, which initially were (1) extremely primitive drawings and (2) clumsily coloured mandalas and became, sequentially, (1-A) more lifelike images of persons and (2-A) neater and more symmetrical and more creative patterns of colouring, from day to day, week to week, month to month. In addition to the patterns of cognitive development evident in comparisons of his successive forms of art, Dan's ability to respond more and more normally in social situations improved. His constant improvement was paralleled by his constant instability, a situation I had come to expect.

Variations in Language

We had decided early in this episode of Dan's illness that it would be a violation of his privacy to videotape or to record him, especially because he had said in the past that he thought such a recording would be unfair. In early July, I had noticed that Daniel's self-talk had slowed down to the point where I could record it by hand if I paid close attention. He was reviewing the usual slate of demons when I was struck by the insightful and poetic statement he made in a more lucid phase: "I am the mouth of the Gate [of hell]; Wardun [perhaps in the sense of gatekeeper, although he also used "war done" meaning "war ended"] is the little demon that I slew." Then he said, "I took over the demon-slaying from Jesus because it would have been too dangerous for Him." Perfect grammar and syntax. Next, as he slipped into the confusion phase of his cycle he mentioned images from Tolkein—Pippin—and from *Star Wars*—the Saarlak Pit. Our friend, Rev. Bern Warren, "became Warren Hastings [Spider]. Not that he's a demon, he became a Morgul son after destroying Pradular." The simplistic word association (of one "Warren" with another "Warren") in this phrase was typical of the non-purposeful associations of some of his chatter—and identical to the mistaken connections with reality all of us make in dreams and that amuse us once we awaken and remember them. More important, this speech pattern is what I thought of as his "washing machine way of speaking."

Using similar referents, he said quite ungrammatically, "He was going to be Frodo like I was sitting beside him there who would be Matt who is Jesus as a result." The last phrase, "as a result" (or "in result"; or "in result: fact"), was one he used very frequently and usually to end a sentence or a group of sentences. It seemed to convey his satisfaction with having sorted out something in his reasoning mind, even if the logic of the syntax was not normal or was opaque to the listener. A few weeks later, he would use the same phrases clearly as the equivalent of "therefore," but still misplaced according to the sequence of normal English syntax. His statement about Frodo made perfect sense to him, although the arrangement of words approached German syntax. However, just as I, a decade before, had sensed very clearly what I *intended to say* under the influence of Zoloft, while *part of my brain produced rapid streams of disconnected words*, Dan's rational connection with the words he was producing was to some extent disconnected. However, in this instance, I could transliterate Daniel's sentence.

> ***He was going to be Frodo like I was sitting beside him there who would be Matt who is Jesus as a result.***

He meant:

He [meaning "I" but referring to himself in the third person because his rational brain is objectively contemplating the image of himself in his right brain paired with Frodo]
was going to be Frodo [a simile made from the pairing of images of himself and Frodo in the right brain; but he has trouble with temporality, so the tense—past progressive—means "was like Frodo" or "intended to be like Frodo" or "tried to be like Frodo"]
like ["like" meaning "in the same sense that"],
I [referring to himself again, but in the normal first person, so the left brain is exercising its normal sense of self in contemplating a pairing of images in his right brain]
was sitting beside him there [meaning "was sitting in for him" in the sense of "representing" or "symbolizing" him]
who would be Matt [the antecedent "him" in the previous phrase now is explained as the "who,"—that would be his friend Matt]
who is Jesus [the second member of the pair, Jesus, completes the second simile: Matt symbolizes Jesus]
as a result [meaning "so I believe" or "consequently"; in this instance he uses two expressions, "as a result" and "like" (after "Frodo"), to affirm (for the satisfaction of his left brain) his having made the connection between the two pairs of associations, i.e., the metaphors, offered by his right brain]

Or, more succinctly in normal syntax:

> ***I have tried to be like Frodo in the same way that Matt represents Jesus.***

The basis of all language is metaphor, although we seldom notice until helping small children learning the language. We grow to manipulate complexities of metaphor at high speed. Daniel's sentence is an expression of "slow-motion" hemispheric integration and demonstrates the metaphorical pairing of ideas made by the right brain under the effort of reason and linguistic dominance made by his impaired right ear/left brain. In this sentence he has linked two such pairings: himself and Frodo, and Matt and Jesus. These sorts of comparisons are the way a child's learning proceeds.

The disconnected pattern of speech used by Daniel is a precise rendering of the way sentences typically are heard by the dyslexic individual, such as my friend Nancy, who heard parts of sentences with alternate ears that produced cacophony in her left brain.[8] As Tomatis said, "the larynx can only produce what the ear can hear";[9] and it is evident from my observations in this instance (as well as in many others) that a parallel notion should be applied to syntax: "the larynx can only produce what the integrated brain can sort." Order in time also is sorted in the ear as sound impulses are conveyed to the brain. Daniel's sentence indicates the trouble he has being precise in his choice of words for expressing temporality. I found significant, also, the way he began to speak of himself in the third person, but switched to the first personal pronoun once his interior concept of subject (produced in the right brain) and predicate (sorted in the left brain) had been satisfied and as he completed the verbalization of his idea. I considered the quick switch of personal pronouns Daniel made to be a sign of improving short-term memory capacity and of improved hemispheric integration. *The foundations of logic and of language depend on hemispheric integration because they are dependent on the distinctive capacities of the dominant and the non-dominant hemispheres; perception (prioritized for certain purposes in the right brain) and thinking and speech (prioritized for most purposes in the left brain) are interior and exterior manifestations of a hemispheric duet.*

The Ability to Make Choices

As Daniel said repeatedly in terms most people would consider delusional, he is God in the sense that he is the master of Heaven and Hell, the hero of his own fantasy, as Frodo is the hero of Tolkein's tale while on his way to his mission in Mordor. But that "fantasy" is part of his rational belief system, essential to his reality, too. Dan makes the choices (albeit under unusual pressure) that lead to suffering or to health, to isolation from friends and family or to communication in a world that is a mixture of problems and pleasures, disappointments and joys. We are god-like in our choices, for they determine how we will impact our environments. Whether we have "freedom" in those choices and how the environments thus created in ever-widening social spheres impact one another are theological and sociological issues too complex to debate here, although I comment on these issues in Chapter 15.

Another three weeks of deterioration passed before Dan picked up the headphones again in mid-August. His behaviour had regressed to a point that one night we were on the verge of increasing his medication or of seeking an alternative one. The next morning Daniel seemed to have noticed his own deterioration or to have answered some inner question or to have consolidated his gains, because he donned the headphones in the morning and over the ensuing days continued to choose to listen. Within about four days he had regained most of the ground he had lost during almost *two months* away from the headphones. His mood lightened, although his self-control would deteriorate later in the day with fatigue. As he resumed the colouring exercises he became more creative with the designs. He told me he was working at staying inside the lines and I noticed his hand tremor. Again, I asked myself, was this tremor the result of the medication or of the lack of co-ordination between his hemispheres?

At Luke's suggestion we added a junior-level jigsaw puzzle (100 pieces) to his "while listening" activities. The next day he got his own breakfast, making muesli from oatmeal, currants, almonds, honey, and milk. His interest in his environment improved: he expressed caring about his dog again after weeks of non-concern, and resumed feeding and watering him; he asked about other family members—Eric had been away for a few days. His verbalizing became calmer, even humorous, rather than a defiant, anxious, or frustrated rant. His voice was more modulated, less harsh. He became insightful to the point where he could explain that certain "demons" represent "greed," "jealousy," or "lying." He told me that the "demon of jealousy" is double-headed: an actor and a cook show host. These roles entail activities at which he imagines, correctly I think, he could excel. That metaphorical image represents an integration of the left and right hemispheres very close to fully rational insight.

Is this healing process under the influence of sound and hemispherically integrative activities the means by which some brain-injured people "regrow" their brains? Is it necessary to know the minutiae of a process when you see that it is evolving in the desirable direction? These are important questions to debate and answer before more fragile and vulnerable people are subjected to potentially damaging electric shock therapy, magnetism (in MRIs[10]), radiation (CaTs, SPECTs,[11] x-rays), to irresponsible surgeries and implants,[12] and to unnecessary pharmaceuticals and harmful behavioural strategies.

Syntax in Schizophrenic Language: Daniel's Schizophrenic Symptoms

Once I had determined the cause of Daniel's schizophrenic symptoms, I turned to the Internet to see whether my discovery was already known. In *Brain* I came across a paper of Drs. Rachel L.C. Mitchell and Tim J. Crow, which provides a summary of the investigations by many researchers of schizophrenic speech patterns, "Right hemisphere language functions and schizophrenia: the forgotten hemisphere?" I was interested to find that schizophrenia has long been considered a disease of laterality, although I would suggest that it might more accurately be considered a disease of non-laterality. I was astonished to find that the awareness of laterality among psychiatrists has not led them to an understanding of the role of the ear, neither in laterality in normal persons nor in the schizophrenic individual.

I have not known many schizophrenics and those I know were or are heavily medicated so that their speech patterns were/are affected. It was helpful to find all of Dan's symptoms identified by Mitchell and Crow in conformity not only with their wide experience, but in the massive literature under their review. Daniel's drug-induced schizophrenic symptoms had been thoroughly established, initially, to the satisfaction of the psychiatrists who saw him; then, extensively in my record-keeping over the years (and in his family physician's notations); and, finally, those symptoms were fully verbalized and displayed in his recent episode of illness, as I have described them here. If Daniel is not spontaneously schizophrenic, he appears to have reproduced in his many episodes of psychosis precisely the condition of the spontaneously schizophrenic brain, as well as several times to have recovered normal left-brain dominance and full mental function. Some of the conditions of Dan's post-psychotic brain function, i.e., the ability to appreciate metaphor and to use language normally, are noted by Mitchell and Crow (in quoting Drury et al., 1998), and have been recognized in other schizophrenic patients experiencing remission.[13]

The Four-Chambered Model of Language Production

The reports of such researchers as Drury into the language patterns of schizophrenics provide the extensive material for Mitchell and Crow's analysis of the function of the right hemisphere of the brain in language production by schizophrenics. Their review "advances the hypothesis that the core deficit in psychosis is *a failure of segregation of right from left hemisphere functions*."[14] This implies that schizophrenics lack normal segregation of the hemispheres (i.e., in their operations) and that both left- and right-brain functions are somehow commingled in the schizophrenic brain. These researchers support a "four-chambered model of the processing of language" in a normal brain and provide a diagram, which is based on the size of parts of the brain, the greater number of neurons in the larger areas suggesting particular kinds of development or activity there. Unfortunately, this model does not incorporate the routes of sound in the brain that is fundamental to understanding the integration of the hemispheres in language production. Applying the same four-chambered model to the schizophrenic brain, they conclude some of the chambers lack appropriate partitions so that they become joined to produce "spillover,"[15] by which they may mean the phenomenon I have termed the "washing machine effect" in Daniel's verbalizations that loosely join words by various kinds of associations with disordered syntax.

One of the difficulties with Mitchell and Crow's thesis is that they make assumptions about the nature of speech production in the normal brain, which they claim:

> . . . involves a transition from engrams that represent the precursors of speech ("thoughts" and "intentions"), which are at least in part located in [the] right dorso-lateral pre-frontal cortex, to . . . Broca's area in the dominant hemisphere, and that the perception of speech occurs in part by activation of remote association in [the] right occipito-temporoparietal cortex (the extraction of "meanings") of the activated primary sensory engrams in and around Wernicke's area . . . .[16]

As I, like most of my readers, bring only a layman's knowledge to neurology, let us examine this description in stages.

Broca's and Wernicke's Speech Aphasias

Paul Broca (1824–80) described loss or defect in the expression of speech: motor aphasia. Carl Wernicke (1848–1904) described loss or defect in the understanding of speech: so-called "sensory aphasia." The areas of the brain related to these functions—the motor production of speech and the understanding of speech, respectively—carry the names of their respective investigators: Broca's area is towards the front of the cortex and Wernicke's area is back of it within the temporal lobe. Wernicke's research built on the previous work of Broca.[17] However, it is important to note, as does the *Oxford Companion to the Mind*, that the term "sensory aphasia" attached to Wernicke is a bit of a misnomer. "This term . . . is by no means totally appropriate as expressive disorders undoubtedly occur in Wernicke's aphasia, but they are disorders *in word usage and choice* rather than disorders in the articulation or expression of speech. In severe cases, indeed, the patient's speech approximates to incomprehensible jargon."[18] Thus, it is clear that in psychosis Broca's area is *not* affected but the ability of Wernicke's area to function is impaired. Whether this inability is a result of disruption or interruption of input to the area, or an intrinsic malfunction of Wernicke's area, is not discussed.

Wernicke went on to develop a flow chart for the production of language in the brain that presumed a movement or flow from "understanding" (in Wernicke's area) to "speech production" (in Broca's area), which his pupil Hugo Liepmann used to investigate apraxia (the inability to make purposeful, skilled movement,[19] related to lesions of the corpus callosum).[20] Possibly, Wernicke's flow chart for language production is the concept that Mitchell and Crow are developing further in their theory because they allude to "the motor engrams in Broca's area" as being "qualitatively different from, although closely related to, the sensory engrams in Wernicke's area."[21] At this point, the researchers acknowledge a developmental relationship between the words that one hears and those that one understands and becomes able to speak; but the implications of that *continuous* relationship between hearing and speaking and of the anomalous development of that relationship during some persons' childhood—which produces dyslexic syndrome, and that can produce psychotic or other uncontrolled expressions of language—are not explored. Their preoccupation is between "understanding" and "speaking"; the element of perception—the mechanisms of "hearing"—has been omitted. The nature of the *input* to Wernicke's area *from the ears* has been omitted from the flow chart.

The Shape of the Brain

To develop their thesis, Mitchell and Crow note anatomical differences in structure and point out that "the precise nature of the bias is of great interest," drawing attention to the "bias" from "right frontal to left occipital."[22] Starting with this anatomical difference in size or volume, they impute certain causes to it. They assign to particular parts of the brain the storage of "engrams"—the "traces" inscribed somewhere in the brain that account for memory—as they ascribe aspects of "thought" to particular parts of the brain.

> Specifically, it is suggested that the process of speech production involves a transition from engrams that represent the precursors of speech ("thought" and "intentions"), which are at least in part located in [the] right dorso-lateral pre-frontal cortex. . . .[23]

Language must develop from experience; engrams are traces of experience in both halves of the brain. Therefore, without doubt, some language will develop from engrams located in the right side of the brain. The authors are not specific, however, about the kinds of engrams available from the right brain or why they may be significant in psychosis.

Traces in the brain that can be "committed to memory" must reside in the particular places where the brain has been altered (1) by sensory experience, (2) by reinforcement of various kinds within the brain as it works with these engrams to classify, reinforce, and otherwise use them conceptually, or (3) where the brain has been changed by conditions that erode engrams, also. Many researchers are loath to assign particular sectors of the brain to memory where such engrams might be "stored." They consider that the engrams of perception lie along the neurological routes of that perception and may be retrieved from those parts of the brain's neurology in the processes of "remembering." A particularly vivid example of this premise is explored by Mario Beauregard and some of the interpretive implications of this approach are expounded by him and Denyse O'Leary. Beauregard scanned the brains of Carmelite nuns who were remembering their states of mystic consciousness to demonstrate the reproduction of that form of consciousness (as they attested to its having recurred).[24] Notable, also, is Beauregard and O'Leary's description of the nuns' sound-deprived lifestyle and mental environment, which would tend to undermine left-hemispheric control to create that tendency to experience the mystic consciousness and ecstasies of the right brain.[25]

The use of "engram" generically rather than specifically by Mitchell and Crow makes one's interpretation of their assignments to particular locations in the brain problematical. However, the engrams of sound (which may provide "anchors" for visual and tactile engrams as well) follow particular routes and have created specialties of function in particular areas of the brain.

A further difficulty is that they assign both "thought" and "intentions" "at least in part" to the "right dorso-lateral pre-frontal cortex." My reading and observations suggest that *distinctive* forms of thought (or thought processes) tend to reside (because they are assigned as a consequence of the loop that strengthens laterality) in the dominant and non-dominant hemispheres and that *intentions*, because they are rational and underlie action (as opposed to "urges" or "impulses" associated with the brain stem and limbic system), reside less in the right than in the left (or dominant) hemisphere. The semantic distinction between "desire" and "intention" is relevant and, I think, distinctly appropriate to only one or the other of the hemispheres, the right and the left, respectively.

The Loci of Perceptions, Feelings, and Thoughts

I am uncertain what these authors mean by "thought." In their view, "thought" (perhaps just for the purposes of discussing language) is originating in the right brain. Does an engram constitute a thought? Do the photons entering the eye constitute a thought? Is a musical note a thought? Must we be able to name objects to have a thought? I remember from when I was less than a year old sensing joy in the smiling expression on my father's face when he came early in the morning to my crib where I was standing, holding the side bar, and burbling with baby laughs. I remember his saying something to my mother about how happy I always was when I awoke; he called me "sunshine," which I did not understand for a long time afterwards. I remember my perceptions, my feelings, and that I did not have the ability to make sense of some of his words. At what point of access to an image or part of an image can we assert, "There is a thought." Surely, it is only when the awareness of image or sound or tactile sensation (received photons or sound waves or stimuli of skin receptors named as nouns such as "the moon") is linked to our reactions—feelings or other perceptual pairings related temporally by verbs ("is rising") and with other associations by prepositions and other modifiers ("in the east")—that we have "thought." Nonetheless, vivid sound impressions or images or sensations can persist in memory until we have words to associate with them so that we can think about or express our thoughts about them. I think it is useful to distinguish "perception" from "thought." It may also be important to distinguish *linked* images, and images linked with feelings, from "thought" until they are endowed with relationship in silent or spoken language.

Certainly, sensory perception and the awareness of temporality, which combine to form language constructs and those associations that build a value system within the varying degrees of health (including mood) of the body across all of its parts, conspire to produce "thought." Sensory experience travels via the thalamus to particular parts of the cortex, "though there is also some mixing, with different sorts of inputs going to the same region."[26] The locations for the engrams of sensory input are, thus, only loosely associated with the specific areas of sensory reception in the cortex; there are surrounding areas of secondary sensory areas, association areas, and, at the perimeter of these, motor areas. "Each secondary area connects with the corresponding area of the other hemisphere, with every other secondary sensory area of its own hemisphere, and with association areas."[27] If you also factor in the preferential screening in the temporal region that prioritizes input from the right ear, which normally affects the motor functions of the right side of the body, these preferences affect the development of the brain in the left hemisphere and perhaps supply it with the preponderance of engrams—and with the facility for mining them wherever they may be.

Thus, I find these researchers' assignment of particular engrams to certain parts of the brain problematical both from the standpoint of sensory reception and localized distribution and from the standpoint of screening for "meaning" by the left (or dominant) hemisphere for the classification and storage of its associations among these engrams. "Meaning" is nuanced in the left hemisphere on the basis of sensory and emotional information provided by the right hemisphere, while the right also contributes to variety in emotional climate and states of consciousness. But the left hemisphere, as long as it is dominant, is the primary locus of "thought." It endows with meaning the perceptions of the senses and their associated emotions; it controls access to the memories of them; it decides what may be predicated of those perceptions and feelings.

Spoken Thoughts and Interrupted Sounds

"The larynx can only produce what the ear can hear," Tomatis has shown. The researchers who speak about how the brain processes sound for speech without taking account of *how the ear perceives sound and how the brain attends to sound differentially coming through the left and right ears* are missing essential factors in the equation. They are left with the idea that the schizophrenic brain is mismanaging the production of speech somewhere between Wernicke's and Broca's areas in the brain. Thus, Mitchell and Crow assert that "the perception of speech occurs in part by activation of remote associations in right occipito-temporoparietal cortex (the extraction of 'meanings') of the activated primary sensory engrams in and around Wernicke's area."[28] In part, yes; but Wernicke's area is not the primary area for the perception of sound. Auditory reception occurs adjacent to Wernicke's area "in the upper part of the temporal lobe"[29] and is, we might note, particularly receptive to high tones[30] that are the most energizing (because of their short wavelength/ higher frequencies) and most important for the perception of nuance in tonality as the ear functions as a "frequency analyzer" in the sound it conveys.[31] Furthermore, in normal, right-eared individuals, meaningful linguistic sound input is discriminated to the left occipito-temporoparietal cortex, i.e., the *left* hemisphere. Sound received in the left ear also is received in the left hemisphere, but via the right brain, fractionally delayed, and is screened by the rational left brain for relative importance. The memory of sounds linked to images the person wants to think or talk about could reside in the right hemisphere. These are the data available for the perception of speech.

The assertion of Mitchell and Crow about the contribution of the *right* occipito-temporoparietal cortex uses normal brain function as its departure point and, therefore, applies (if it applies) only as long as the movement of "sound bites" between the hemispheres is flowing normally between the hemispheres to bring that sensory data into the area of sound interpretation in the left cortex. Normally, speech being processed through the right occipito-temporoparietal cortex has come from the *left* ear and the right hemisphere is putting an emotive spin on it. Without dominance by the left brain, part of that sound is arriving in the interpretive (left) part of the brain disordered; hence, only in a specialized sense does "understanding," as they claim,[32] proceed from the right brain. And that specialized sense is that "understanding" is enhanced *only* while the meaningful part of the spoken sound concurrently is being attended to by the left hemisphere (i.e., sound via the right ear) when the left hemisphere is dominant. Thus, the right hemisphere can only be drawn on selectively for the engrams that pertain to other perceivable qualities of sound that refine meaning and provide a subtext that enhances left-brain understanding when it is *not* dominant.

Put another way: understanding is sequential and the result of the left brain's ability to order neurological events in time—and temporality is established *in the ear* as it receives sound waves; "frequency" means "discriminated by time intervals." Understanding is a matter of the left brain's attending first (as a matter of primacy) to the grammar and syntax of meaning, and of noticing the nuancing qualities of the sound only secondarily. The associations made by the holistic tendencies of the right brain become "understanding" only as they are accepted into the templates of the left brain to facilitate or enlarge its categorizing use of language. The problematical behaviour (not only of language) of schizophrenics was acknowledged by Crow to be temporal in his discussion of the efforts of psychiatry to categorize schizophrenic symptoms. He points out that the temporal lobe in schizophrenics develops abnormal structure (determined through post-mortem examination).[33] Overlooked at that time was the possibility that these structural changes (diminished size) were brought about by abnormal (too little) input from the defective right ears of schizophrenics.

Thus, procedural problems in speech derive not from Wernicke's area, but from what is fed into Wernicke's area from the upper part of the temporal lobe from a faulty ear. And what gets to the temporal lobe from the ears will be ordered or disordered according to the dominance or non-dominance of the left brain. Under conditions of right-ear dysfunction and lack of hemispheric integration, the right brain has opportunities to feed its "understandings" to Wernicke's area to produce "disorders of choice, rather than disorder in the articulation."[34] In the absence of hemispheric integration (when the corpus callosum becomes improperly enervated), Wernicke's area proceeds (through no intrinsic "dysfunction") to relay misinformation caused by interruption (every two minutes) in the normal directions of hemispheric communication (integration), causing a form of apraxia. Thus, at intervals, the perception of sound and the production of sound are disordered in both dyslexic and psychotic individuals—and in those affected by other conditions considered "mental," such as bipolar disorder, Tourette's syndrome, or even stuttering. This is why the Listening Centre interventions with autistic children (i.e., pre-speech schizophrenics) have had considerable success—dyslexic syndrome, psychosis, and a number of other mental disabilities or instabilities appear all to be related to auditory-processing deficits and to fall on specific places along a comprehensive continuum of ear-related forms of mental function that includes the range of normal.

Modulated Speech
Mitchell and Crow refer to voice production qualities—"including speech volume, pitch, tone, intonation, vocal quality and rate or rhythm of speech"—that are deficient in schizophrenics[35] and to schizophrenics' unsatisfied need (for verbal response constructs) to perceive changes in tone of the speakers' voices ("emotional prosody")[36] that help to nuance meaning. Yet, missing from their analysis is an explanation of how the reception and perception (or lack thereof) of sound from the ears function in relation to the neurological routes of stimulus (or non-stimulus) and response (or lack of response) within the brain. The particular need of the reception centre of the temporal lobe for high-frequency sound is not mentioned. Mitchell and Crow come very close to affirming the Tomatis Effect when they observe that people can only speak what they hear; but they do not integrate that concept into their other observations of the qualities of schizophrenics' speech. The essential information from Tomatis that the larynx can produce only the sounds perceived by the ear points directly, in schizophrenics' monotonous vocalizations, to a functional failure of the middle and inner ear, which also affects the ability of the left brain to maintain dominance and the "direction of traffic" in the corpus callosum. We must conclude, then, that rather than a failure of segregation[37] of the language perception and production areas of the brain, the dyslexic/psychotic/schizophrenic condition reflects a failure of *integration of the hemispheres* caused by a failure of the right ear to provide the temporal region with consistently ordered input to support the roles in speech production of Wernicke's and Broca's areas. Mitchell and Crow conclude by suggesting that "Future studies of abnormal lateralization of left hemisphere language functions need to take account of the consequences of a failure of lateralization of language functions to the right as well as the left hemisphere."[38] Indeed, this study has been the crux of Daniel's story.

When Lateralization Fails

"The failure of lateralization"—which is the failure of sound energy to adequately reach the left hemisphere from the right ear—does affect the function of both hemispheres; however, in respect to dominance, the effect is not so much concurrent as consecutive. In other words, the left brain is not able to maintain dominance because the right ear is not supplying it with the necessary energy to keep it dominant. And the left ear (assuming it is functional) does provide the right-brain-then-left-brain with enough energy to keep the left brain operating somewhat rationally at two-minute intervals, but without synchronicity and in a manner very strongly influenced by the emotionally charged perceptions of the right brain. This situation produces forms of thought or of understanding in the left brain that has many of the features of those intervals of right-dominance. The left brain cannot help but come under the influence of its disordered, emotional partner. An unhealthy cybernetic is perpetuated when the left hemisphere cannot dominate. The "periodicity of fluctuation" probably varies a bit from individual to individual and perhaps for reasons lying in the hydraulics (movement in the fluids) within the brain and of which Dr. Upledger writes.

In a schizophrenic whose symptoms have been suppressed with medications these fluctuations and symptoms may be much obscured. But perhaps not entirely, as these incidents may indicate. Dalton, a layman uninformed about hemispheric dominance, with his wife, who is a priest, visited my acquaintance Jim, who has been medicated for schizophrenia for almost 40 years. I have often spoken to Jim in passing, but have never had the opportunity to spend time with him. Jim was hospitalized with a broken collarbone. The priest and her husband are new to our community, but upon the death of Jim's mother Gwendolyn while he was still in the hospital, they felt he might welcome a visit. Gwen was 96 years old and had been her son's faithful caregiver since his diagnosis when he was 27. When I asked Dalton at Gwendolyn's funeral how he had found Jim, he replied, "He has his moments," by which, he explained, he meant that Jim could communicate more normally with his visitors at intervals. It had not occurred to him that the intervals might be regular. My attention was caught more recently when I inquired of a colleague how his dyslexic syndrome daughter is, and he used precisely the same expression: "she has her moments," implying she fluctuates between good ones and not-so-good ones. Most likely, regularly.

Clearly, in Daniel's brain, the right hemisphere had its opportunity of dominant lateralization every two minutes. Those were the intervals in which his ability to produce coherent language was most lacking, demonstrating Tomatis's observation that temporality and order in cognition proceed from the sequence of sound waves moving through the middle and inner ears; the disruption of that function will produce disordered hearing, cognition, and syntax (through no fault of Wernicke's area). When Dan's left brain resumed its interval of dominance it did so with greater control of language, but under a continuing strong influence from the right brain. The right brain persuaded the left brain of some considerable nonsense. But not only of nonsense: the right brain "put two and two together" correctly to some extent and with improving rationality as his ear was stimulated so his hemispheres regained integration with dominance to the left. Dan's right ear gradually collected better input for his left hemisphere, which produced better cognizance. Over a period of months, as the right ear gradually regained its function for receiving input from the external world, the corpus callosum recovered its normal operation, and the left brain became less influenced by the dreams and visions and metaphors of the right brain, and more attuned (yes, attuned!) to the influences of the external environment, i.e., the "real" world of rationalized sensory impressions.

Short-term Memory, Attention, and Left-Brain Dominance

As Daniel's right ear–left brain connection grew stronger, his short-term memory improved. Short-term memory is presumably what Mitchell and Crow mean by the "scratchpad" that they assign to the right hemisphere.[39] However, Dan's short-term memory cannot reside "in the right brain" because he uses short-term memory to hold impressions while waiting out the confusing ascendancy of the right brain. I think we all do! Eventually, his short-term memory perfectly bridges the disparate function of his right hemisphere so that communication between both hemispheres under the condition of left-brain dominance can proceed seamlessly. Thus, both short-term memory and the ability to attend are aspects of left-brain dominance and depend on a well-tuned right ear.

The implications of my findings for long-term memory are less clear. Persons who can most easily access their right hemispheres may be those with weaker left-brain dominance, such as left-handed persons. If memory is stored where sensory perception is received, the ease of access in lefties to both their hemispheres may advantage their retrieval of experiences, i.e., enhance their access to memory—and along that continuum of reduced dominance some may be *unable* to forget and *unable* to still the voice of memory. Such a condition reminds me of people I have known with "photographic" memories.

Mood and Muscle: Myopathy

In my dyslexic syndrome friend Nancy, the same sort of struggle as Dan's to order ideas into words of conventional syntax was arduous until she lost her hearing in her left ear; then, she gained the ability to express her ideas without effort in normal syntax—and her mood improved. The relationship between mood and cognitive function in Daniel has been mentioned frequently here. My observations of him and of the dozen or so people I have known suffering from schizo-affective disorders, i.e., bipolarity with or without psychosis, and those experiencing some forms of depression are that these mental disturbances are related to varying degrees of hemispheric coordination (or lack thereof). I also have noticed in those individuals that mood disturbance does not precede ear dysfunction, but that ear dysfunction precedes mood disturbance. Here are a few examples.

A woman with a tendency to "mood swings" was comfortably employed prior to treatment for breast cancer and was able to continue to perform remarkably well in that job during much of her treatment, including surgery from which she rebounded, chemotherapy, and radiation. However, one of her medications was a steroid prescribed to control the nausea often experienced with chemotherapy. "Oh, didn't someone tell you that your meds could affect your moods?" one of her doctors asked after the fact. (Whatever happened to "informed consent"?) She, too, had grown up with a bipolar father and might have decided nausea was the preferable alternative to bipolarity. The steroids left her with dys-integration so intense that she was unable to maintain her employment, even when "stabilized" on medication. Steroids can negatively affect muscle strength and muscle mass and can cause "myopathy," a general term for problems in muscle.[40] One is prompted by this scenario to ask whether a person already experiencing some difficulty with the muscles of the middle–inner ear and hemispheric stability could be very adversely affected by a steroid medication for any purpose. Another friend who recently underwent chemotherapy with steroids had hearing deficits from childhood and appeared for a week or two to be losing hemispheric control under the influence of her medication. She feels that focused listening throughout her chemotherapy helped her to tolerate its side effects. She tells me of a woman in her support group who lost her hearing as a result of chemotherapy. Yet again, a woman with elevated bipolarity is taking medication that causes dizziness and affects the production of saliva, which symptoms indicate the medication could be affecting her middle and inner ear. The risks to hearing and to mental stability of cancer patients undergoing chemotherapy conceivably might be reduced if substitutes for steroids in the protocol could be devised and/or if focused listening became part of the therapy. These women's experiences suggest yet another area for research and investigation.

A hopeful treatment not widely known, apparently, is that "some cases of bipolar illness may be ameliorated by nutritional supplementation"—chelated vitamins and minerals.[41] Conventional wisdom associates extra nutrients with an improved immune response. This study recalled to me the brochure I had read in the hospital waiting room that suggested some researchers thought the schizophrenic's immune system might somehow be a cause of schizophrenia. Since the health of the middle and inner ear affects the body's immune response, as Tomatis demonstrated with his monks and as I experienced, any connection found between the function of the middle and inner ear and schizophrenia likely would affirm the intuitions of those researchers, however, not as a cause a priori but as a coincidental symptom.

Metaphor in Schizophrenic Language

As a student of English language and literature, writer of poetry, introspective artist, and student of behaviour more generally, I was knowledgeable about metaphor and symbol *in* language and art and about language and art *as* metaphor and symbol. Daniel was going to show me, with his hemispheres integrating in slow motion, just how that works for those of us who operate at higher speeds and more efficient levels of integration. As seems often the case, the breakthrough came under pressure.

Richard and I were feeling stressed. We were variously unhappy or worried about our household of needy people. Richard struggled still with CFS symptoms intermittently. Apart from these worries, the only thing I had to complain about was skin problems related to the cancer surgery and skin graft in the middle of my face done eight months before. Richard began to be very insistent about our taking a few days off and going to the cottage with Dan, even if we had to take some proofreading with us. I, too, longed for a break, but I felt deep conflict. Dan would be destabilized both going to and coming from a brief time at the cottage. If moving Dan through the landscape to the cottage would destabilize him, his recovery would be impeded. While I believed that he would recover from a setback, I did not know how to weigh one loss against the other gain, especially for the several people involved. At length I acceded to Richard's wishes and he, I, Dan, and my sister made the brief trip.

We were glad to see our friends at the marina, whose praises for Luke and family and for Eric delighted me, even as they emphasized the contrast with Dan, who walked away while she and I were talking—doubtless because he knew updates about him also would be discussed. We carried editorial work with us, my sister proofread an early draft of my manuscript, and the rain poured down all three days, ensuring that work was accomplished. I was not mistaken in how the change would affect Dan, although I could not have guessed how profoundly; two months later, at the end of October, he would illustrate the complex Creation Myth that he had formulated on his perceptions during those few days. Immediately obvious was that he became more confused, less able to overcome his confusion with reason. He tried to cope with the change in environment by spending a few hours each day chanting to a monotonous tune on his guitar. The heavy rainfall had brought out a colourful profusion of mushrooms. Mushrooms had one major meaning for Dan and he was assaulted by their prevalence. Before long, each of us represented a different kind of mushroom to him. Images of wilderness animals became part of his pantheon: bears and wolves and rabbits. His toenails and fingernails had become very long (I had forgotten to buy new nail clippers for him), so he identified with "bear." While the three days of heavy rain didn't interfere with the older adults' enjoyment of the change of scene, Dan was much more limited than usual in what he might do: no swimming, hiking, or fishing. His manic symptoms increased at the cottage.

When we returned on Friday afternoon, as I had predicted, Dan was further destabilized. His self-talk had increased; he was attaching to people he knew the images of those bears, wolves, rabbits, and mushrooms. By Saturday morning, I was much more on edge about Dan's mental condition and had lost most of the benefits of the holiday. As Dan's dog Lox leapt all over the sunroom and onto my work and scratched me every time Dan went out for a cigarette, I remarked to Richard on the need (from my perspective) of having Lox neutered. Dan immediately flared that we might as well neuter Daniel—his identification with his dog was intense. Over the next half-hour he struggled with the concept of who should be in control, setting Lox in a variety of other relationships to himself: Lox as God (God's creature, but also what Richard humorously called the "ultimate symptom" of dyslexic-psychotic misapprehension: "d-o-g" equated with "g-o-d"); Lox as son (Dan's creature); then, he verbalized God in a variety of other forms: God as earwig, God as vulture, God as black bear, God as heron (we'd seen a great blue heron swooping around our boat at the lake the day before); and he spent further time testing other family relationships in reference to the God–animal relationships he was trying to sift and to arrange hierarchically.

Dan must have noticed his own inability to reason as he wanted to, because he took up the headphones. While he was engaged in focused listening and mandala-colouring, he became calmer and quiet for a while, then observed, "April is my pet rabbit." This caught my attention because I recalled that his gift many years ago of his pet rabbit Betty had been April's idea; his association between April and Betty was not as hard to explain as his association of the family members with mushrooms. Next, to my surprise, he made a string of remarks that revealed his reasoning process that led him to the "April is my pet rabbit" metaphor. He again employed the connective term he used in reasoning, "in result."

> ***April is my pet rabbit.***
> ***Jesus, He's alive in result.*** [meaning "Although I cannot see Him, I believe Jesus is alive."];
> ***April, she's alive in result.*** [meaning "Although I never see her anymore, I believe April is alive."]
> ***I ate mushrooms and I'm alive in result.*** [meaning "I ate mushrooms that made me feel like I had died, nevertheless I believe I am alive." Notice the fragility of his concept of his own existence. Notice also the role of his belief system in his reasoning process.]
> ***And April is my Betty.***

Notice that Dan has progressed from the pairing of objects, as in the Dan/Frodo = Matt/Jesus example, to the pairing of activities, that is, pairs of objects of which various actions or states of being are predicated. Dan here provides another example of how language develops to employ increasingly complex metaphors.

Dan then came to some sort of grudging acknowledgement of my problem in trying to control a naturally hyperactive golden retriever bonded to him and excessively interested in the female dogs—Eric's aging Tequila and Dan's spayed boxer Feather—by asking me, "*When* will you take Lox to the vet?" His emotional sense of losses that connect his lesbian sister long-absent, his long-lost pet rabbit that she gave him, and his inability to control the pet he now loves and who may for that reason be castrated and rendered impotent are related notions attached to emotions that run very deep and flow together. At dinnertime, I realized that my threat of the morning had not disappeared from his anxieties when he remarked candidly and accurately that the fried scallops we had eaten looked like (as he expressed it, "were") his dog's testicles. I decided to postpone taking action on Lox—or on serving scallops again!—hoping that Daniel would soon be well enough to take full responsibility for his pet. The fact that he was trying to analyze his own reasoning process provided a basis for optimism.

As Dan mused about associations between animals and people, he was also reviewing what he remembered of *Animal Farm*—and the Napoleon with whom he identified this time was not the French Emperor, but the pig from George Orwell's novel. He knew Betty had been purchased from a couple who farmed rabbits. He considers himself in one respect a farmer—in fact, he continued to weed his garden through the summer, to pick the beans, and to watch the progress of the tomatoes, peppers, and corn. His verbalization suggested to me that he was groping for the "authority (dominance) meaning" of his farmer role in the context of his family.

When Eric, who had returned late the night before, arrived in the sunroom, I noticed that Daniel's emotions shifted as he felt a rising tension. Luke's appearance a few minutes later levered Dan's emotions up another few notches and reduced his self-control. Intending to add a positive note, I explained to Luke the improvement in Dan's reasoning processes that I had just noticed. Dan reacted the way he had to my talking to our house guest about him and to my speaking with my friend at the marina—he removed his headphones and left. I realized too late that he had felt demeaned or betrayed and that once again I had been condescending and tactless.

Integration Disorder in a Range of Mental Syndromes

I have referred several times to the condition of lack of hemispheric dominance in Daniel and his inability to control his verbalizations as a dream-like state similar to that most of us experience only when sleeping (perhaps narcolepsy also is related to loss of hemispheric dominance) and as revealing associations that when he was mentally stable he would not normally want to reveal. In that state of consciousness, his right brain is poorly under the control of his left brain, although most forms of *procedural memory* (deeply ingrained activities, such as being able to walk, eat at the table, take care of minimal hygiene, and so on) still serve him through *both* parts of the cycle. From observing his word associations more or less continuously for many months, it is clear to me that the kinds of knowledge I retrieve at will (notice the relationship of left-dominance implicit in that phrase) from my right brain he is forced to contemplate willy-nilly—and he must somehow come to terms with this fluctuation almost continuously. Only in sleep or in particular kinds of waking, contemplative mental activity (such as daydreaming, artistic creation, meditation and other types of prayer) do most people rest from the matrix of dominations schooled into the left brain, although left-handed and other right-brain-leaning people feel less dominated and may resist the strictures of those dominations—a "sinister" state of mind from the standpoint of the norm in domineering societies. However, experiences arousing intense emotion, whether viewed as positive or negative, cannot so easily be subdued by the dominant hemisphere or repressed into the non-dominant hemisphere—by any of us. Similarly, those who are left-lateralized and processing sound through their right hemispheres on the way to the left hemisphere are more likely to have propensities for a broader range of "sub-conscious" imagery, including the macabre; again, the influence considered by others—quite accurately in terms of laterality—to be "sinister."

The abnormally lateralized brain (constantly cycling left, right, left, right over an interval of about four minutes in Daniel's brain) struggles to articulate associations made in the right brain that the left hemisphere cannot consistently order syntactically. By means of its tendency to create holistic understandings of its perceptions, the right brain sculpts its own empire of "subconscious" knowledge during its intervals of opportunity. These are the sorts of associations familiar to all of us in imagination and dreams and that are famously expressed in the art of all cultures and throughout history, and they are strongly influenced by the limbic system, i.e., the basic drives and emotions.

This mental process occurs in the normal brain as well, but usually under a much greater (although sometimes lesser) degree of control because the left brain—via the healthy right ear—has been parented, socialized, schooled, and acculturated to particular codes, mores, folkways, taboos, and beliefs that define particular controls. (Or that militate against controls. Societies continue to debate and negotiate the borders of control for individual, institutional, and state behaviours—from parenting issues to sexual issues to gun control and the waging of warfare.) We learn how to shut down the dominance of our left hemispheres by relaxing our (right) stapedius musle, although we are just now learning that we are accomplishing precisely that when we allow ourselves to sleep, to relax at a undemanding physical activity, to explore the imaginitive side "of our nature" (i.e., brain), to daydream, to "sink into a reverie," or otherwise give the left brain some respite.

Our decision base will depend on our ability (or our perceived ability) to counter or shape the impulses arising through the right brain. For some individuals, some codes become extremely entrenched in the dominant hemisphere through rituals of church, nation, or secret society; or through threats, abuse, deprivation, or other assaults on body or mind. Most forms of socialization incorporate repetitions of language, taboos concerning the divulgence of certain aspects of language (e.g., the silence of the Confessional; the tenets of Masonry; the secrecy of agencies for spying; the confidentiality of the medical profession; and so on). Of course, the normal brain is subject to assaults on the right ear: loud sound, viral and bacterial infections, chemicals in medications and street drugs, and physical assaults such as beating, sports concussions, automobile accidents, and so on. Impairment of the audio-processing abilities of the right ear usually are not investigated under such circumstances, but can have profound effects on the integrative relationship of the two hemispheres.

Although trends in thinking, such as existentialism and structuralism, have sought for universals that could be considered to determine the form given to thought and hence to speech and language and the social structures reliant on human communication, it seems likely that the amorphous quality of the working of the right brain is anarchic and defies definition. Its function is to offer fluid associations to its hemispheric twin for rational processing, and some of those associations will be fragmented by memory overlay and deterioration and will be distorted by the circumstances in the limbic (emotional) brain that give context to the sensory perception and to the memory of it. This divergence or specialization in mental capacities is an aspect of *nature*. The effort of the rational left brain to protect and defend its partner's observant, emotive, but non-rational fluidity of sensory input *as a form of rationality* is futile and produces endless tautologies and circular reasoning and renders meaning in language useless precisely because the basic interaction of the hemispheres is "circular" in the sense of a loop. Reason necessarily objectifies those right-brained perceptions as data. We see this interaction in slow motion in Daniel in the way he rebuilds imagistic associations (right brain) through acculturated rational language (left brain) during his psychotic episodes. However, even though the same process is happening faster, we can introspect to see the same hemispheric interplay as we engage in problem-solving situations where the repeated activities of reasoning, "picturing solutions," and contemplating the reasonable outcomes of those imagined solutions eventually lead to a rational solution. We can contemplate this process because it also takes place more slowly than the learned patterns of observation and responses that dominate in our routine lives. When we were young, those processes occurred slowly, sometimes painfully, until they became routine and could proceed without our careful attention.

Individuals absorbing the teaching of their culture usually come increasingly to comply with the rules, trivial and great. This response of the left brain to what it hears (and otherwise perceives) is an aspect of *nurture*. This process reinforces the connection between the attentive right ear and the left brain, producing ever-increasing self-control related to the expectations of the particular social milieu and culture. Tomatis observes this effect of ear-attunement when he writes, "As we move towards effective listening uncontrolled behaviour occurs less often."[42] Where the social standards are conflicting or shifting and the messages mixed, the individual will be less strongly reinforced in those social concepts. With the failure of the ear to perform its function, whether from infancy and childhood or from some other place along the life course, that learning and compliance will be compromised and either will take place problematically or will start to break down. The implications of that breakdown for the society will depend on how the members of the society understand and decide to deal with the disabled individual. In fact, the culture will decide whether or not that process should be viewed as a danger (to isolate), a disability (to remediate), a misfortune (that no one knows how to fix), a crime (to punish), a problem (to solve), or an advantage with superior social collateral (like the ancient Greeks, who believed psychosis put a person in closer touch with the gods).

One can see where *on the spectrum* of left- and right-brain control certain areas of behaviour fall on a negotiable border. For example, the relative strength of the left brain over the right brain in an individual will define the relationship between reason and self-control (left brain), on the one hand, and fear and desire (right brain), on the other hand, to produce (1) the necessary or unavoidable act (according to belief) in his or her decision-making or (2) a sense of indecision or compromise (based on feelings) negotiated before the act. We can see, thus, that hemispheric dominance is a matter of degree, even in those whose behaviour is regarded as falling within the range of normal. Overall, balance may be strong or less strong, and so may come under duress to produce explosive losses of self-control, many instances of which have been cited in this story.

The person who is able to come to understand his or her formative experiences (residing in the right brain, which is the subconscious) becomes better able to understand his or her motivations. Whether or not that knowledge will or can be turned into more rational decision-making will depend on the entrenched quality of the patterns of habitual or addictive behaviour and the person's left-brain *physiological* stability for enacting left-brain motivations in reason and in belief for changing those behaviours. It appears to me that recognizing the rational and rationalizing work of the left brain as a distinct process that continuously builds and reforms structures for thought as based on the perceptions of a changing environment—an environment that can be adapted to according to the plasticity of the brain—obviates the search for a neuro-anthropology other than the one we have found.

Both nature and nurture must be considered in understanding how hemispheric dominance and interaction occur. We all recognize a wide range in the degree of self-control people demonstrate, but we do not always understand what causes those differences. Among my experiences in which I have lost my hemispheric dominance is the one I described in Chapter 2 over the condition of Dan's tattered jeans. I also became hysterical in my early teens after my sister and I were released from an amusement park ride in which centrifugal force had thrown me against her so that she could not breathe; I was aware that I was killing her (or tending to) under the control of a force I was helpless to offset. When I discovered my father's body after he had died in his sleep I behaved rationally as I called Richard, the police, and interacted with the family; then, I began to weep. As people arrived and I conversed with them, still quite rationally, my tears continued uncontrollably, a situation that persisted for an hour. Any occasion of prolonged spontaneous, unfettered emotion—tears, laughter, or raging—may indicate some measure of loss of hemispheric control. The frequency of such losses in an idividual, measured against a large population, will determine whether or not the label of "bipolarity" will be applied. The release of energy in such an expression of emotion appears to allow hemispheric dominance to be re-established with a change in mood, perhaps by resetting the electrical charge in the brain's circuitry. But very frequent episodes will impact personality and the decisions a person feels able to take.

From the changes taking place in Daniel's degree of hemispheric dominance during recovery that affected his degree of self-control, parallels can be drawn with others who demonstrate similar sorts of behaviour, but without the background of severe and continuous lack of dominance that defines psychosis. In such persons we can postulate a comparable (to that stage of Dan's recovery of ear function) auditory processing deficit that has compromised the strength of dominance of the left brain. As anyone could observe Daniel's "short fuse" in recovery, the buildup of emotional pressure, called "kindling," would contribute to a loss of self-control and the sudden hemispheric switches demonstrated in an uncontrollable outburst of emotional speech. Such rage, wailing or keening, and other loss of control over the right brain are the conditions observed in bipolarity, with and without psychosis (as we saw both of those conditions in Dan). Actors (and others) learn to manage more extreme expressions of emotion, usually without losing rationality, although Method actors enter into considerably greater risk.

I have lived for most of my life with or close to people with diagnosed (and undiagnosed) bipolarity. That buildup of emotional energy—as events that might seem trivial to others grate on the individual having a less stable condition of hemispheric dominance—becomes not only easy to notice, but predictable in terms of the sorts of irritants that kindle in the particular individual. Such irritants are not neutral, although they can include the physical stresses to which any person might react; they are likely to touch "sore points" that impel the individual towards a hemispheric shift. The sudden explosion of emotion into anger or fury (which may be prolonged, not transitory) expressed in "pressured speech" or abusive language, and more rarely in violence, may also be followed by a kind of regret or remorse. Or, these outbursts may become so habitual to the individual that rationalizations and justifications for his or her behaviour, strategies for controlling the environment, punishing the self, blaming and punishing others, and so on become entrenched, and tertiary behavioural syndromes, such as borderline personality disorder, become established. Such behaviours that grow up around bipolarity constitute its syndrome.

It should be mentioned that the person experiencing a burst of right-brained expression may have little or no idea how he/she appears in that state and may have little or no memory of the event, as one awakening from a dream may retain a memory of it for a short period of time. It may be much more difficult for the targets of such behaviour to forget the event(s).

These syndromes I have discussed—dyslexic, chronic fatigue, bipolarity, schizophrenic, and the patterns of substance abuse—are not discrete illnesses, such as tuberculosis, which can be ascribed to a particular bacillus, or polio, which is caused by a virus. As Dr. Jay Goldstein's Checklist in Appendix A demonstrates in reference to chronic fatigue syndrome (ME/CIFIDS), they are referred to as "mental problems," but they are identifiable by a cluster of other physiological symptoms associated with them, such as the digestive problems frequently present with autism. In this regard, Tomatis's description of the relationship between the ear and the viscera can hardly be over-emphasized:

> Let's begin with a closer look at the right vagus nerve. It is most important to remember that one of its branches innervates [is the nerve that activates] both the lower part of the external auditory canal and the tympanic membrane. This is its only external point of emergence but it plays an enormous role. It is also important to emphasize that the vagus nerve extends a sensory nerve fibre onto the muscle of the stirrup, which receives its motor activation from the facial nerve.
>
> The vagus nerve working in association with the glossopharyngeal nerve, innervates the eustachian tube and the pharynx. It also supplies sensory innervation to the larynx through the upper laryngeal nerve and provides motor function to the larynx through the lower laryngeal nerve, also called the recurrent nerve. This name is derived from the loop that this nerve makes around the right subclavian artery.
>
> The right vagus nerve then innervates the bronchi, the esophagus, and the coronary arteries. From there it descends the diaphragm near the pylorus, generously innervates the lesser curvature of the stomach, and fuses with the left vagus nerve behind it.[43]

Once Tomatis has traced the route of influence of the right vagus nerve he outlines a sphere of influence proceeding from the tympanic membrane (the eardrum) that stimulates:

the pharynx
the sensory larynx
the motor larynx
the bronchi
the coronary arteries
the stomach
all the viscera, and finally
the gall bladder.[44]

I needed to look no further for an explanation of why the medications that affected Dan's ears had caused so much vomiting, or why chronic fatigue syndrome had affected Richard's digestive system, or why Daniel's damaged ears affected his voice production (making him sound robot-like), or why those deprived of sound stimulation might develop irritable bowel syndrome. The effects of dysfunction in the ear are far-reaching indeed. However, while Dr. Goldstein would develop chemical protocols for altering CFS symptoms,[45] Dr. Tomatis used music for the audio-processing deficiencies that produce those symptoms.

You may also have noticed how the symptoms of one of these syndromes overlap with the symptoms of another cluster. For example, vertigo (vestibular dysfunction) and difficulty focusing on mental tasks (cochlear dysfunction) in chronic fatigue syndrome are related to the lack of balance (vestibular dysfunction) and inability to attend (cochlear dysfunction) in dyslexic syndrome, just as those ear dysfunctions in full-blown psychosis present as a severe lack of co-ordination (vestubular dysfunction) and the loss of rationality (cochlear dysfunction). The overlapping of one set of symptoms that seem attached to a central problem—difficulty learning the language in dyslexic syndrome—onto part of another set of symptoms that seem attached to a different central problem—disordered speech in psychosis—has made these sets of conditions and behaviours very confusing for doctors to sort, a situation that may be exacerbated by medical specialties. However, when you locate them relative to ear function and listening ability (dominance or non-dominance on a spectrum, as in Figures 13.1 and 13.2), the cause-and-effect relationships are much easier to comprehend. And these "mental" disorders can be seen in terms of symptoms that are alike (or strongly related) in kind, but not in degree of severity.

I have described in some detail the parallels between dyslexic syndrome and schizophrenia—between the status of listening (i.e., the functional state of the ears) and its effects on speech. When we examine the range of types of speech production and non-production on Figures 13.1 and 13.2, we become aware of the importance of listening and of speech in all of our interior (mental and bodily) and exterior (family and social) worlds. As I began to grasp how and why the status of the mid–inner ear impacts hemispheric integration, the spectrum of "mental" disorders arrayed itself. Thus, I noticed in a few weeks that Daniel was passing through a stage of bipolarity (with psychosis) and that it was an indication he was nearing the end of the episode of schizophreniform mental illness. He had become calmer from day to day and more able to accept the alternating dominance of his hemispheres; while his intervals of disordered speech continued, they were less disordered and he was sometimes happy and occasionally humorous or responsive to humour.

Then, an unusual situation developed. Eric, employed in Luke's construction business, decided to buy a property between our house and Lainey's. At first we kept this evolving situation from Dan, fearing it would be stressful. As the day of the closing approached, however, all the packing and rushing around to lawyers, bankers, and realtors became evident to Daniel. When I told him about Eric's new enterprise, the surge of positive feeling towards his brother in his delighted, "That's wonderful!" was followed swiftly by a surge of jealousy, "How come *I* can't . . ." that flared into a noisy rage—although he couldn't kick in the bottom panel of the storm door the way he had a few years before during such a phase because recently the whole door had fallen off its hinges. He had enough self-control to take his rant out into the yard. His loss of verbal control had displayed precisely as bipolar behaviour. He calmed down after a walk in the yard and ten minutes later apologized to me for his outburst. For the next few days he remained more labile, more likely to become upset. This incident, taken together with similar but less dramatic ones, was an indication, I thought, that Dan's corpus callosum was working better as a bridge between his hemispheres (its "directionality" had been re-established), so that his left hemisphere had a higher level of control over the right; but the level of dominance was not so strong that it could withstand a surge of intense emotion. Daniel's overall improvement in mood led me to consider reducing his medication. Another bipolar event occurred the evening he discovered Sarah and Tom were going to move out of their upstairs apartment and into Alex's house; Dan flared into anger and remained more stressed for a few hours before he resumed his former level of self-control. His level of self-control fell fairly within the range of normal.

A few days before, while I was with my friend Faith snooping around our favourite charity store, I had bought him a huge, stuffed lion, which to me was a contrapuntal symbol of strength to the "dyslexic" rhino I had taken to him during his first hospitalization. The day Sarah and Tom moved, I found him in the living room hugging the lion and looking the way he had

Figure 13.1 The Proposed Continuum of Ear-Related Hemispheric Affects on Behaviour

Listening Ear	**Right** (with normal bi-aural hear ing, left ear trails)	**Both** (with fully bi-aural hearing)	**Left** (abnormal bi-aural with loss in right ear)	**Left** (monaural)
Sensitivity to high frequen-cies of sound*	Low—Mid—High	Low—Mid—High	Low—Mid—High	Low—Mid—High
Dominant Hemisphere	Left hemiphere	Both; alternating	Left hemisphere, with switches	Right hemisphere
Prevailing Mood	Energetic and calm	Swings	Energetic to manic with depressions	Less energy, depressions
Stability of Mood	High	Labile	Unpredictably labile	Emotive; labile
Attainable self-control	Very high	Low to moderate	High, with sudden losses	Low to moderate
Learning of Control: acculturation, socialization	Proceeds normally	Learning problems	Proceeds with some difficulty	Proceeds with more difficulty

*"Sensitivity" is a reminder to the reader that dominance has a range: it is subject both to physiological ability to process high-frequency sound and to a variety of modes of social learning and reinforcement. Hence, the concomitant range of symptoms and behaviours that will be observable in each of the categories in the column beneath Listening Ear also have a range subject to a degree of unconscious and semi-conscious control as well as to a significant degree of deliberate control. The first four conditions (I-IV) are strongly tied to physiology: nature (which may nevertheless be modified). The last two (V, VI) vary also according to nurture: social influences in family, community, the expectations of the umbrella culture, and so on.

Figure 13.2 The Proposed Spectrum of Mental Conditions and Types of Speech Production Related to Listening Ear and Consequent Degree of Hemispheric Integration

Normal Listening: Right ear dominant, left ear trails	**Bi-aural Listening:** Neither ear dominant	**Left-Eared Listening:** Left ear dominant, right ear trails	**Left-Eared Listening**: Little Input from right ear
LEFT hemisphere dominant	**NEITHER hemisphere strongly dominant**	**LEFT hemisphere dominant: more influenced by right; sudden switches to right dominance**	**RIGHT hemisphere dominant: some influence from left**
Mood: Energetic and calm	*Mood:* Swings	*Mood:* Energetic to manic; depressions	*Mood:* Less energetic with depressions
Stability: High	*Stability:* Labile	*Stability:* Unpredictable	*Stability:* Labile
Control: Very high	*Control:* Low to moderate	*Control:* High, with sudden losses	*Control:* Low to moderate
Learning: Normal	*Learning:* Problematical, memory may be enhanced	*Learning:* Social learning somewhat difficult; memory enhanced	*Learning:* Impeded by emotionality
Speech: Normal for syntax, prosody	*Speech:* Intermittently disordered; pressured speech	*Speech:* Normal to rapid, kindling to bursts of emotive, disordered, or pressured speech	*Speech:* Emotive; reluctant speech

Each panel represents a *range* of hemispheric integration or non-integration from low to high; some spillover occurs in some categories.

as a child on the day Luke and Sarah had returned to school leaving him at home. I had not appreciated until that moment how their mere presence in the house had ameliorated his sense of frustration in having had to return to living in the parental home. Three weeks later, after he had regained left-hemispheric dominance, his occasional outbursts persisted: clearly the condition of bipolarity but no longer accompanied by psychosis.

Integration Disorder and the Spectrum of Ear-Related Mental Illnesses

For a very long time lateralization has been recognized as an issue in dyslexic syndrome. In a sense, lateralization also is an issue in Daniel's psychotic integration disorder, and schizophrenia has been regarded as a disease of (non-)laterality. Because I have seen all the symptoms both of dyslexic syndrome and of schizophrenia during Daniel's episodes of psychosis, it seems to me that his similar routes to healing from both dyslexic syndrome and from psychosis may be relevant to those who experience schizophrenia as spontaneous psychosis.[46] I would hypothesize that fragile hemispheric dominance creates an underlying problem so that when such a person develops a severe problem in the middle and inner ear a mental crisis may be precipitated. No one heretofore has been looking for a middle- or inner-ear problem when a person shows signs of mental illness; and so the downward spiral of medications that worsen the condition of the ear while modifying the mental landscape commences. For persons strongly habituated to their lack of hemispheric dominance, I would imagine such crises are less likely to occur.

The array of symptoms in these two syndromes—dyslexia and psychosis—which have been shown in Daniel to be related to middle- and inner-ear dysfunction, suggests that not only those who suffer from fibromyalgia/chronic fatigue syndrome, dyslexic syndrome, attention deficit disorder, attention deficit hyperactivity disorder, and psychosis through substance abuse can benefit from some version of the Tomatis Method. The same strategy might also be applicable to those who suffer from Asperger's disease, Tourette's, autism, spontaneous schizophrenia, schizo-affective disorders (mood disturbances) such as bipolarity (I and II), and those suffering from sound-deprivation depression.[47] Among the authors I have referred to in this book, Dr. Joseph Huggins and Drs. Rachel Mitchell and Tim Crow see schizophrenia and schizo-affective and affective psychoses and the related disorders on the same continuum or within the same spectrum.[48] I think further investigation of the spectrum of dysfunction in the middle and inner ear will more precisely define the relative positions of these disorders along the continuum. The possibility of Tourette's syndrome (Daniel's inability during a phase of his recovery process to control vulgar language and his tendency to perseverate with vulgar language at about the same stage being a shared symptom of that disorder) and such other disorders as epilepsy and Asperger's syndrome falling along this continuum is incontrovertible.[49] The Tomatis Method of intervention is already being used with a high degree of success with autistic children, as I have noted, as are similar interventions with sound for autism.[50] Surely, sufficient evidence of the relationship between ear function and brain function has been found to excite more rigorous investigation into these treatments for these and other disorders. I expand on this continuum of ear-related mental disorders below and in Figures 13.1, 13.2, and 14.1.

Supporting Evidence from Audiometry

One of the important means of learning about the ear is through audiometry, which measures "cochlear and brainstem evoked activity and electrical stimulation of the eighth [cranial] nerve. Electrical signals, either derived from real or simulated activity inside the auditory system, . . . form one essential link between both the diagnostic and the rehabilitative aspect."[51] More simply, sound stimulation applied through or near the ear can be sensed along its pathway to the temporal cortex through an electrode placed adjacent to the brain stem in the scalp. (The methods of audiometry usually are among the less invasive means of learning about the activity of the brain, not only in regard to hearing but in electroencephalograms [EEGs] that measure differences in electrical potential between various points on the skull; however, some audiometric measures are taken through holes drilled in the skull with the electrodes set on or into the brain.) Since the route of sound, both via the vestibule and via the cochlea, is through the brain stem on its way to the cortex, measures of electrical activity taken near the brain stem reveal aspects of its function or dysfunction, and (by its known associations) with other parts of the brain.

In the ascending, that is, from the ear to the cortex, pathway of sound, Tomatis places the (superior) olivary nucleus, "located in the mid-lateral tegmental region of the pons, which receives fibres from the ventral cochlear nucleus"[52] with the neural feedback system he calls the "vestibular integrator." He places the thalamus with the "cochlear integrator" and mentions the red nucleus (in the upper portion of the mid-brain, extending into the subthalamus[53]), as part of the neural network that joins the vestibular and cochlear integrative systems. His schematic demonstrates the *highly integrated circuitry of these two routes of sound between the ear, the brain-*

stem, and cortex.[54] The cortex, as we have seen from Tomatis's study of audio-vocal feedback, responds to the information fed to it from the ear, for instance, by altering the larynx to change the vocal emission—the voice. Thus, the voice is able to produce what the ear can hear. If the ear *cannot* perceive sound energy properly, the information fed to the cortex from the ear will not be normal; the brain will respond, but abnormally, and a problematical sort of cybernetic will be established. The return or "descending" route from cortex to ear, that is, the route of the potentially abnormal response of the cortex, is the neural pathway under examination in the following study.

A recent audiometric study in France of dyslexic children focused on: "[t]he medial olivocochlear (MOC) system, an inhibitory pathway functioning under central [cortical] control."[55] The authors point out that the cortex appears to influence the sounds produced by certain hairs in the cochlea: "The MOC inhibitory fibres originate from a part of the superior olivary complex and project onto the outer hair cells [of the cochlea], forming the last step of a descending auditory pathway which originates in the cortex . . . (Huffman and Henson, 1990)."[56]

If the MOC system is part of a feedback system (as it appears to be from the changes it can undergo), only the second half of the loop is described by these researchers. The authors of the study start with an assumption of "control" in the brain, thus inadvertently also assigning a priori the problem to the brain. They speak of the sound being produced by the hairs in the cochlea, whereas sound in the first instance is produced externally and transmitted through the cilia in the cochlea after it has passed through the middle ear. Thus, significant parts of the ear are not brought into the cybernetic (loop). "In humans, there are several lines of functional evidence for cortical influence on the cochlear micromechanical properties reflected by evoked otoacoustic emissions, which are sounds thought to be generated by the outer hair cells . . .(Kemp, 2002)."[57] Except that, again, the sounds are not generated so much as transmitted by these cells. However, since the relationship between the ear and the brain is cybernetic, the cilia–cortex relationship could just as well be expressed in terms of the ear's cochlear emissions influencing the temporal cortex—the ear functioning as the "microchip" that affects the "control panel" of the body, the brain—the cortical influence on the cilia (and on the brainstem) describing the second half of a loop. The investigators postulate that the neural messages from the brain are mediated in the MOC fibres: "Such an effect has been demonstrated in cortically resected [perhaps referring specifically to "dual-brain" surgeries] patients . . .(Khalfa *et al.*, 2001) and by electrical cortical stimulation in epileptic patients . . . (Perrot *et al.*, 2006)."[58] In other words, both surgical severing of the loop and artificial stimulation of the brain can affect this pathway. (Khalfa also finds autistic children laterality deficient in the left hemisphere, a condition whose cause she cannot assign, but assumes to arise from "more central auditory processing alterations.")[59]

The authors also note that alterations in the MOC system reflect laterality and the condition of schizophrenia—which I have suggested should be called a disease of *non*-laterality—that may be defined as the absence of hemispheric dominance and which, in a situation of sustained psychosis, would produce an absence of laterality. "Moreover, the MOC fibres which originate from the ipsilateral [on one side] superior olivary complex (uncrossed MOC system) present a pattern of functional asymmetry influenced by handedness (Khalfa *et al.*, 1998)—a pattern found to be absent in schizophrenic patients (Veuillet *et al.*, 2001)."[60] While the authors go on to source the cortex for changes in the MOC, I consider that this information importantly supports Tomatis's contention that the condition of the ear defines laterality and my contention that improved ear function brought about through sound stimulation not only alters the dyslexic condition but also ends the symptoms of schizophrenia and at the same time establishes or improves laterality (the skills of the right hand).

The study concludes that "some auditory system processing mechanisms are impaired in children with dyslexia and that audiovisual training can diminish these deficits."[61] Undoubtedly, the connections between the visual and the auditory cortex allow for a degree of remediation, but the more direct means of altering the problem of dyslexia, reflected in the MOC, would be to alter the middle-inner ear so that it has the ability to process sound normally. Similarly, with the other subjects under these researchers' consideration: autism and schizophrenia. We have seen the evidence for the efficacy of this approach in Daniel's listening intervention and several times in his focused listening.

Patterns of Recovery

In the course of a single day at any time in his illness, Dan's drawings might recapitulate all the stages from which he had recovered while also illustrating a freshly attained level of complexity. This recapitulative activity became clearer to me about halfway through this episode. Daniel's uncontrolled verbalizing, as it became cognitively more and more complex during his recovery, also would recapitulate during the course of a day many of his prior levels of cognitive function. This range of expression would make some family members think Dan was not improving or had stalled at a plateau. However, I noticed that whenever Mountain Man would "sight the crest of a ridge" and attempt it, he would then fall back to a lower level as if to consolidate his gains. Doidge notes such a pattern of recovery for stroke victims and cites findings to indicate that plateaus in neurological recovery do not signify a maximum has been reached, but perhaps merely a level of consolidation.[62] That sort of plateau may have been what happened in a major way in June when Dan experienced hours of left-hemispheric dominance and slid back into non-dominance. Or, that brief reprieve may have been interrupted by his having found some marijuana and used it, so that he reverted to a lesser level of brain function. Occasional incidents of improved self-control happened very clearly in lesser ways, as when, for example, he sat absolutely quietly for 20 minutes in Dr. A.'s office waiting for the doctor to arrive to renew his prescription. While he did not walk out of the office retaining that level of verbal self-control, over the next week he was able oftener to keep quiet, his self-talk became much quieter, and his ability to reach for other goals—such as restraining Lox's running all over the neighbourhood by reinforcing the fence with additional boards—was enhanced. That pattern of two steps forward and one back characterizes a great deal of human learning.

Doidge's writing on recovery from stroke also helped me to revise my strategies for helping Dan by showing me the sorts of activities that could help the brain to re-establish certain kinds of learning. The repetitive physical tasks that employ both hands and call on fine motor control, whether in the kitchen cooking and cleaning or in the yard splitting logs and picking up sticks for kindling or in the field swinging a golf club, require the co-ordination of both hands, both sides of the body, and both brains. Here is another field where therapies for recovery could be developed and explored.

The Time Factor in Recovery

Daniel's intervals of psychosis, that is, the length of time it has taken his ears to recover from damage by drugs (street and prescription) and/or from loud sound, do not appear to depend simply on his previous history of psychosis. In other words, he does not appear to have a progressive disease ("once a schizophrenic always a schizophrenic") that simply deteriorates from bad to worse, or that improves gradually over time. His most severe symptoms occurred during his first psychotic break; thereafter his psychotic breaks showed greater levels of self-control until the last. Another point to consider in the time factor is that Dan's abuse of hallucinogens leading up to this episode was much more extensive than anything he had done for a very long while and was most similar to the drugging that had precipitated his first psychotic break. It appears that Daniel's psychosis is the product of his addictive use of drugs, his prescription medication, noxious sound, and the resiliency of his ear mechanisms and neurology to resist those insults.

The severity of symptoms at the outset of spontaneous schizophrenia is perceived among those diagnosing schizophrenia as "the acute phase." I think it may simply mean that persons experiencing psychosis initially are terrified and upset, but become accustomed to their symptoms of loss of hemispheric control and to the accompanying social dislocation implied in their diagnoses. Plus, of course, medical interventions (chemical, behavioural, and psychological) will have some ameliorative effect on acute symptoms. Dan's episode in 2003 lasted from 8 February to 26 October (eight and a half months) and was complicated by two head injuries (the car accident the previous December and the snowboarding accident in February). The 2005 episode lasted from November 2004 to mid-January 2006 (14 months). The episode that began in late December 2007 showed a brief remission in June (at the six-month point), and ended 26 November 2008 (11 months). However, the recent episode clearly involved multiple injuries to the ear and brain: marijuana and hallucinogen damage, Risperdal damage, and sound damage. The recent episode also included more focused efforts to effect change.

The familiar patterns of improving cognition, mood, and self-control that had eventuated in previous episodes became evident in mid-October; we agreed to reduce Daniel to half-medication, i.e., the same dose but every second day. Within 10 days Richard had noticed how much less stable Dan was on the day following his evening dose of ¼ of 0.25 mg Risperdal. At first I did not trust Richard's observation, but within four days I could see the same pattern. It made me wonder how much sooner Dan might have improved *without medication or on a different one* but under focused listening.

Ten days after Daniel stopped taking medication, he regained left-hemisphere dominance. In mid-November a flu virus struck two of the family households and Dan was the last in ours to come down with it. For a day and a half he slept and the sound of his voice was silenced. Richard kept the radio in the adjoining room tuned to the classical music station. On the morning of 26 November, Dan awakened fully rational, as I could tell in his observations about the differences between the decision-making capacity of his dog and humans and how that related to God—"d-o-g" was no longer being misapprehended as "G-o-d." He had complete control of his speech. He played his guitar, singing quietly the original lyrics to tunes of The

Pogues and whistling in tune at intervals. An hour later he was lecturing me on the severity of "this" psychotic episode compared with others, the differences between the inciting drugs, and what he thought about comments I had made regarding the increasing severity of his symptoms from one such episode to the next. He complained about being "stuck here" and became restless about his future. He had only fragmentary memories of the previous 11 months, but felt as though he remembered everything, as any of us may feel on awakening from a dream. His ability to focus optically gave his interactions with people a sense of "presence." I realized his prevalent nystagmus that came and went with alterations in hemispheric dominance probably were the "eyes darting around" of his first recorded psychotic episode—and likely the R.E.M. (rapid eye movements) recorded of persons in the state of dreaming.

Obviously, many questions have not been answered, such as the time it usually takes in a range of representative individuals for the stapedius muscle to recover spontaneously without medication of any kind; and for the cilia to recover without medication or with supportive therapy; and for neurology subjected to hallucinogens, marijuana, alcohol, and tobacco to return to normal. What would be the optimal use of sound for recovery and what integrative activities would most help? How could Daniel's setting for recovery be transposed into an urban setting? How can appropriate levels of tranquility be achieved through pharmaceuticals (and which are the best for the purpose?) without negatively affecting the ears and the neurological processes involved in listening? How do age and general health affect the recovery process? What supportive therapies and counselling would expedite not only neurological healing but the psychological and social ills attendant on these particular ear-related disabilities? What changes in community attitudes would smooth the way for re-entry into the mainstream?

The end of psychosis is the beginning of another odyssey of continuing sound stimulation, reinforcing of left-brain strengths, and recovering of important threads for learning, self-esteem, and responsible patterns of living. Left-hemispheric dominance is only as strong as one decides to make it or is able to decide to make it. What we now know about the plasticity of the brain opens the possibilities for recovery and wholeness exponentially.

Antipsychotic Medications and Compliance

Who studies how comfortable or uncomfortable the patient feels with his or her psychotic symptoms? I garner a great deal about how uncomfortable other people feel about the patient. I hear such complaints with sympathy because I know how it feels to live in a household with someone who is fully psychotic and minimally medicated; with people who are depressed and isolating; with bipolar people who are prone to sudden outbursts of vitriolic emotion; with people who try to organize the activities of those around them to accommodate their own conditions of fragile mental comfort; with people who resist opportunities to alter their condition of comfort for the comfort or well-being of anyone but themselves. We all know how some of these living conditions feel to some extent: Gerald May[63] describes this as the human condition, the tendency to addiction built into our nervous systems. For those with easily altered hemispheric dominance, the need felt to shore up that dominance with certain kinds of routines explains some of these addictive behaviours.

Persons researching and treating mental disorders with antipsychotic drugs remark on the tendencies of their patients not to comply with treatment and, therefore, they also assess drugs in relation to how consistently they are taken: compliance. I have seen a tendency, in the studies I have read, for researchers to assess patients with value judgements along the lines of their compliance or non-compliance. These concerns have become so widely recognized that they have gotten into television portrayals of bipolar characters who are presented as unwilling to take their medications because they too much enjoy the "high" of the manic episode of their cycle and do not want it blunted (e.g., the Canadian television series *Traders* in its final season). Another assumption in the medical profession regarding non-compliance is that the patient is rebelling against the extrapyramidal effects (neural degeneration outside of the pyramidal tracts of the central nervous system and basal ganglion, such as tremors, tardive dyskinesia, chorea, athetosis, dystonias, and Parkinsonian symptoms) caused by the medication, which can be so severe as to be life-threatening.

While both of these assumptions have validity, medications affect both hemispheres of the brain, causing both the less functional and the better functioning hemispheres to be suppressed in their activities. One can easily see why persons striving to maintain what mental facility they retain would be reluctant to sacrifice that ability for the reduction in symptoms originating from their other hemisphere, if they can tolerate their own psychosis. Not without reason, many people today who have passed through mental hospitals and psychiatric wards also have chosen to forgo medication; some consider themselves "psychiatric survivors" and in some instances have organized to further the cause of patients' rights. Apparently many do feel that they can more easily tolerate their unruly hemisphere than the even greater loss of their reason and self-control. Even the tiny amount of medication

Daniel took made him feel "numb in the brain," and affected how he saw things "all trippy and weird," thereby contributing to hallucinatory images. In fact, it was remarkable how much more of the "real Dan" returned when he dropped back to half-meds, still more, on one-third, and yet more on none—even though he was not free of psychosis and had not yet attained left-hemispheric dominance. His comments on his own mental process as he continued to manage the images from his right brain could be amusing: "I've been orthodosed"—meaning "by the medication into orthodox beliefs"—and "I finished that one [hallucinatory image] off. It's rather like a chessboard and I'm Bobby Fischer."

14

The Cerebral Society

Cerebral Dominance Institutionalized

Dan, on one-third meds, was in great form this morning, starting his listening session in some confusion but much more insightful and perceptive in his left-brain phase and generally calm and in an upbeat mood as he listened to Bach through his headphones and made drawings. As I had not attempted to discuss hemisphericity with him, his observations expressed remarkable perceptions of his divided brain:

"I died and went into the abyss, but I had pleasure in it and went directly on to the left."

"I've slain things by sending them away."

"I'm not trying to control others; I am trying to control myself."

"I am not all of these people; I am more of a twin."

Then, without my noticing, he turned the volume high on a selection he likes and our day became one of the more difficult kind, his non-stop chatter running the gamut of his right-brain images with a greater range of repulsive memory than usual.

Late this October afternoon the sun broke through the overcast turning the broad landscape of lawns, fields, and trees, with their pale, wet leaves and grasses still tossing in the breeze of the rainstorm, into springtime greens. Every drop of rain coating the eleven windows in the sunroom became a bead of pale gold; the room was frosted with a filigree of shimmering luminosity, an enchanted place from which to view the magical landscape outside. Dan came out of his room muttering and was silenced by the beauty. Then he spoke quietly, with awe, of how the light had transformed everything we could see. Richard came in a few minutes later when the raindrop jewels on the glass were evaporating, the sun setting behind clouds, and the garden fading to the dun colours of late autumn. Even then, the magic lingered to delight him. In that brief interval Lothlorien had come and gone; but the peace and the sense of distant elven voices remains.

Re-Creating the Non-Dominant Hemisphere in Society

In society we create roles and groups mirroring the dominant, rational, left brain and ensure that our children are schooled in these from the cradle, through kindergarten to high school graduation, and we laud them if they continue their left-brain development through graduate school and specialties. Society also creates roles and groups mirroring the kinds of expression and non-expression for which we use our non-dominant hemisphere. We expect them to listen and not to speak—or to speak in a carefully prescribed way. We expect our friends to keep secrets. We expect family members not to embarrass one another in public. We expect employees to be "discreet" about "privileged" information. We have confessors and insist that they obey the rules of a dominant rationality, usually an institution such as church or a profession with a defined ethos such as psychiatry or psychology. Many are familiar with the advantages and disadvantages of such a role through the practice of formal confession. In the confessional the expectations are ritualized and rigidly supported by the institutional church.

Some people are uncomfortable with confidences. "Don't tell me anything you want kept secret," a friend remarked; she has difficulty controlling the verbalization of her "stream of consciousness." Where the notion of confession has been institutionalized in the role of a confessor, one finds that some priests are more comfortable than others in their role. What happens when the confidante, priest, psychologist, psychiatrist, lawyer, guru, or other confessor finds her/himself burdened by confidences?

With another of those slow-motion Gestalt shifts of my mental framework that accompanies new and fundamental knowledge, the importance granted to confidentiality in our culture has come into a new perspective in my understanding of social behaviour as a reflection of the way individuals use their non-dominant hemisphere. I recalled the time a priest had prefaced his public comments about something I had said to him with the words, "As you told me the last time you were at confession . . ." and how profoundly betrayed I had felt and how I had judged his lapse to be the result of mental instability. Now, I understood the nature of the priest's "nervous breakdown," perhaps exacerbated under a pressure he endured in bending his ear to the confidences of his parishioners. Perhaps for the same reasons, following my confession to a different priest the confessor then made his confession to me. I remember times during my teens when I discovered some people had expected me to lie, if necessary, to shield something said in confidence about someone else who was to be "kept in the dark." This sort of "requirement" coming from a family member to apply within that small circle can be extremely stressful, depending on the degree of trust among its members and the nature of the secrets. How much more difficult for a pastor already hemispherically stressed to carry the right hemispheres of an entire parish?

Knowing how the loss of hemispheric control in others can affect me, I am aware also that the decisions of others made interpersonally or in private can and do affect not only me, but the larger society as well. What happens to people who confide in priests or psychiatrists or counsellors who purport to have superior knowledge about the mind, or the soul, or the psyche and find themselves not to be the recipients of knowledge, wisdom, and insight but the victims of ignorance, mistakes, manipulation, or of greed, desire, and a hunger for power? As long as the functions of the two hemispheres of the brain remain mysterious, opportunities for self-deceit and for charlatanism prevail within institutions and in society more generally. Freud and his followers, some ministers of religion, and a long and various troupe of meddlers in brain processes have been able to take advantage of persons who do not understand and do not know how to take care of both their brains.

If socially constructed roles and groups that mirror the function of the hemispheres can work either positively or negatively, does knowledge of how people tend to use their two hemispheres remove the need for or value of social roles that mirror the subconscious? At present, we live in a society where people who have access to their subconscious and people who emphasize self-control over the subconscious tend to assume opposing positions. Some are in a pitched battle to subdue one another, each of whom represents to the other the aspect of brain dominance or of brain liberation that the other finds lacking or abhors. Silencing by various methods plays a part in this conflict. Thus, some subjects are declared taboo, by an individual in a family, or members of the family towards others, or by social groups, or by even larger segments of the society. In a remarkable number of situations our "freedom of speech" is compromised by the unwillingness of others to listen and the willingness of others to be silenced. As I came to see where such impulses originate, a ray of light beamed all the way back to those years when April wrapped herself in a selective silence, when my husband declined to discuss certain issues with me, and when I found my welcome to express my opinions on a number of subjects curtailed. All social groups develop these patterns of limiting and promoting certain kinds of speech. Taboos of various kinds fostered in the media become "politically correct" and many trade their own desire for privacy, for a host of reasons, in the same coin. Where one occasionally finds someone expressing a dissenting opinion, he/she is ridiculed or treated much more harshly; ostracism can precipitate a range of depressions, even suicide. In some societies dissenters are put to death.

My scientist father and musical, artistic mother symbolized hemispheric polarization to me as I was growing up—of course, I did not have that name for their differences—because each argued from a fully rationalized point of view about very different ways of "knowing" that included their religious beliefs. Similar kinds of polarization are seen when science is divided from religion, when one religion claims greater rationality than or superior insights to another, when the "practical minds" of finance line up against "the bleeding hearts" of the socially conscious, when "rational enlightenment" meets "faith." Humans also create institutions that will seek to unify these dichotomous elements, such as the United Nations to take the most prominent global example. We also find inter-denominational and inter-faith groups; the International Red Cross; Doctors without Borders; and other organizations designed to bridge gaps considered by some to be unbridgeable.

The "Mental" Outsiders in Society

A different dynamic comes into play when the conflict is joined by someone who lacks both the ability to rationalize his belief system and to rationalize or to control the expression of his subconscious. Most of the polarized "rational" social groups feel stressed by the "unteachable child," the "gay person," the "addict," and the "insane"; they are face to face with an uncontrolled or uncontrollable right brain and subconscious. These rational groups suddenly come into a kind of agreement and operate under broad social mores, taboos, folkways, and notions of how the brain works or should work. When someone like Daniel fails to be able to function within even broad terms of social acceptability, or when men and women choose to reject for their own reasons those broad social values, the various belief systems in the surrounding conflicted groups become active and whatever is hidden in the non-dominant side of each person's brain may become a factor in how he or she will think and behave towards the dyslexic, homosexual, substance-addicted, bipolar, or schizophrenic individual. In other words, the *irrationality* of the supposedly rational members of the society comes into play as they turn on the persons who have lost, lack, or have rejected particular forms of rationality and self-control. They want to treat that person the way they treat their own non-dominant hemisphere: by ignoring, repressing, or silencing. Ironically, the persons who consider themselves to be rational may force irrational, unreasonable, and "right-brained" solutions upon these less-controlled persons, or they may explode into a loss of hemispheric control on their own account. The targeted "uncontrollables" are marginalized, ignored, isolated, subdued, shunned, drugged into silence, put to death, or otherwise partitioned from integrated socialization.

A decade passed before I could see that the courtroom judgement releasing April's mother into society had been accepted by her as an opportunity to become a balanced person; she managed to rebuild her life. You will recall from Chapter 1 that April and her friends were assaulted by other students and the citizenry with bricks that happened to be available, an American incident of attempted stoning. It appears that Americans who consider some Middle Easterners worthy cannon fodder on account of their medieval laws do themselves resort to similar tactics outside the law when they encounter a state of mind they cannot comprehend. No less damaging to the groups of marginalized people is that because they are relatively few in numbers and vulnerable mentally, they become objects of study in ways that can be harmful to them or even deadly. Some sorts of damaging interactions have been demonstrated towards Daniel, within our family as well as in the surrounding society, all his life. To a lesser degree, but no less truly, his sister Sarah has suffered with disabling conditions that were almost within the range of normal—colic, shyness, fatigue—yet almost overlooked for being fairly common. Only when her fragility in those directions became full-blown chronic fatigue syndrome did she receive the specialized attention that perhaps she, too, had needed from birth.

In some sense, each of us is a danger to oneself and to others. Whether or not another person experiencing psychosis actually is "a danger to him/herself and others"—as I was told by hospital staff that Daniel was during his early episodes of psychosis—will depend a great deal on the underlying personality and belief system of the individual, the degree of impairment, and the surroundings and how people in the environment define "danger" and what sort of strategies they have learned for coping with psychosis. It is important for others to realize that quite horrible nonsense being spoken by a psychotic person usually is completely divorced from intentionality and that he/she cannot be held morally responsible for it. Similarly, bipolar outbursts ought not to be considered representative of a person's thinking on a subject because they are not reasoned. All of which is *not* to say psychotic or bipolar persons never are a danger to themselves and others; the suicide rate among schizophrenics is greater than that in the general population and some psychotic persons have posed grave dangers to others. These conditions and others outside the range of normal are not benign. The caution I raised in the Introduction remains important.

After the Subconscious Is Revealed

Exactly what do we understand by "vulnerability" to words? Especially to words spoken by a person without the capacity to control and shape his or her words in socially acceptable ways? One of the frequently repeated bits of folklore in my childhood was, "Sticks and stones can break my bones, but words can never hurt me." Of course, words *can* hurt. Children may not be able to defend themselves from aggressive or offensive language, so they are taught gradually how to do that and some of the folklore helps. Hence, by the time the person is grown, we expect a certain level of detachment from what he or she hears. Then, a man shouts in vulgar language at a woman in his apartment building. The second time he does this, he follows her to her apartment and pounds on the door. She is shaken and calls in the police. The abusive tenant has a reputation "for doing such things occasionally." The woman no longer feels safe.

How is someone going to be able to find out what accounts for the man's alarming behaviour? At first, she wondered if he suffers from Tourette's syndrome. Nowadays, in the case of Tourette's syndrome, what used to be considered a disorder that humiliated the patient or amused others has become understood as a malfunction of the nervous system and is no longer viewed as intentional or as appropriate for ridicule. We can view a stream of obscenity and invective from such a person with knowledge that leads to objectivity, tolerance, and compassion. The second time, however, his words were followed by a threatening act—a greater loss of control. Is he drinking and acting out some childhood wounding under his lessening self-control? Perhaps he is a drug addict experiencing temporary psychosis? We are less tolerant of addicts and of those who enter a state of continuing loss of hemispheric control. Is he a schizophrenic who has chosen not to take his prescribed medication? The commonly held belief at present is that schizophrenics should take their medication and they are given little sympathy if someone discovers they have not. We need to learn more about all of these losses of self-control, how they may be ear-related, how they affect verbalizations, how they are related to other losses of control; and we need to disseminate that learning widely.

Why do some adults with high levels of socialization feel themselves vulnerable to words, misunderstand intended meanings easily, and withdraw from hearing them, wrapping themselves in a blanket of silence, isolating themselves from the possibility of hearing certain kinds of sound, certain kinds of language? Why do others feel strong and resilient to sound, seeking opportunities to speak, to listen, to make music with voice, breath, and hands, and to create and listen to music? We have learned that a person's sense of self, system of beliefs, socialization, and acculturation are profoundly attached to hemispheric dominance—for most people in the left hemisphere of the brain—and that the process is directly related to the functional level of the right ear. Depending on the person's right-ear function, hemispheric dominance may or may not be strongly established.

Depending on the person's age and her or his socialization, the person's belief system—a fundamental aspect of the reasoning process—may be robust or fragile. If hemispheric dominance is not yet strongly established, the person will be more vulnerable to impulses from the non-dominant hemisphere: to emotions, impulses, mood swings, even to the possibility of a sudden loss of (shift in) dominance. Under such circumstances, the person, especially a young person, may be highly suggestible. If the person's belief system is fragile, the ability of the left hemisphere to maintain dominance over the interior stresses of mood, emotion, and primal urges will be diminished. The interplay of just these two variables generates a considerable range of possible behaviours. If you add interior stresses arising from poor ear function (e.g., a selection of the ailments listed in Appendix A) and external stresses arising from isolation (internalized as loneliness), little education (which reduces the reinforcement of more-or-less rational ideas in the left hemisphere), and impoverishment (internalized as fear and hopelessness), the pressure on the brain towards "the downward spiral of depression" is inevitable and the individual becomes even more vulnerable on many fronts. Intervention in any of the stresses may ameliorate the person's situation, but correcting the function of the ear can provide the health, energy, and elevated mood for the person to more effectively address and alter some of the external circumstances. For the person who has made unfortunate decisions with far-reaching consequences (such as addictions) while in a state of vulnerability, a great deal of assistance may be needed to support recovery, even after healing has been effected in the ear. Organizations such as Alcoholics Annonymous and other 12-Step programs, some churches, and psychiatric survivors' self-help groups make very long-term commitments to their members.

When children cause one another harm or when mutually consenting adults cause one another harm, we feel less threatened. We more easily maintain the objective viewpoint that releases us to some extent from our tendencies to identify with the victims. We can view not only our own right-brained tendencies (rational *introspection*), but the right-brained tendencies of others. We can learn objectivity and empathy. When the problematical behaviours are exhibited by adults, people tend to fell less sympathetic and look for someone to blame that "the problem" hasn't been dealt with sooner. Mothers seem to be the handiest targets. I sympathized with my frightened friend listening to the pounding on her apartment door but was also concerned for the perpetrator—can this man be helped? Can we more objectively seek an abstraction—a model of the ideal—that will be fair and helpful to both adults unhappy in their social situation? With objectivity we can look for a solution to losses of self-control for children and for adults that will pertain to that model of the ideal and will alleviate harm.

We are not agreed, however, on the model. Ideals vary and come down to matters of belief—among sociologists, psychologists, and psychiatrists, as among all others. Where knowledge can replace or fuse with belief, we may be able to develop firmer rational grounds for improving on the models or ideals or beliefs to which we teach. An understanding of the causes of hemispheric integration disorder and of how we may find healing for a mind hemispherically unbalanced should help us to revise our present models.

Acculturation in Parenting

Parents try to teach that they did not invent the rules they are passing on to their children but that the code is attached to a particular history; and few parents pretend to follow perfectly the precepts they teach. Nevertheless, children who are bonded to their parents in both hemispheres of their brains from childhood, and from before birth to their mothers, cannot but look to their parents as some kind of "ideal" and those two fields of learning—about parents and about values—coalesce in a child's mind. Daniel's illness has allowed us to look into his subconscious as easily as we have watched undersea life in a tropical pool. In psychosis, Daniel's right-brain talk often expresses associations between absolute ideals and Richard or me and between absolute depravity and himself. This way of associating the parents and their beliefs with absolute goodness is natural to the child and reflects some of the earliest work of the right brain. Daniel's verbalization does not mean that we taught him that we are wonderful and that he is depraved, but the teaching of values leaves the child with that tendency. That tendency, once hemispheric dominance is re-established, should be answered reasonably by the left brain. And in Dan's case, it is; he becomes appropriately independant of his parents, insofar as he can, once he has regained left dominance.

The primary conditioning of the right brain in associating primary attachment with values does not mean that parents should cease to teach values, but that they should include an avenue in reason by which the child comes to independence and maturity. Rationality in the left brain—and maturity in the seat of belief—entails separating the beloved teachers from the ideals; keeping the teaching that proves to be valid and legitimate, rejecting or holding in abeyance what seems not to be true to experienced reality, and recognizing the parent as an imperfectly human, if sincere, teacher. The results of that same sort of left-hemispheric evaluation process from generation to generation over thousands of years have produced the tenets of the major religions and the cultures that make up the world we have today. Increasing specialization of the hemispheres and dominance in the left hemisphere appear to be fundamental to that process of social evolution and to the intellectual direction of the human species.

Social Mores Not Linked to Hemispheric Control

Sometimes, behaviour *not* controlled by the left brain is rationalized in the society and enshrined in its laws; thus did the Napoleonic Code, which persisted in Louisianan law, view the passionate retaliatory response of the betrayed wife and mitigate the penalty for murder. In various ways our relatively affluent society increasingly has "rationalized" individualism and freedom of expression for the impulses of the right brain. When individuals need to "hang together" to survive, as in the Western world during World War II, we develop a broader tolerance for some kinds of variance. The reverse situation of affluence may also generate latitude and tolerance, as we recall of the 1960s. Thus, the social definition of "excess" most often comes through hindsight, which is the study of history. How will we manage our perceived differences now that the planet is at risk?

Investigating social roles and groups and expectations is a fairly straightforward matter where people express their ideas and beliefs, but we must note, also, the effects on others of the refusal to speak, the withdrawal of words, or falsity in the words spoken. From perjury to shunning, from ghettoizing to withholding evidence, from rationalization to mute disapproval, we are well aware of the cruelty, malicious damage, pain, and deadly force of words not expressed. The polar trait of the pressured speech of psychosis is the hyper-controlled speech of catatonia, which also may make a person a danger to her/himself and to others. We are less attuned to silence; but we can and do see the behaviours that fall between these two extremes and that tend to be more disapproved socially as they approach the poles. What are or should be the limits to silence and to privacy in society more generally? I have a friend, who wants to continue writing and to publish, who is being asked, following the dissolution of her abusive marriage, to guarantee she will not speak or write of the legal settlement terminating her litigation with her former husband. Her attorney says this is a standard agreement. Has she merely lost a compelling story, potential income, or freedom of speech? Some questions have been decided (perhaps temporarily) for us in legislation and in the legal system; but others remain very much at issue and deserve vigilant attention.

The Psychology of Hemispheric Options

"But that's just another way of not letting go of them, isn't it? Just like pushing things you don't want to think about into your subconscious. But they are still there and you can still revert to them rather than going forward into a healthier life without them. Right?" In the midst of his last psychotic episode, in June of 2008, I had spoken thus to Dan about his choices of music. Is that not one of the ways all of us tend to use our right (or non-dominant) hemisphere—as a recycling bin that doesn't get emptied? As a hiding place for experience and thought we prefer not to examine with reason or deal with or admit into our belief system? As a place to repress things we would rather not have heard, or seen, or experienced? As a place where we keep private our uncomplimentary opinions of others? As a place in the brain where memory is more difficult to access? As a retreat into pleasures where the retreat itself is rationalized by the left brain as healthy, necessary, unavoidable, or intrinsically justifiable? I am not suggesting that the non-dominant hemisphere cannot function otherwise, just that dominant laterality makes those uses of the non-dominant hemisphere easy *because the corollary of hemispheric dominance is that it focuses the attention elsewhere.*[1] The uses for differing hemispheric functions are central to our understanding not only of so-called aberrant psychology, but of normal psychology as well.

Seeing Ourselves in Others

The right hemisphere of spontaneity and creativity—of art, dreams, the imagination, and the subconscious—is the twin of our ordered left domain of reasoned language undergirded by belief and trained in self-control; we abide with both brains according to kinds of awareness and beliefs that should more easily allow us to recognize both halves of our selves in others. Furthermore, our degree of self-control, in which we may take considerable pride, may reside largely in the comfort and peace of our surroundings rather than in some intrinsic strength. Our rural location has contributed to some kinds of peace.

Urban social pressures create mental stress. The "slow burn" in a highly competitive business that erupts into "road rage" on the commute or a spousal crisis at home is a lesser degree of the kindling that incites the more severe forms of mental illness. Such a setting may impact more severely those with unstable hemisphericity. Moving along the spectrum, kindling, as in epilepsy and bipolarity, must also be considered as an aspect of the forces impinging on those suffering the more consistent forces of hemispheric alternation in psychosis. In psychosis, the rational left brain is subject to the usual psychological impacts of previous experience and of the surrounding social situation. In other words, the mentally ill (unless very heavily drugged) are subject to the same environmental and emotional pressures as someone more stable. I have mentioned some of Daniel's reactions while he was psychotic to stresses I created by failing to respect the degree to which he did have self-consciousness. I thoughtlessly talked about him in front of others and he reacted with understandable embarrassment and withdrawal. I knew this normal human reaction to humiliation was true of Elizabeth's son, who might have come to his mother's funeral (despite his hospitalization) but for others he dreaded meeting who had taken a critical view of his (even heavily medicated) schizophrenic symptoms. In segregating people with hemispheric integration issues we are hardly doing better for our ear-damaged mental patients today than doctors of 75 years ago; but I believe we may have the knowledge now to cure and support healing, rather than merely subdue symptoms.

As the famous writer Charles Dickens, sensitized by his own dyslexic experiences, taught in all of his writing, we cannot assess persons' character according to rigid divisions between categories of acceptable or unacceptable behaviour alone, but must also take into account those environmental stresses (often inherent in social class) that propel them towards more profound kinds of loss of control. Assessing persons in terms of social pressures is not to remove all issues of responsibility, but to better understand what are the limits of individuals' capacities for various kinds of stress before their hemispheric dominance collapses—and to be able to evaluate what degree of mental control they may still maintain via the left hemisphere. The renaming of "schizophrenia" from "Seishin Bunretsu Byo" ("mind-split-disease"), with the new term of "Togo Shitcho Sho" ("integration disorder") by the Japanese Society of Psychiatry and Neurology in 2004 was a step in the direction of removing social stigma from people who suffer varying degrees of hemispheric non-integration.[2] The change of nomenclature for schizophrenia in Japan suggests that the broader social context of mental (dys)function should be opened for study. In the United Kingdom the Campaign for Abolition of the Schizophrenia Label (2006) was working to achieve similar goals as the Japanese and for the same reasons.[3]

Lack of Hemispheric Dominance and the "Evolution" of Societies

If we consider other social contexts in which lack of hemispheric dominance occurs, I think we will find ourselves looking at the evolution of social history in new ways. The associations Daniel was making in his right brain between mushrooms and wild animals and the people in his family circle during his eighth month of recovery find parallels in the animism of the ancient Sumerians[4] and of Aboriginal peoples around the world where they have until recently remained relatively unaffected by the advancing "civilized" and industrialized world. Integrating our new awareness of the mode of function of the hemispheres of the brain will influence our view of problems that have perplexed us for a long time, some of which Julian Jaynes explores in *The Origin of Consciousness in the Breakdown of the Bicameral Mind*[5] in a highly speculative thesis. Readers of Jaynes will have found in my exploration of Daniel's illness the tools for separating fact from speculation in his thesis of the evolution of civilizations and of some forms of consciousness. I have explicated the process of language acquisition and manipulation in dyslexic syndrome/psychosis (including the activity of metaphorical pairings in the right brain of which the left brain may predicate conditions and actions) in a way that bears out some of his observations about the metaphorical basis of language. Insofar as I have contributed answers to questions about how and why various forms of consciousness occur, Jaynesian theorists may want to revisit some of their definitions of consciousness.

I think, also, that persons theorizing about the origins of language need to consider non-linguistic forms of communication simultaneously. The distinctions between hemispheric dominance (and social domination) and *integration* in the two halves of the brain are important for postulating not only changes in consciousness of humans over time, but for understanding differences in "ways of knowing" by persons who use both halves of their brain today but—by choice—differently. Other issues, such as the inherent pains of impoverishment, have less to do with choice. Does a simpler language lead to simplistic thinking, or do people find other ways of communicating complex and nuanced ideas? Knowledge of hemispheric integration raises many questions of this kind: Is it possible that North American Native peoples' extreme vulnerability to the introduction of alcoholic beverages into their cultures refers to a lack of insistence on left-hemispheric dominance and concomitant forms of control in their acculturation of their children? First Nations infants experience less sound stimulation from their elders than other ethnic groups.[6] It is obvious that quiet children reduce the cues for predatory animals, prey, or hostile human groups. How does muted hemispheric dominance work *positively* in hunting and gathering cultures to preserve life? Is there a simple relationship between needing to be attuned to the environment with *both* ears that has marked some "primitive" societies and made their transition into contemporary mainline cultures problematic? Daniel expressed the longing to become a part of Aboriginal culture from the age of about 10 when he began to be aware of the ways in which he was different from other members of his family. What did he intuit that we have yet to learn?

The history of social evolution from ancient times to the present can be seen as a "progression" of increasing exercise of left-hemispheric dominance (and of social expectations dependent on that dominance) in the cultures of our earth, as Richard Gregory unintentionally implies in his "logical" hierarchy of the modes of thought (regarding their environment) that characterized societies starting with the Sumerians' animism, mentioned above, and proceeding through the Greeks' pantheon to the Jews' monotheism.[7] However, any concept of "progression" neglects some of the elements of human emotion (right-brained) and human activity (left-brained intentionality) that flourish beyond language but that incorporate the motivations of both hemispheres.

The sprig of lilac in my baby sister's hand locates the sunny day in April. Other "snaps" taken by Dad on his 1930s Kodak camera with the sliding, black accordion box to adjust focal length place the year as 1944 and my age at almost three. "You explain to her what we want," Mummy said to me. "You know what she means; you can talk to her." Lainey was 14 months old and I understood exactly what my mother meant. They wanted to pose her in a spot in the backyard in bright sunlight, but I knew that felt too bright and hot to her. Both parents and my young Aunt Lucille had tried to manoeuvre her into position without success. Although my sister's command of English was slight and mine was precocious, she and I communicated in a language private to ourselves: body language, grunts, gestures, occasional words. We knew what the other was thinking and bonds of trust forming on our neural networks placed us closer to one another than either of us could be to the adults towering above us. At the level of long blades of spring green grass and the dark earth under the back porch where a lattice cut incoming sunlight into bright diamonds we shared the exploration of our limited world with our developing neurology. Despite my efforts, the developed photo of Lainey in her pale blue, knitted coat showed her Kewpie doll knot of enviably blonde hair topping a frowning face. Already, my parents were thinking of her as stubborn and resistant to their schooling of her behaviour. Recently, she handed me a blossoming twig and at the first whiff I recognized choke cherry. "Who else will know?" she asked unnecessarily. Its bittersweet scent evoked for both of us the bent choke cherry tree we had crawled upon and bounced on as "riders" and played under and around with neighbours near the house we'd moved to when I was eight and she was six. To make the effort of turning that shared experience into further language might or might not enhance the sharing of memories that binds us as sisters who experienced together the learning universal to children and unique to us. Humans know and communicate also "by glance and wordlessness."

Similarly, the superficial accoutrements of daily life, from dress to tools to manners, may vary in complexity from place to place or time to time without much affecting what most people value: not so much the "stuff" but rather its metaphorical/symbolic value. One may contend that at every point in history, the human experience changes along a fairly narrow range, despite rhetoric about the "achievements" of industrial and technological change. Whether or not people are in general more self-controlled now than were the ancients is by no means clear. The proliferation of languages has not stopped warfare. The body of "useful knowledge" to be passed from generation to generation remains a matter of considerable debate. The creeds may change but the fundamental process of belief determines the knowledge structure of the language and how individuals will use it. I think directionality in the integration of the hemispheres is built into our forms of consciousness and into our social institutions, but remains flexible to change. If humankind were to experience an apocalypse, who would be the fittest to survive—the right-eared, learned scholar or the binaural Aboriginal? We have too small a slice of human history under examination to prove much about "evolution." Even the purposes to which our elaborate languages are turned today are as often "psychotic" in the sense of right-brained and hedonistic as they are rational. We remain creatures with "more than one way of knowing."

The efficiency of the right ear–left brain route observed by Tomatis has implications not only for the fields of study I have emphasized in telling Daniel's story—medicine and psychiatric practice, psychology, and education in the teaching of reading and other subjects of instruction—but impinges also on the studies of anthropologists, sociologists, historians, students of literature and art, philosophers, lawyers, and theologians of all of the world's religions. My new perspective is reinforced as I see Daniel trying to identify who he is by reaching out to the extensive genealogies preserved by my father that map Daniel's association with one English family's involvement in European society back through historical eras to A.D. 800. How did hemispheric dominance affect the choices of family members across the generations to alter the development of the society—the women who became nuns; the men who became priests; the women who married into other wealthy houses; the men who chose mates outside the expectations of the society, like the heir who married a Spanish woman; the warriors who preserved the dominant status of their duchies or nations and of their family from generation to generation? How did the access of those warriors to the raw, unfettered emotions and blunt metaphors required for battle with swords mesh with the beliefs and rationalizations provided by left brains little schooled in language or in the complexities of thought that had become sequestered in monasteries, but that arose in the population more generally with the Renaissance? How do the strategists of contemporary warfare manage to control the brains of today's young warriors, whose schooling towards left-hemisphere dominance likely outstrips that of historical kings? Does a brain educated to certain rational values revert under pressure to contrary impulses associated with the non-dominant hemisphere? Why do soldiers return from battle and murder their own wives? I think we can answer those questions, or come closer to answers, by way of our new understanding of hemispheric dominance, listening capacity, and the role of belief in managing integration.

We have a new focus for research into our reasons for behaving as we do and into the great variety of cultural heritages of the peoples of the earth. It is difficult to think of a discipline in the traditional academic forum of the humanities and social sciences for which an understanding of hemispheric function and dysfunction as ear-related has little significance. The Socratic tenet of self-knowledge adheres: one needs to understand how she or he became who she or he is at every age and in every age—including this age—to be able to theorize about how the rest of us arrived at this point in human history. I hope I have contributed to those forms of inquiry.

The Spectrum of Mental Disorders and Social Tolerance

Daniel was diagnosed as schizophrenic, an opinion at least not refuted by any of the eight psychiatrists who saw him in the initial stages of his illness. Is Daniel schizophrenic? According to the definitions first given to me by our family physician, a diagnosis of schizophrenia cannot be made at present because all of his psychotic experiences have been associated with his use of marijuana and hallucinogenic drugs. According to other definitions of schizophrenia, he must not be, because he has recovered after each episode of illness to a greater measure of mental function and self-sufficiency—but his recovery may be attributable to the extremely low doses of medication he has been given over recent years by comparison with other symptomatic persons who have been studied. Daniel reacts very negatively to the suggestion that he may be schizophrenic; he accepts more easily that he has drug-induced psychosis. He feels/thinks/believes that the condition induced by drugs is justifiable, although he regrets that he cannot maintain a kind of control over psychosis that would make its effects temporary. Ac-

Figure 14.1: A Proposed Model of the Continuum of Listening Abilities, Degree of Hemispheric Integration, and Related Behaviours

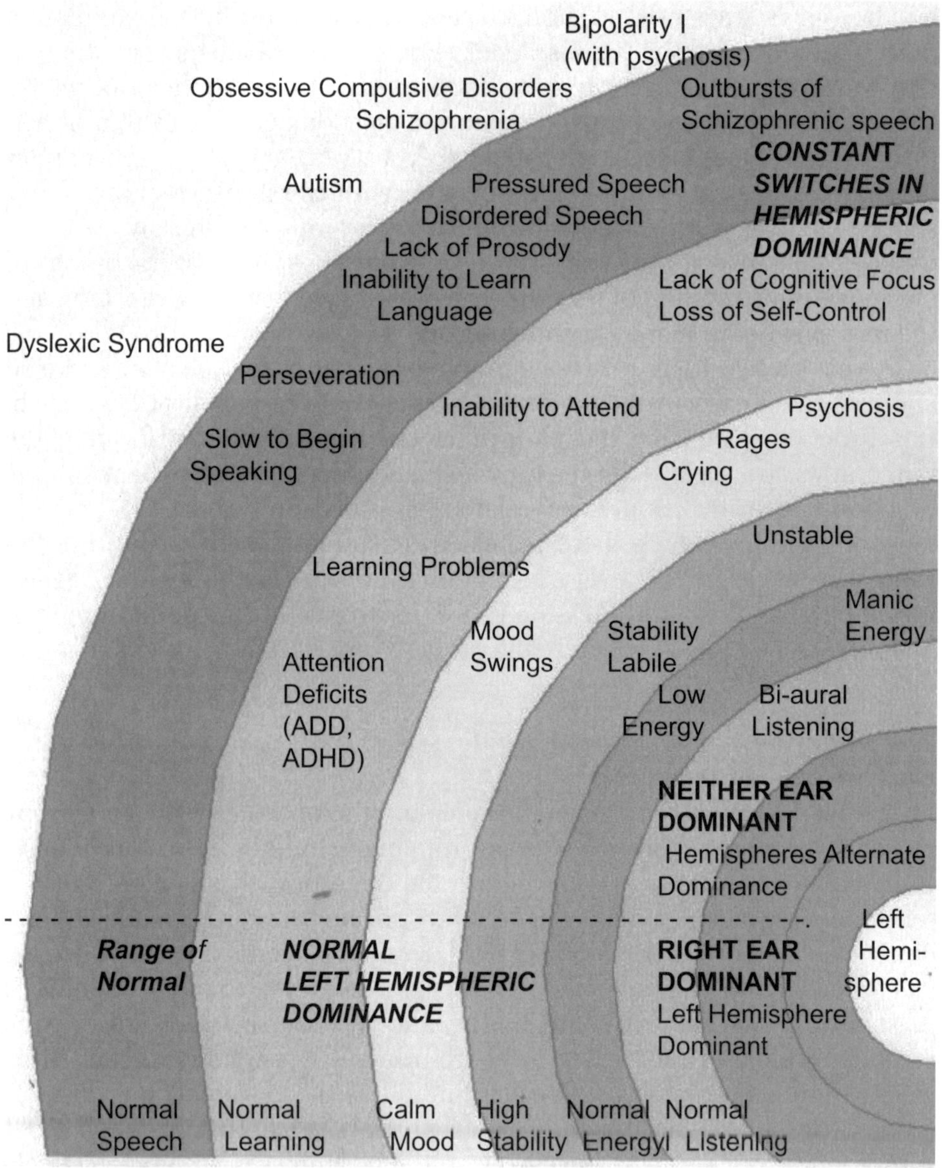

Note that persons whose behaviour generally falls within the range of normal experience varying degrees of dominance and a considerable loss of dominance under such conditions as fatigue, sleep, illness, physical and emotional stress, intoxication, and so on.

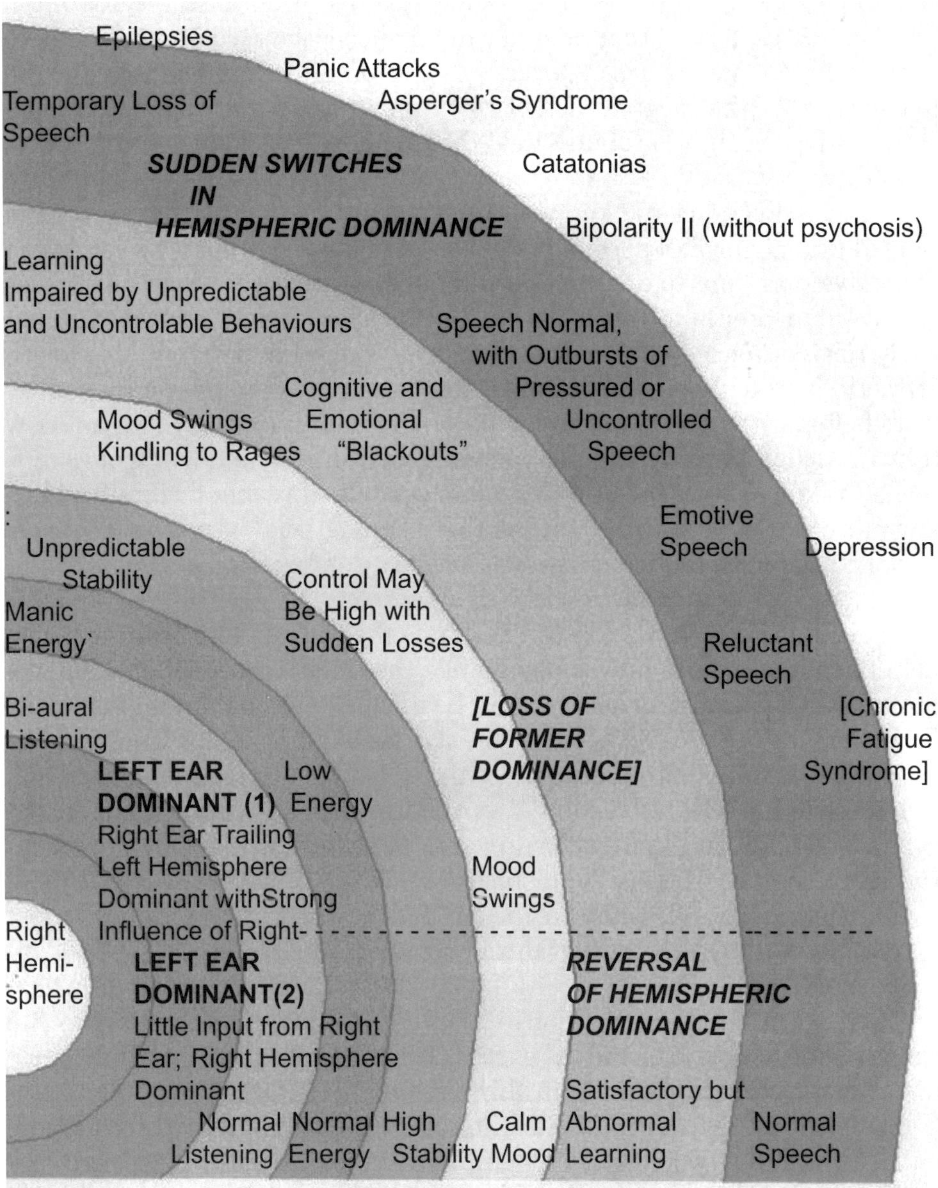

Note that a complete reversal of hemispheric dominance produces behaviours that approach the range of normal.

cording to other definitions, Dan's periods of psychosis all have included the symptoms that define schizophrenia,[8] therefore, he is schizophrenic. We can agree with the Japanese that Daniel has had intervals of "integration disorder." We can add that many factors have aggravated and continue to contribute to his episodes of severe integration disorder. And we can stop to consider to what degree each of us is "integrated" and what spot we occupy on the continuum (Figure 14.1) that includes everyone heretofore and presently.

Daniel is an addict and knows it. He has affirmed that for him marijuana is a gateway to alcohol and hallucinogens. He has said that if he can break his addiction to marijuana that he will have control over other substances and avoid further bouts of psychosis. I believe his addictions are neurologically related not only to reinforced pathways of addiction, but also to the attractions of feeling right-brained, to which his hemispheric instability from before birth also conditioned him—a view of his brain that I am now trying to teach him. He has not yet succeeded in staying away from marijuana for longer than the duration of an episode of psychotic illness plus a few more weeks or months. I think that keeping Dan on pharmaceuticals once he has recovered from psychosis is a way of fostering his addiction to drugs.

Learning about how we listen and how that affects our brain will also help us to understand how someone like the artist Vincent Van Gogh, referred to in the previous chapter, can be diagnosed as having so many different mental disorders, because he slid through several of them as he ascended the more extreme bands of these syndromes related to ear function towards his self-destruction. We can introspect and see in ourselves the range of moods and tendencies towards behaviours that show up as extremes in persons who lose hemispheric control. As I have described in several chapters, Daniel, while in the process of recovering from his most extreme symptoms, also moves through a range of classifications of behaviour—bipolarity I and II; Tourette's; depression; lability—to perfectly normal self-control and even superior self-control judged according to the range of normal reactions to life events. That other hemispheric-imbalance disorders such as autism fall in this ear-function-related continuum has been shown by successful application of the Tomatis Method for autistic children and in considerable contemporary research that I have cited. My early suspicion, based on Dan's seizures both from street and from prescription drugs, that epilepsy belongs in this spectrum is confirmed in Doidge's discussion of Merzenish's work.[9] The potential for healing these mental problems through stimulation of the middle and inner ear should be investigated.

Individuals may be able to deal with extremes of behaviour, but societies vary in what they consider tolerable, even in so narrow a behavioural phenomenon as speech. A broad range of hemispheric control in individuals allows us to understand why certain sorts of events happen in social institutions and among social groups differently acculturated along ethnic, national, religious, or philosophical lines. A quick look at history shows us that psychotic behaviour has been institutionalized in bacchanals and art forms, subcultures and armies. Today, we can more easily pinpoint the individual who is "out of control" in a rigidly structured society or social institution and within such a setting can more easily insist on some form of uniformity, even if that is detrimental to the individual or to the relatively small group of individuals proclaimed "mental." However, within that rigid society so much emphasis on uniformity remains that even when one does find drug compliance from a schizophrenic that will bring him/her in some ways closer to the norm, one also finds the brutally damaging results of heavy medication, so that schizophrenic people nonetheless may be severely ostracized.

How the society treats those people it has marginalized will to some degree determine the willingness of those persons to co-operate towards the social project. The majority tends to define itself as "the society" and does not consider its attitudes asocial, even when they impinge negatively on the minorities. Even among young dyslexic people, who feel unable for a long list of reasons to meet the expectations of a society whose structures they no longer respect, one also finds a high degree of self-medicating through pot, alcohol, and other distractions, which is a way of hiding their shame, confusion, guilt, and misery while it also further removes them from socialization. The negotiating of social boundaries among individuals and between groups could be facilitated by an understanding of what lies at the root of these anti-social or asocial behaviours. As long as the majority remains remote from its minorities it, too, is practising asocial behaviours. The usefulness of members of the society to one another could be facilitated by finding better modes of treatment for persons with diminished hemispheric dominance than the forms of separation and the pharmaceuticals on which societies increasingly depend, and by exercising a wider tolerance for those suffering in ways that they and others heretofore have been unable to comprehend.

Putting Music to Work

The therapy that brought Daniel back into hemispheric integration seems almost too simple. But we are just beginning to understand how music, or sound frequency, affects physiology in the same way that we are just beginning to imagine through the physics of string theory that all substance is composed of vibrations. The logistics for testing the Tomatis Method and related methodologies for the benefit of others experiencing similar disabilities and for implementing therapies found to be useful in other ailments will take some imaginative retooling. However, the costs involved would be vastly less than those entailed in the merely controlling and supportive therapies presently in use. If persons suffering hemispheric imbalance can be brought into normal brain function and maintained there through focused listening and holistic manual exercises, the prognosis for many will be more hopeful, the costs to society less, and the benefits to all greatly enhanced. I am fully aware that Daniel is one case. But as his dyslexic syndrome healing was merely representative of thousands of young people who today are experiencing similar healing of their learning disabilities, the possibility that his experience of healing from psychosis also may be representative of a much larger cohort that could be healed in the same way deserves consideration and research. I hope that the matrix of Daniel's disabling conditions, their correlative behaviours, the forms of intervention, modes of recovery, and his familial context will provide some guidelines for those who embark on this exciting new phase of research on behalf of those suffering from ear-related mental illnesses.

I remember the overwhelming majority show of hands at the family group therapy session at the general hospital when I asked that roomful of schizophrenics how many of them had experienced difficulties in school. I think of the daughters of two of my friends whose social and physical lives have been full of pain and upheaval; they are dyslexic syndrome children inadequately diagnosed, poorly supported, saddened and marginalized so that they became drug-addicted, which has compounded their pain and that of their mothers on one hand and of their children and partners on the other. I think of the children and adults and addicts suffering the effects of "shock and awe" from bombs during warfare. I think of countless people affected following automobile accidents and viral infections with the symptoms of chronic fatigue syndrome and/or fibromyalgia. I see the products of noisy, drug-indulged partying on the streets of our village and sometimes hear about their incarceration in the Ontario hospital or jails remote from their homes and communities. How many of these people might be living very different lives if the schools had included programs managed by teachers cognizant of the effects of singing and listening, programs designed to exercise those tiny muscles of the middle ear to facilitate left-hemispheric dominance and the integration of the cerebral hemispheres? Indeed, how many schools today have organized music programs as an integral and important part of the curriculum, as was the norm when people of my age were schoolchildren? Even in mental hospitals, the means of healing that have worked time and again for Daniel might be applied with little cost and very widely. Ambient music therapies in hospitals have been mentioned in Heslet's work and their use is spreading across the world. Focused music therapies designed specifically for the deficits of particular aspects of damage might be developed very easily and at relatively little cost.

The need is evident for talking therapies that help individuals who have experienced hemispheric imbalance to see themselves as valued persons on a universal spectrum in their communities and the wider culture. To develop these applications in psychological and religious counselling is beyond the scope of this book. However, where communities gather to make and to attend to healing music they are enhancing their own well-being while they are bringing well-being to others.

The Listening Centre

> I think that Daniel would not have had the motivation to wrestle with his addictions and that he would not have been open to his family's supportive efforts if he had not been given the experience of fundamental normalcy in his central nervous system through his program of sound stimulation at The Listening Centre.
>
> Daniel's repeated healings have made possible his continuing will to live and have provided a foundation for the restoration of caring relationships among the members of his family.
>
> My healing from chronic fatigue syndrome symptoms through the listening programs provided by The Listening Centre and through related singing and humming and listening disciplines demonstrates the health-promoting effects of sound stimulation to the middle and inner ear.

More than 12 years following Daniel's first psychotic break it is clear to me that these affirmations, written in the midst of this story, are more certain than ever. Daniel's recoveries from psychotic episodes induced by street drugs, prescription drugs, head injuries, and alcohol would have been greatly complicated or perhaps impossible had he remained in his natal condition of dyslexic syndrome. If the Listening Centre intervention had been postponed, it is likely that Daniel's drug addictions would have been left to the "expertise" of some of the same medical personnel whose prescriptions and lack of knowledge at the very least complicated his early mental illness and impeded his recovery. As The Listening Centre is not geared to treating some forms of "mental" illness, Dan might never have received the platform of "normal," i.e., "right-eared," neurology that linked him to the "normalized" world and gradually built trust and hope into a reality he had perceived through the confusion and unhappiness of dyslexic syndrome.

Focused listening inspired by the Tomatis Method as well as listening interventions I obtained through The Listening Centre have been essential to my maintaining my health and, by extension, to the survival of the rest of the family as Richard and I have worked not only to support Daniel's recovery but also to support the struggles towards independence of the other young people in the family whose illness, addictions, educational deficits, and economic times have kept them semi-dependent much longer than had been the norm in recent generations in North American society.

Looking Forward

Certainly, Dan knows that work and family and friends are essential to a satisfying life, but that if sobriety does not undergird those social relationships they cannot keep him from self-destructive behaviour. His challenge will be to find sobriety apart from his deep impulse to trust and to imitate those he loves and admires but who feel they can afford to socialize with substances that would erode his self-control and damage his ear and brain. The friend whose drinking triggered the collapse of the building project of two summers ago told me proudly last summer of his eight months of sobriety; he and his wife (who has kicked a cocaine habit) took in hand one of those lads I used to drive home in a drunken stupor—the one who's mum was hospitalized—and he, too, was clean and sober and working. Hope for Dan resounds from the success of others, but cannot be seen by him as a success he can depend on. For Daniel, too, the redeeming possibilities of honourable work are at hand. But treachery lies everywhere, because it lies within. Daniel and others facing similar challenges need not only music, but a strong belief system and a community of compassionate and informed people with some insight into their own right hemispheres. Daniel's family, of differing world views and religious outlooks, has survived his mental illness to remain concerned and compassionate.

Mountain Man's further climbing and walking and running will outlive my reporting ability. He may be able to reintegrate with "mainstream" society, but as Erving Goffman pointed out in *Stigma*[10] over 40 years ago and as John Hoffman and Susan Froemke reiterate today in *Why Can't They Just Stop?*,[11] people who have suffered mental illness, alcoholism, drug addiction, poverty, or educational success below the norm tend to be sidelined, if not completely ostracized, by those lacking such experiences, or who lack insight into their own similar experiences. I hope this book helps people to see they, too, occupy the same continuum: a range of degrees of ear function and of hemispheric dominance that affects the function of their own cerebral hemispheres and their own capacities for contributing to the social project.

The editor of the *Oxford Companion to the Mind* also turned his interest in mental processes to the historical development of thought about the brain. Richard L. Gregory's observations at the end of his survey, which spans more than two millennia of perspectives on the central nervous system, reveals what he did and did not know about hemispheric integration:

> Brains, we believe, construct predictive hypotheses of aspects of the world which are generally useful for survival. We now appreciate that most brain hypotheses, and especially perceptual hypotheses, are largely at variance with the realities of physics. . . . Sometimes, the fantasy traveller returns to bring gifts back to our world. The Creative mind is Pandora's Box: sometimes the gifts are good and sometimes they bring disaster.

I think I have shown that the perceptual hypotheses integrated by listening are less "free-floating" than Gregory realized, so that Pandora's realm can be more firmly located in the brain's right hemisphere. However, Gregory intuits the "tolerances" of integrative hemispheric functioning:

> Perhaps where tolerance limits are set provides the basis of personality, as also the structure of societies. If so, surely the setting of the limits within which we are free of the physical world and yet able to survive within it, distinguishes sanity from madness—and due tolerance is necessary for us to observe and understand—let alone to communicate.[12]

Societies construct their "tolerances" of the border between sanity and madness. And we need to exercise "due tolerance" to observe those borders. The application of compassion (right brain) and objectivity (left brain) is essential in humanity's ongoing efforts to understand, to communicate, and to survive. The abilities of both halves of the brain must be integrated to achieve these goals.

15

The Nuclear Family

Microcosm and Macrocosm

Only in his mid-seventies did my father acknowledge to me a mental component to his problems, although I had heard about the "nervous breakdown" he had in college, the strange highs and lows of his courtship, and knew all too well the rest of his history for which I had had a ringside seat. He and Mother came to our house to ask our permission to move into my sister's house nearby. Permission? Richard and I were caught off guard in confusion and amazement. Who were we to deny them the right to live where they pleased? They proceeded to enthuse about a radio program they had heard assuring the listeners that at least 25 per cent of persons have "mental problems." Some unknown broadcaster had pulled the thorn of the stigma from which both must have shrunk all their married lives. They had come, at last, to feel they were "normally abnormal," in Erich Fromm's phrase that Richard used with Daniel to help him find his place in the spectrum of brain function.

It is hard to convey the shock to one's own mental balance to hear someone, who has asserted his mental superiority to all and sundry during all of one's life, suddenly implying that he has known all along that something between his ears was amiss. My mother seemed almost to have been admitting to co-dependency, at the same time. We had the presence of mind to assure them that we not only loved them but would be thrilled to have them near and to be able to help them in any way we might. They hastened to assure us that they didn't need help, but just wanted to know how we felt about their buying Lainey's house and living on the same road with us. They would be moving from the anonymity of the city to "the sort of community where everyone knows everyone else and we don't want to make life difficult for you."

Psychological Sociology

The study of sociology has evolved significantly since my undergraduate introduction to the subject in the 1960s. My editorial work has kept me abreast of some of these changes, yet my feeling as a young woman that the study of society was not keeping up with the changes in society remains strong. As then, I think that most of the paradigms and methodologies used by sociologists for examining groups of people have tended to miss the mark in two directions, one micro, as they generalize about the individuals in the society to focus on the social group, the other macro, as they apply horizontal (e.g., economic) or vertical (e.g., hierarchy of organizational) models to the society as a whole. However, I am sympathetic to the perspective of Georg Simmel, who took an eclectic approach to his sociological inquiries,[1] which started with the individual and, in his view, ought never to lose sight of that person/those persons.

> Society and the social structures that comprise it are, in actuality, only the sum of individual interactions. The appropriate locus of sociology, then, shifts away from the study of macro structures, such as economy and class, to the study of the processes underlying the interaction of individuals.[2]

The late nineteenth- and early twentieth-century urban setting, of modest size and slower pace than we see today, in which he observed personal interaction seemed to him ideal for such study and reflection. He was not enthusiastic about what was to be gained by studying "structures such as economy and class," but remained sanguine about "the study of the processes underlying the interaction of individuals."[3] As interest in the psychological nature of the individuals who form groups has grown, sociologists have addressed micro issues. Sociologists who concern themselves with emerging global issues of population, the migration of peoples, the issues of work and trade, and the geographical influences on the workers and traders address the macro issues. Meanwhile, the problematic aspects of the ways people interact with one another have escalated exponentially to the point where our group home—our "planet without bulwarks, out in space"[4]—has come under threat of annihilation and humans as a macro society are in danger of extinction as a direct result of the way they interact with each other in their space. Might the extinct members of the Heaven's Gate cult in their Nikes typify the human race? The methodologies of sociology, like the soon-to-be-fuel-less engines of the Industrial Revolution that spawned them, have provided templates for social understanding inadequate to the societies that meanwhile have evolved. As we have found the source of our ecological problems in the reckless squandering of our natural resources and have faced the need for radical measures in the reduction of hydrocarbons in our atmosphere, we as desperately need to understand the sources of our social problems in the individuals who create the society after their own image so that we can face the need for radical measures in the reduction of social tensions that prevent co-operation in the social project.

My hope for understanding society, or what some call "the human condition," began with a need to understand the people I knew best. When I eschewed behaviourism in the psychology department at my university for studies of English language and literature, I was choosing an avenue of study that I had hoped would open to me knowledge of how the individual mind and psyche worked upon the *other*: one-on-one, one-on-others, in ever-widening concentric circles. I was not disappointed. Learning to analyze the writing of one writer valued by the society and comparing with it one's own thinking and writing are ways of studying the workings of one's own mind and psyche in relation to one's own family and the larger society: psychology and sociology combined, both in the writer's historical context and in one's contemporary social setting.

The study of English in the context of a liberal education is layered. Not all writers have been received generously by their society; the student considers varieties of reactions. The rise of specialties—the critics of broad reading and educated taste, the publishers of increasingly greedy purpose, the historians who give context to the ever-more-remote early writers—produced multiple facets for viewing the original writers, their ideas, their manners and styles, the milieu of their personal lives and times. Yet each sentence, whether of an original work of art or of a critic of that work, is the product of an individual brain. And individuals live in societies. The greater one's acquaintance with individual minds, the better-equipped one has become for understanding the next sentence and the next writer—and one's next personal encounter in one's contemporary social context. Writers simply are articulate members of the population expressing thoughts and feelings commonly or uncommonly held—to distinguish which of these categories is the case becomes part of the expertise of the diligent scholar.

The study of the language and the literature of English-speaking people grew from such monolithic exercises as those of Samuel Johnson, who was disposed to write a definitive lexicon[5] of the English language and a compendious summary of its poets,[6] to English studies that dare to fragment works of art and examine them microscopically in conjunction with psychology, sociology, religion, philosophy, manners, customs, history, exploration, anthropology, politics, art, music, theatre, folklore, and the literatures of other peoples of the earth. The diversity of English language and literature inquiry we have seen proliferate in the last half-century was inevitable. My impulse, however, was to synthesize. I chose an area of study that would straddle three disciplines: English poetry, visual art, and psychology. When my formal studies were interrupted, my desire to continue to explore certain issues and to measure my learning against "real life" continued. Despite experiencing some of the emotional equivalents of warfare as my life unfolded, I have tried to maintain, along with my wholehearted engagement with the people around me, the objectivity and honesty of the scholar.

The study of any of the disciplines mentioned in the preceding paragraph is the study of persons; even the sciences that focus on the earth and its plants, animals, and minerals are studies conducted by persons in social contexts with social consequences. Consider a member of what sociologists call "the nuclear family" in which I was raised: my father was a chemical engineer, in his day the equivalent of today's engineering physicist. He was one of the designers of the first nuclear power plant in Canada, at Chalk River, Ontario. If he were alive today he would disapprove of that facility, which is in the process of being repaired for the twenty-first century's demand for medical isotopes, because he came later in life to appreciate the problems to succeeding generations inherent in the disposal of nuclear wastes and he spoke out against the use of nuclear energy for electrical power also for that reason. During World War II he advised the Canadian government on issues of atomic energy, including the armaments being developed in the United States, on the understanding that their development was to be of the nature of a "deterrent." When both governments betrayed his trust and the U.S. bombed Hiroshima and Nagasaki, he profoundly regretted his part in their departure from what he conceived of as "honourable" war that hoped to exclude civilians from theatres of conflict. He left the National Research Council for private industry. Still, we look back on his achievements for Canadian Industries Limited (CIL); Standard Chemical Company; and General Abrasives and Nilock Chemicals in New York State with their explosives, DDT, chlorine bleach, and other pollutants to the household and landscape with a jaundiced eye. The earth is not a better place to live in as the results of a great many chemical engineering efforts. Many of us no longer view warfare as "honourable" along any dimension.

My father also loved the earth, the wilderness of the Muskoka area where he built a cottage and canoed and where he laboured for years to preserve the quality of the water in the adjoining lake. He devoted himself to church projects, both building and teaching. As a social being his influence was considerable. He never understood or came to terms with his unusual brain and the effect it had on his moods and health or on the lives of others. One could say my father's life, both for its ignorance and for its knowledge, had enormous impact on a small circle of individuals and on society more generally—even globally. My father's life makes Simmel's point: that a study of the interaction of an individual with other individuals would better allow us to understand the society as a whole than knowing that he was the scion of an ancient English house brought up in a blue-collar Toronto family in the first third of the twentieth century, which included the Great Depression that preceded World War II.

Our children recoil in horror from the effects of their forebears' unquestioning embrace of the engines of war, the means of travel, the products and by-products of industry, the ecological costs of energy consumption, and from the unimaginable havoc certain to follow upon plant and animal modification. The souls motivated by a spirit of inquiry into the material world that characterized the few generations before mine and that march on in droves today have failed to ask the important questions before getting up in the morning and going to work: What will be the result of my place in the social scheme today? What is my primary responsibility to the members of the society of which I am a member? These are the sorts of questions many sociologists have avoided. The crumbling of social institutions is nevertheless providing some answers.

First, do no harm. Most of us would have to stay in bed for the day and for many days afterwards working out our own salvation with fear and trembling to achieve that tenet of Hippocrates. A society may be evaluated, some believe, in terms of how it treats the most vulnerable of its members: children, the sick, the aged, the marginalized, and the mentally ill. The population of the earth and the expansion of some very anti-social expressions of cultural beliefs are growing exponentially. Is there any way out, short of mass suicide, which seems more or less to be what the peoples of the earth are attempting? The depletion of energy sources will answer other questions for us. The means of flying away from issues, whether on holiday or to find more of some resource to exploit or to expand markets, is diminishing. Along with the oil resources, "globalization" may have peaked. Lives focused on smaller communities such as Simmel enjoyed or on the even smaller communities of my grandparents may be thrust upon us. The institutions—governmental, commercial, educational, and charitable—on which we have come to rely are shrinking, redefining their mandates, and vanishing. How, then, will we solve the issues of order, commerce, education, and mutual aid that persist in the human condition? How will we find healing for children, like Daniel, born with learning disabilities? Will we swallow the pills doled out by the pharmaceutical companies, the new lords of coping and well-being? Will we continue to support forms of education intent on biological destruction? Will we continue to waste the planet's resources by running away from our problems to hot vacation spots? Or will we sit still and listen?

Daniel was viewed by his grandfather as a problem on my hands as, indeed, Daniel's problems had strongly affected every member of the family. My father, however, failed to realize that he had been a problem on his wife's hands, his parents' hands, and the hands of his children. Despite its mathematical and scientific brilliance, Dad's bipolar brain caused pain in every direction: within his natal family, his nuclear family, his office families, his church families, his communities' families, and the human family. It may not be too late to discover how his brain function affected his learning, his work, his decision-making in the atomic age, his voluntary contributions to his churches and his communities, because his shifts in mood and personality may not have differed essentially from the bi-hemispheric cycling of his dyslexic syndrome, drugged, or psychotic grandson. A few weeks ago I found a diary he kept sporadically as a child that ends with an account of a serious illness. He was just 16 and suffering the usual symptoms of scarlet fever, a staphylococcus infection of the throat that easily affects the ears through the Eustachian tubes. His symptoms included a very stiff neck for which his doctor at first prescribed hot linseed oil poultices. A few days later, his doctor lanced his neck just behind and below his dominant right ear. Under the doctor's instructions, my grandmother daily opened and probed the ensuing secondary infection at the incision, eventually producing a suppurating crater at least an inch deep.[7] For the rest of his tumultuous life he would look back to that time of illness as a turning point in his formerly robust health.

I have tried to present Daniel's story in the context of my journey in inductive reasoning. That scientific mode of thinking, inculcated in me by my father, led me to what I grasped first intuitively about Tomatis's discoveries—on the basis of many life experiences—and understood more fully later—on the basis of my own careful observations—namely, a principle of the kind J.S. Mill calls "a uniformity of Nature," in my case concerning people's behaviour and the ability to listen.

> We must first observe that there is a princple implied in the very statement of what Induction is: an assumption with regard to the course

> of nature and the order of the universe, namely that there are such things in nature as parallel cases; that what happens once will, under a sufficient degree of similarity of circumstances, happen again, and not only again, but as often as the same circumstances recur. This, I say, is an assumption involved in every case of Induction and, if we consult the actual course of nature, we find that the assumption is warranted. The universe, so far as known to us, is so constituted that whatever is true in any one case is true in all cases of a certain description; the only difficulty is to find what description.[8]

It is for that reason my description of Daniel's singular case merits consideration.

As the knowledge that George III of England[9] suffered from porphyria (and was very nearly killed by his doctors) allows us to reinterpret the events of his reign, similarly, the knowledge gained from Daniel's hemispheric instability allows us to reassess the behaviour and decision-making of people, past and present, affected along that part of the continuum of hemispheric instability. If my insights into Daniel's induced alternating hemispheric dominance open opportunities for healing to others who are born with dyslexic syndrome and bipolar and autistic and schizophrenic predispositions, not only individuals will be healed, but society will change. If people who have come away from an illness, an accident, an assaultive sound, or an assaultive psychological event with chronic fatigue syndrome or impaired listening skills can find healing through a Tomatis Method program, the Mozart Effect®, the laughter regimen recommended by Norman Cousins, or by their own humming and singing and laughing, society will change. If people who have experienced despair or isolating depression can be given hope through listening retraining, society will change. If people who have damaged their ears and their brains with street drugs can be brought gently back to normalcy with minimal pharmaceutical intervention and wise use of music, society will change.

And all of us, from Toronto to Nagasaki, will feel better about getting out of bed in the morning.

Appendix A

*Checklist of **ME/CFIDS** [Chronic Fatigue Syndrome] Symptoms*

The following is a checklist, "Looking in the Mirror," by Dr. Jay A. Goldstein, Anaheim Hills, California, who uses it in his preliminary consultation with patients who may be suffering from ME/CFIDS. This does not constitute a comprehensive diagnostic procedure and patients should consult their physicians to rule out other diseases that produce similar symptoms and to undergo appropriate medical tests to confirm their diagnoses.

Symptom Patients with Symptom (%)

1. Fatigue 95%
   (usually made worse by physical exercise)
2. Cognitive function problems 80%
   attention deficit disorder (ADD)
   calculation difficulties
   memory disturbance
   spatial disorientation
   frequently saying the wrong word
3. Psychological problems 80%
   depression
   panic attacks
   personality changes (usually a worsening of a previously mild tendency)
   emotional lability (mood swings)
   psychosis (1%)
4. Other nervous system problems 75%
   sleep disturbance
   changes in visual acuity
   seizures
   numb or tingling feelings
   disequilibrium
   light-headedness, feeling "spaced out"
   frequent unusual nightmares
   difficulty moving tongue to speak
   ringing in the ears
   paralysis
   severe muscular weakness
   blackouts
   intolerance of bright lights
   intolerance of alcohol
   alteration of taste, smell, hearing
5. Recurrent flu-like illnesses 75%
   often with chronic sore throat
6. Painful lymph nodes (especially on side of neck and under the arms) 60%
7. Severe nasal and other allergies, often worsening of other previous mild prob lems 40%
8. Weight change, usually gain 70%
9. Muscle and joint aches with tender "trigger points" or fibromyalgia 65%
10. Abdominal pain, diarrhea, nausea, irritable bowel syndrome 50%
11. Low fevers or feeling hot often 70%
12. Night sweats 40%
13. Heart palpitations 40%
14. Severe premenstrual syndrome (PMS) in women 70%
15. Rash of herpes simplex or shingles 20%
16. Uncomfortable or recurrent urination —pain in prostate of men 20%
17. Other symptoms seen in less than 10% of patients:
    - rashes
    - hair loss
    - impotence
    - chest pain
    - dry eyes and mouth
    - cough
    - TMJ syndrome
    - mitral valve prolapse
    - frequent canker sores
    - cold hands and feet
    - serious rhythm disturbances of the heart
    - carpal tunnel syndrome
    - pyriform muscle syndrome causing sciatica
    - thyroid inflammation
    - various cancers (a rare occurrence)

Appendix B

Counting Activity Showing Intervals of Lucidity and Confusion, October 2003

Time	Interval	Cognitive Mode	Counting Activity, Comments
9:22	0.5	Unfocused.	False starts.
9:23		Focused	Starts "1, 2, 3, . . . 49"
9:24	2.0	Focused	"50, 51, . . . 50, [pause] . . . 93" False start from 50, recovers without mistakes.
9:25		Unfocused.	Focused with effort, recovers focus.
9:26	2.0	Unfocused, slightly focused.	"49, 60, . . . 64 uh . . .78, 80, . . . 93,
9:27		Very focused.	"97, . . . 100; 1, . . . 39 very easy and relaxed, pause 40 . . . ." Completes task, and begins again.
9:28	2.0	Focused.	Focused with effort. Focused, losing focus.
9:29	1.0	Unfocused, then, largely focused.	
	0.5		Jumps to 28, long pause; starts over at "1, 2, 3, . . . 8," (Needs a new start.)
9:30	0.5	Unfocused.	
	0.5	Largely focused.	[Switches again to very quiet and fast mode, but clear] "1, 2, 3, 4, . . . 95." Again, in order to focus Dan begins the task again.
9:31	1.0	Unfocused.	"96, 99, 91, . . .100." Very slow incorrect counting.
9:32	1.0	Focused, then unfocused.	101, 102, [slowly]. . . 109. Dan continues past 100, contrary to instructions.
9:33		Unfocused, then focused.	Pause ". . . 119; [complete stop], Very good, now what?" He continues after a long pause.
	1.0	Fairly focused.	Me: "Do you want to start again at 1?" Dan, "[long pause] . . .1, 2, 3 . . .
9:34	0.5	Focused.	[and stopped]. . . 31." Time ends.

Appendix C

Reading Task Showing Intervals of Lucidity and Confusion, January 2005

Time	Interval	Attention	Comments
12:20:40-56		Pause	
12:21	1.0	Focused.	Reading fairly clearly, loudly, forcefully. Defoe's style is archaic and poses some challenge to his reading ability.
12:22	1.0	Focused.	Has hard time over the meaning of the word "physic"; long pause. I prompt him to begin reading again.
12:23	½	Losing focus.	Continues and corrects himself on "cured"; pause. I say, "Out loud."
12:23:38-53	½	Pause continues.	Becomes delusional. Expresses frustration and anger: "That's what frustrates myself." Explains that he is "literally beside himself."
12:24:56	1.0		Continues delusional. He can "see others around and from the dead."
12:25:08	½	Refocusing.	Long pause; appears to be reading silently.
12:25:26	½	Focused.	Long pause; appears to be reading silently.
12:25:54	½	Focused.	Pauses over the word "Bibles" says, "which I mentioned."
12:26:11	½	Focused.	Restarts the sentence, reading clearly.
12:26:31	¾	Unfocused.	Pauses after the word "table."
12:27:15			Pause continues. I say, "If you can, start again, do."
12:27:35	¼		Able to focus on me but not on task. Dan repeats my words, "If you can start again, do."
12:28	½	Unfocused.	He appears to be reading silently.

Time	Interval	Attention	Comments
12:28:22	¾	Focused.	Starts again reading clearly. A fatigue factor or compensation for a fatigue factor seems about to begin.
12:29:0	¾		Slight pause; continues.
12:29:06	¾	Different focus?	Having trouble. Perhaps having difficulty with the words. Perhaps finding the subject and expression of feelings interesting or else hard to process emotionally as he reads.
12:29:51	¾	Focused.	Recovers and continues reading quite well.
12:29:55		Loses focus.	Dan stops and recapitulates the sentence about tobacco and then runs it into a commentary on himself. His commentary on himself rambles a little. He seems to identify with Defoe's smoking.
12:30:36	¼	Very focused.	Begins reading very quickly. There is a change of mood in his reading style. Manic?
12:30:50	½	Unfocused.	Stops reading aloud. May be reading silently.
12:31:18	½	Focused.	Starts reading aloud again, "Call on me in the. . ."
12:31:40	¼	Unfocused.	". . . day of trouble . . . me." Stops, looking ahead of him, not at me. Well-focused to end of sentence.
12:32	½	Unfocused.	Glances at me and tries to find his place in the text. Loses place in text.
12:32:31	½	Unfocused.	Expresses anger directed towards the task or the writer. Mood either frustrated or angry.
12:32:49			Dan says, "It could be something *I* wrote . . . . [Mood changes to almost humorous.]

12:33 Dan adds, ". . . It's so silly." His eyes move over the page for a few seconds, then he goes into the other room. A minute later he comments: "I liked that. [Pause] Disturbed and all this shit." I think he means he is pleased that he could read as well as he did in consideration of his overall mental ("disturbed") condition.

Daniel was reading from Daniel Defoe's *Robinson Crusoe*, a selection chosen randomly. The pattern of focus here matches very closely the testing of a counting activity that was done during Dan's previous episode of mental illness (Appendix B), including the fatigue "fibrillation" after 5 minutes of trying to concentrate on a task.

Appendix D

A Demonstration of Masking

On 18 April 2008 at about 10:30 in the morning, I noticed that when I spoke to Daniel while he was wearing the headphones for focused listening that he removed the left earpiece to attend with his *left ear*. I asked him to reverse the situation: to replace the left headphone and offer the right ear. "Can you hear me with your right ear?"

"Yes, I can hear you, but the in-between parts are too loud." His response made me suspect that he was experiencing the phenomenon of masking as a result of misfiring neurons due to sound damage in the cilia of his right ear. I wondered how I could test my idea.

A recent issue of *National Geographic* (May 2008) was at hand. Daniel was looking at the photographs, which I did not comment on. I imagined he did not have the ability to interpret and certainly not the ability to read the magazine. He had not been able to read for almost four months. To my amazement, he began to read aloud and managed to read the entire caption under a photo of striped tents. My eyes filled with joyful tears and I praised his reading as I recognized this leap forward in his recovery. He noticed my reaction and showed his own pleasure with a shy, proud smile.

A few moments later, Dan started to read the accompanying article aloud. When he paused I suggested he continue to read whenever he felt able and I would keep track of his reading the way I had once before; I was thinking of timing his cognitive cycles. He responded immediately by picking up the headphones *as though he needed them in order to read*. I realized then that the headphone sound was in some way counteracting the masking condition inherent in his right ear. Perhaps the sound was "tuning" the stapedius muscle so that the damaged neurons would fire more consistently. Perhaps also the music makes neural connections across the corpus collosum to facilitate inter-hemispheric communication while the music is playing.

Because I was sitting at my computer I clicked onto its clock, more accurate than others I had used in previous timed tests, to keep track of Daniel's efforts to read.

Time/Interval		Attention	Comments
10:40:00	1.0	Unfocused.	I prompt him again, “Whenever you feel like reading, you can go ahead.” He removes the left earpiece in order to hear better and looks at me blankly, still confused.
10:41– 10:42:32	2	Unfocused.	[2 min. of confusion] Dan starts reading from the magazine and comments on the picture in a disjointed way.
10:42:32	1	Focused.	From the motion of his eyes back and forth and down the page Dan appears to be reading normally to himself, silently.
10:43:00		Unfocused.	He flips through the pages of the magazine, commenting “Hmmm. Interesting.” I think he has become confused and is covering his embarrassment at not being able to perform in the way he had earlier. [starts period of confusion]
10:43:45?	2		Near the end of this minute he starts to recover. [completes 2 min. confusion]
10:44:00		Focused.	He begins to read aloud again. He interrupts his reading to “rationalize” the confu sion of a minute previous. [begins focus]
10:44:43	2	Focused.	He returns to full rationality and focus as he slowly turns pages to look at photos and again comments, “Hmmm. Interesting.” This time, I had the impression he actually was able to “read,” i.e., to interpret, the photos. [completes about 2 minutes of focus]
10:44:54		Unfocused.	[begins interval of confusion] Starts to lose rationality and begins to reminisce about a neighbour who was his friend in childhood.
10:46:48		Fully confused.	Dan says, “The Torture. God does it automatically. I put them out forever. My own father holding my face. I don’t get much pleasure at all, just a friend.”

10:46–10:47			Richard intrudes for a full minute; anxious as to whether I am doing the proofreading he has passed on to me. I explain briefly what I am doing. Presumably Dan's confusion abated at the beginning of this minute as the lucid, focused interval commenced.
10:48 –10:49:30	2	Focused.	Dan resumes his focus on the magazine. "Look at the disgusting alligators." His cognition is rational and he continues reading all of the captions associated with alligators or crocodiles. [completes about 2 minutes of lucidity]
10:49:30	1.5	Unfocused.	Dan loses his focus and starts flipping through the magazine with the chatter typical of his confused phase. This marks the beginning of a fatigue factor ("fibrillation") between cognitive modes.
10:50:30		Focused.	Return to rationality. Dan ends his phase of lucidity with the comment, "There's no map with this one."
10:50:45?	1/4	Unfocused.	Dan enters phase of rapid cycling [confused rationalizing, then 15 seconds of focus].
10:51	1	Unfocused.	
10:52		Focused.	Dan returns to lucid phase with his dis covery of an article showing the animals of "Russia or other [similar] countries." I could not tell how long this rapid cycling interval lasted because he stopped the reading aloud activity for a period of 7 minutes.

For the next seven minutes Daniel read silently to himself or looked at pictures. I continued to attend to my computer. At 10:59, something he saw in the magazine reminded him of an incident four years ago when he had killed a treed porcupine (that Eric's dog, Tequila, would not leave alone) with his hunting bow and arrows . His possession of such weaponry already had raised an issue with Richard and me and everyone in the family was aware of our strong disapproval.

The porcupine killing, because Richard had requested it, created a serious rift between us, and I considered what occurred to be extremely ill-advised, given Dan's wavering mental condition. So much that I had schooled into Dan throughout his childhood and that I had protected him from in later years with a view to his tendencies to instability—and to what could eventuate during a psychotic episode to a brain experienced in any way to violence—was compromised in that incident. That incident resurfaces in what Daniel said next.

10:59	2.75	Focused	Dan, clearly, "I can put animals out of their misery. Look what I did to that violent porcupine." Starts to read. Stops to talk about porcupines and quills. "You are probably just missing your mate. A legendary thing that I did with my bow . . . and [this is] not that one I shot but it's cool looking, eh?" And he showed me a picture of an animal that resembles a porcupine, but is not.
11:01:45		Unfocused.	Dan starts confused talk.
11:02: 20	2	Unfocused.	Puts the *National Geographic* aside, stretches. [2 min. of confusion]
11:03		Focused.	Dan talks fairly coherently about me, angels, healing plants, and flowers.
11:03:30	1.5	Focused.	Dan removes headphones and walks away. [1.5 min. or more of lucidity]. Presumably the masking effect of cilia damage returns as he removes the headphones.

From 10:41 to 10:49:30, a period of 8½ minutes, his cycling was consistent, probably because his return to lucidity was supported by the listening activity. When Daniel returns to reading after a significant rest, his lucid interval is slightly longer; then his brain appears to resume its usual 4-minute cycle, likely due to the masking effect of the cilia damage in his right ear.

The alternation between confused and lucid cognitive states follows very much the same pattern of previously timed tests of cognitive cycling.

Notes

Introduction

1. John Vacca, ed., *The World's 20 Greatest Unsolved Problems* (Upper Saddle River, NJ: Pearson Prentice-Hall, 2005), chs 13, 14.
2. Gerald May, *Addiction and Grace* (San Francisco: Harper Row, 1988).
3. *And the Band Played On*, 1993, directed by Roger Spottiswoode, with Matthew Modine, Alan Alda, Richard Gere, and Angelica Huston. This film is based on the important book by Randy Shilts, *And the Band Played On: Politics, People, and the AIDS Epidemic* (New York: St. Martin's Press, 1987).
4. Demonstrated in the tragic story of James Lonnee, who was killed in jail; see *Toronto Star*, 22 Sept. 1996, B1.

Chapter 2
Starflower Tree: *Chronic Fatigue Syndrome*

1. Hillary Johnson, "The forgotten epidemic," *Rolling Stone*, 21 Mar. 1996.
2. *Lorenzo's Oil*, 1992, directed by George Miller, with Susan Sarandon, Nick Nolte, Zach O'Malley Greenburg.
3. Terry Wotherspoon, *The Sociology of Education in Canada: Critical Perspectives*, 3rd edn (Toronto: Oxford University Press, 2009), 226–7.
4. *National Longitudinal Survey of Children and Youth* (Ottawa, 1996).
5. Edith Joyce Smith, "Does it make you happy that I tell you of a kindness?" *Toronto Star*, 27 Aug. 1985.
6. *Toronto Star*, n.d.
7. Joey Slinger, *Toronto Star*, *c.* 1996–7.
8. Bruce Blackadar, "American dream goes sour, and we're not far behind," *Toronto Star*, 21 Jan. 1996, F6.
9. Dennis Forcese, *The Canadian Class Structure*, 4th edn (Toronto: McGraw-Hill Ryerson, 1997), 133, quoted in Wotherspoon, *Sociology of Education in Canada*, 256.
10. I regret that I am unable to give credit to the original author and would appreciate any information that would allow me to do so in future editions of this work. The book was loaned to me and I did not keep a reference, but it reflected the point of view of Lehtinen (see Chapter 3, note 2) and what Dr. Harold Levinson refers to as the traditional "dyslexia stemming from brain injury" stance: "this flawed 100-year-old brain-damage theory has led absolutely nowhere in so far as medical ways of diagnosing, treating, and explaining the dyslexic syndrome as expressed as 'dyslexia stemming from brain injury'." <www.dyslexiaonline.com/information/whatis.html>.
11. Harold Levinson, M.D., *Smart but Feeling Dumb: The Challenging New Research on Dyslexia—And How It May Help You* (New York: Warner Publications, 1994).
12. Harold Levinson, M.D., *A Solution to the Riddle—Dyslexia* (New York: Springer-Verlag, 1980)
13. Johnson, "The forgotten epidemic," 21 Mar. 1996
14. Hillary Johnson, *Osler's Web: Inside the Labyrinth of the Chronic Fatigue Syndrome Epidemic* (New York: Crown Publishers, Random House, 1996).
15. Byron Hyde, Jay Goldstein, and Paul Levine, *A Symposium: The Clinical and Scientific Basis of Myalgic Encephalitis/Chronic Fatigue Syndrome* (Ottawa: Nightingale Foundation, 1992).

Chapter 3
Dan's Road: *Dyslexic Syndrome*

1. *The Gods Must Be Crazy*, 1980, directed by James Uys, with Marius Weyess, Sandra Pensloo, N!xau.
2. Laura E. Lehtinen, "Have you ever known a perceptually handicapped student?" from the New Jersey Association for Brain Injured Children (mid-1980s). Note that although originating in New Jersey this theory of dyslexic syndrome was being promulgated in Ontario pedagogy.
3. Ibid.
4. Anita Griffiths, *Teaching the Dyslexic Child* (Novato, Calif.: Academic Therapy Publications, 1978).
5. Personal interview, June 1995.

Chapter 4
Light Dawns: *The Tomatis Method*

1. "Mom says 'miracle' treatment healed her daughter's autism," *Toronto Star*, 1 Feb. 1991, B1; Annabel Stehli, *The Sound of a Miracle: A Child's Triumph over Autism* (New York: Doubleday, 1991), cited in Don Campbell, *The Mozart Effect: Tapping the Power of Music to Heal the Body, Strengthen the Mind, and Unlock the Creative Spirit* (New York: Avon Books, 1997), 234–5. Stehli worked with another French otolaryngologist, Guy Berard, employing a listening apparatus for work parallelling that of the Tomatis Method listening centres. See <www.georgianainstitute.org/quest.htm#BVST>. On this website, the description of the theory as to how this strategy works is too vague to compare with the extensively developed Tomatis theory. But see below, Chapter 11, note 20. Another story of autism healed through early childhood intervention is *Awakening Ashley* by Sharon Ruben. See "Awakening Ashley" at: <www.awakeningashley.com>. See also Valerie Dejean, at: <http://216.194.201.205/SI /valeriedejean.org/>. Dejean appears to be having a high level of success in treating children having autism. See also Chapter 12, note 8.
2. Paul Madaule, *When Listening Comes Alive* (Norval, Ont.: Moulin Publishing, 1993).
3. The Tomatis Method at: <www.tomatis.com/English/Articles/Biography. html>. "In recognition of his discoveries, he was named Knight of Public Health of France in 1951. Later, he was also awarded the gold medal for Scientific Research (Brussels, 1958)."
4. Dr. Harold Levinson at: <www.dyslexiaonline.com/information/treatment. html>. "The anti-motion sickness antihistamines have the greatest chance of helping the academic and non-ADD part of the dyslexic syndrome." Dr. Levinson was the originator of the use of Ritalin, a stimulant, for ADD, which was tried but not found helpful with our son. Daniel's use of street drugs might have been confounding the effect of the prescription medication. The fact that Ritalin itself becomes a street drug, a pharmaceutical to sell to and barter with others, as it did for Daniel, only compounds the problem.
5. Nicholas Regush, "The listening cure," *Equinox Magazine* (May 1997): 5. Accessed at the Listening Centre site: <www.listeningcentre.com>. See also Alfred A. Tomatis, *The Ear and the Voice,* trans. Roberta Prada (Lanham, Md.: Scarecrow Press, 2005).
6. Tomatis, *The Ear and the Voice*, 84: "The cochlea stimulates virtually the entire brain." The complexities of the passage of sound from the ear through the brain to the hemispheres is reduced here to the elements relevant to the problems under discussion. The cortical-cochlear feedback loop, minus the role of the middle ear, is mentioned by Oliver Sacks, *Musicophilia: Tales of Music and the Brain* (New York: Vantage, 2008), 146.
7. Alfred A. Tomatis, *The Conscious Ear: My Life of Transformation through Listening*

(Barrytown, NY: Station Hill Press, 1991), 50–1; Tomatis, *The Ear and the Voice*, 75. The great complexities of the passage of sound from the ear through the brain to the hemispheres (see note 6) are reduced here to the elements relevant to the problems under discussion.

8. Madaule, *When Listening Comes Alive*, 75.
9. Pierre Sollier, *Listening for Wellness: An Introduction to the Tomatis Method* (Walnut Creek, Calif.: Mozart Center Press, 2005), 272–4.
10. Madaule, *When Listening Comes Alive*, 70.
11. Richard L. Gregory, ed., *The Oxford Companion to the Mind* (London: Oxford University Press, 1987), 108. "Territories of the left temporal and parietal lobes, which are essential for language understanding in the great majority of adult humans . . . are already asymmetric in a human fetus of twenty-four weeks gestational age—four months before birth. . . . the human brain is set to acquire some kind of language long before it hears a single word." Except that Tomatis thinks hearing occurs this early in utero. Tomatis, *The Conscious Ear*, 127, 131.
12. Madaule, *When Listening Comes Alive*, 71; emphasis mine.
13. Tomatis, *The Conscious Ear*, 131, 208–9; Madaule, *When Listening Comes Alive*, 75; Sollier, *Listening for Wellness*, 274–7.
14. Tomatis, *The Ear and the Voice*, 84.
15. Chris McManus, *Right Hand, Left Hand: The Origins of Asymmetry in Brains, Bodies, Atoms and Cultures* (Cambridge, Mass.: Harvard University Press, 2002), 156.
16. Madaule, *When Listening Comes Alive*, 75.
17. Tomatis, *The Ear and the Voice*, 26.
18. The neurology of hearing in reference to the Tomatis Method and to music more generally also is explored in Campbell, *The Mozart Effect*, 18–21.
19. Thus, Tomatis was able to talk about Paul Madaule's "*dyslateralité*: 'mixed dominance.'" Madaule, *When Listening Comes Alive*, 8, 165–6. See also Campbell, *The Mozart Effect*, 50. Campbell emphasizes the effects of sound crossing the corpus callosum en route from right ear to left brain in the developing brain and in the brains of musicians. Ibid., 192–3.
20. Gregory, ed., *Oxford Companion to the Mind*. Differences between the functions of the hemispheres are augmented in the article on "Split Brain and the Mind," 740–6.
21. Wikipedia, at: <en.wikipedia.org/wiki/Laterality>:
"It turns out that the hemisphere of speech is the hemisphere of action in general and that the command hemisphere is located either in the right or the left hemisphere (never in both). Around eighty percent of people are left hemispheric for speech and the remainder right hemispheric. Ninety percent of right handers are left hemispheric for speech but only 50 percent of left handers are right hemispheric for speech (the remainder are left hemispheric). The reaction time of the neurally dominant side of the body (the side opposite to the major hemisphere or the command center, as just defined) is shorter than that of the opposite side by an interval equal to the interhemispheric transfer time (IHTT). Thus, one in five persons has a handedness that is the opposite for which they are wired (per laterality of command center or brainedness, as determined by reaction time study mentioned above)." I. Derakhshan, MD, Neurologist.
22. Tomatis, *The Ear and the Voice*, 21, 52, ch. 9.
23. For example, Lynn Payer's study, *Medicine and Culture: Varieties of Treatment in the United States, England, West Germany, and France* (New York: Holt, 1988) showed that the same symptoms were diagnosed quite differently by doctors in each of several countries. As Clarke writes: "Payer's work suggests, in broad strokes, something of the

relationship between culture and medical practice." Juanne Nancarrow Clarke, *Health, Illness, and Medicine in Canada*, 4th edn (Toronto: Oxford University Press, 2004), 210.

24. <www.tomatis.com>.
25. Terry Wotherspoon, "Education," in Lorne Tepperman, Patrizia Albanese, and James Curtis, eds, *Sociology* (Toronto: Oxford University Press, 2008), 252–8. "The shift in the gender balance of educational attainment has drawn attention to other aspects of education. Findings from numerous surveys that girls have begun to outperform boys on a number of indicators have generated controversy over suggestions that gender inequality has reversed to the point that the education system is now 'failing' boys. For example, the comprehensive Programme for International Student Assessment (PISA), conducted in 2000 and 2003 to compare student performance in core areas, has posed a concern for policy-makers that girls in Canada and at least 30 other participating nations consistently demonstrate stronger test outcomes in reading . . . ."
26. A morphological study comparing male and female brains—Theodore Rabinowicz, Dorothy E. Dean, Jean McDonald-Comber Petetot, and Gabrielle M. de Courten-Myers, "Gender differences in the human cerebral cortex: More neurons in males; more processes in females," *Journal of Child Neurology* 14, 2 (1999): 98–107—found that: "Estimates of neuronal numbers—the product of neuronal thickness times density—were 13% higher in males than in females, with mean male-to-female ratios of 1.13 in both hemispheres. The data provide morphologic evidence of considerable cerebral cortical dimorphism with the demonstration of significantly higher neuronal densities and neuronal number estimates in males, though with similar mean cortical thickness, implying a reciprocal increase in neuropil/neuronal processes [where synaptic junctions are formed] in the female cortex. Thus, the differences between male and female performance on intellectual tasks apparently does not relate to a greater number of cortical neurons in the female brain." At: <jcn.sagepub.com/cgi/content/abstract/14/2/98>.

    But perhaps to synaptic junctions.
27. Wikipedia, at: <en.wikipedia.org/wiki/Geschwind-Galaburda_Hypothesis>: "The Geschwind-Galaburda hypothesis was proposed by Norman Geschwind and Albert Galaburda to explain biology of gender in cognition abilities by relating them to lateralization of brain function. The basic idea is that differences in maturation rates between the cerebral hemispheres is mediated by circulating androgen levels, and that sexual maturation acts to fix the hemispheres at different relative stages of development after puberty. According to the theory, male brains mature later than females, and the left hemisphere matures later than the right."
28. A number of factors creating gender inequality in the classroom that negatively impacts boys, including girl-oriented activities, standardized tests skewed towards female performance, and social class factors (that females may have more flexible strategies for overcoming) are discussed by Terry Wotherspoon, *The Sociology of Education in Canada: Critical Perspectives*, 2nd edn (Toronto: Oxford University Press, 2004), 102–4.
29. Campbell, *The Mozart Effect*, 192.
30. Nancy Joan Gall, personal communication, 3 Sept. 2008.
31. Such as Griffiths, *Teaching the Dyslexic Child*. The date (1978) and dated tone of Griffiths's work should not distract the reader from the soundness of her remedial approach to the teaching of dyslexic syndrome children. Her multi-sensory, non-judgemental, positive-reinforcement method for helping dyslexic children to make viable steps towards reading and writing represents the very best that could be done for dyslexic syndrome children prior to the discoveries of Levinson and Tomatis—and that still may be

done by those unable to access or unaware of the strategies now available for healing the dyslexic syndrome ear. Most of what has been published since is a variation on the same sorts of methods Griffiths used, including methods that attempt to exploit the creative resources of the teacher, such as Jeffery Freed and Laurie Parsons, *Right- Brained Children in a Left-Brained World: Unlocking the Potential of Your ADD Child* (New York: Simon & Schuster, 1997). A similar situation seems to prevail in the efforts at remediation by audiology for "auditory processing disorder [or deficit] (APD)"; see also Chapter 11, note 34.

32. Cited on Dr. Harold Levinson's website at: <www.dyslexiaonline.com>. Accessed 4 Apr. 2008.
33. Griffiths, *Teaching the Dyslexic Child*, 1.
34. Levinson, *Smart but Feeling Dumb*, and Levinson, *A Solution to the Riddle—Dyslexia*. Dr. Levinson's more comprehensive and updated analysis and treatment for dyslexic syndrome is described on his website and in his more recent publication *The Discovery of Cerebellar-Vestibular Syndromes and Therapies: A Solution to the Riddle—Dyslexia* (Stonebridge, 2000).
35. Nancy Joan Gall, personal communication, 2006.
36. Madaule, *When Listening Comes Alive*, 50–1.
37. Ibid., 60–1. See also Campbell, *The Mozart Effect*, passim.
38. Madaule, *When Listening Comes Alive*, 62–3.
39. Tomatis, *The Ear and the Voice*, ch. 9.
40. Personal Communication, 9 Jan. 2009, per e-mail of Morana Petrofski, Co-Director of The Listening Centre, Toronto.
41. Madaule, *When Listening Comes Alive*, 37, 41.
42. Paul Madaule, "The Dyslexified World," translated from French by Dr. Paula Kacher, copyright 1987, The Listening Centre Press. "L'Univers Dyslexie" was presented at the Fifth International Congress of Audio-Psycho-Phonology (Tomatis Method) in Toronto in 1978.

Chapter 5
Light and Darkness: *Healing and Psychosis*

1. "The pulses of equal intensity and the 'all-or-none' response of nerve[s] were described by E.D. Adrian in *The Basis of Sensation* in 1928." Richard L. Gregory, *Mind in Science* (Harmondsworth, Middlesex: Penguin Books, 1981), 204–5.
2. Ibid., 203.
3. Tomatis, *The Ear and the Voice*, 57–9.
4. This description of Daniel's listening intervention was kindly confirmed for me by The Listening Centre Co-Director Morana Petrofski. Personal communication by e-mail, 9 Jan. 2009.
5. Regush, "The listening cure," 5.
6. This explanation was kindly confirmed for me by The Listening Centre Co-Director, Morana Petrofski. Personal communication by e-mail, 9 Jan. 2009. On Gregorian chant, see Madaule, *When Listening Comes Alive*, 62–3.
7. The Listening Centre Co-Director Morana Petrofski. Personal communication by e-mail, 9 Jan. 2009.
8. *The Empire Strikes Back*, 1997 Special Edition theatre re-release, directed by George Lucas, with Harrison Ford, Mark Hamill, Carrie Fisher, and Alec Guinness; voiceover by James Earl Jones.
9. Madaule, *When Listening Comes Alive*, 23, 72–3.

10. This explanation was kindly confirmed for me by The Listening Centre Co-Director, Morana Petrofski. Personal communication by e-mail, 9 Jan. 2009.
11. Ibid.
12. Tomatis, *The Conscious Ear*, chs 9 and 10.

Chapter 6
The Lion's Den: *Daniel among Psychiatrists*

1. *Physicians' Desk Reference*, 50th edn (Montvale, NJ: Medical Economics Company, 1996), 317, 1301–5, for Risperdal (rispiradone); 1575–8, for Haldol (halperidol); 1621–2, for Cogentin (related to atropine, antispasmodic, to offset effects of neuroleptics); 2523–6, for Thorazine (chlorpromazine); 2700–2, for Ativan (lorazepam); 2355–8, for Etrafon (perphenizine plus amitriptyline hydrochloride); 2839–40, for Elavil (amitriptyline hydrochloride); Colase (laxative to offset effects of neuroleptics). How many of these medications actually were used on Daniel and at what dosages, I do not know. I was told that he received Haldol, Cogentin, lorazepam, Risperdal at dosages of at least 6 mg, and Colase. I filled prescriptions for Haldol, which our family doctor quickly countermanded; Risperdal, which our family doctor gradually reduced in dosage; Cogentin, which our family doctor also gradually withdrew; and Colase, which Daniel decided he did not need.
2. *The Snake Pit*, 1948, directed by Anatole Litvak, with Olivia de Havilland, Mark Stephens, Leo Genn.
3. *David and Lisa*, 1963, directed by Frank Perry, with Keir Dulea, Janet Margolin, Howard Da Silva.
4. *One Flew Over the Cuckoo's Nest*, 1975, directed by Milos Forman, with Jack Nicholson, Louise Fletcher.
5. *Awakenings*, 1990, directed by Penny Marshall, with Robert de Niro, Robin Williams, Julie Kavner. Based on Oliver Sacks's memoir; see <www.oliversacks.com>.
6. Erving Goffman, *Asylums: Essays on the Social Situation of Mental Patients and Other Inmates* (Garden City, NY: Anchor Books, 1961). The level of medication in use for the patients in Daniel's environments largely obviated the sort of interactions Goffman studied; not entirely, however.
7. Ibid., 1.
8. For example, Juanne Nancarrow Clarke, *Health, Illness, and Medicine in Canada* (Toronto: Oxford University Press, various editions, 1996–2009), and, especially, a remarkable history whose author is a psychiatric survivor: Geoffrey Reaume, *Remembrance of Patients Past: Patient Life at the Toronto Hospital for the Insane, 1870–1940* (Toronto: Oxford University Press, 2000).
9. See Chapter 8, note 6.
10. Gregory, ed., *Oxford Companion to the Mind*, 71.
11. Ibid., 76.
12. Ibid.
13. Thelma Wheatley, *My Sad Is All Gone* (Lancaster, Ohio: Lucky Press, 2005), 253. This inspiring story appeared late in my journey as an interesting confirmation of my impression that Daniel could improve as successfully on less medication, i.e., by using Risperdal at a low dosage, although I had no way of assessing the relative amounts of chemicals in Dan's system in the way Dr. Huggins could for Julian and had relied instead on my observations of changes in Dan's behaviour.
14. Ibid., 149–66.

Chapter 7
Out of the Pit: *Daniel's Diagnosis*

1. Wotherspoon, "Education," 244–75.
2. On the uselessness of IQ testing, see Gregory, *Mind in Science*, 326–
3. Griffiths, *Teaching the Dyslexic Child*, 66–7.
4. Ibid., 18.
5. See Johnson, *Osler's Web*.
6. One of the brochures in the ward waiting room produced by the government of Ontario stated that schizophrenia is suspected by some researchers to be a result of malfunction of the immune system.
7. Paul Madaule, "The Dyslexified World," an excerpt from Tim Gilmor, Paul Madaule, and Thompson, *About the Tomatis Method*, trans. from the French by Dr. Paula Kacher (Toronto: Listening Centre Press, 1987), 8.
8. One of these was Joanne Greenberg, *I Never Promised You a Rose Garden* (New York: New American Library, 1964), a novel about a schizophrenic young woman who is helped by an empathetic woman psychiatrist. The debate concerning the role of empathetic psychiatry in schizophrenia continues, for example, among the philosophers, psychiatrists, and published writers who produce book reviews for MentalHelp.net, at:<www.mentalhelp. net/poc/view_doc.php?id=439&type=book&cn=7>. Lara Jefferson's *These Are My Sisters: A Journal from the Inside of Insanity* (Garden City, NY: Anchor Press/Doubleday, 1975) is a record of her insanity by an inmate of a Midwest institution who died in the late 1940s; the author's name is pseudonymous. It reflects the high intelligence of a person whose brain nevertheless frequently loses its control (although the staff often call on her for assistance with other patients, suggesting that her loss of control is either sporadic or at predictable intervals) and describes the conditions in a mental hospital in the first half of the twentieth century. On this, see also Reaume, *Remembrance of Patients Past*.

   The nearest city library has since culled several other schizophrenia and learning disabilities titles I read of which I did not keep a record.
9. This was copied from a psychiatric text, which was removed from the library after I had consulted it. I regret that I am unable to give credit to the original author and would appreciate any information that would allow me to do so in future editions of this work.
10. Getting information from his mother proved difficult. Only after I became her caregiver four years later was I able to learn from family members and from her (through the distorted lens of her Alzheimer's disease) that her strategies of denial and intense, superficial cheerfulness combined effectively to veil the reality of her own depression. She eschewed psychological discussions, but through an occasional comment would reveal considerable psychological understanding. I think she regarded depression as a weakness and proved herself an extremely strong-willed and stoical individual in masking it. She was a devoted listener to classical music and considered daily access to that activity essential to her well-being.
11. *Physicians' Desk Reference*, 317, 1301–5.
12. See note 8, this chapter.
13. I have been unable to locate this author, whose very useful book also was culled from the public library. I regret that I am unable to give credit to the original author and would appreciate any information that would allow me to do so in future editions of this work.
14. See note 12, this chapter.
15. Ibid.
16. The Tomatis Method, at: <www.tomatis.com>.

Chapter 8
Mountain Man in Spider's Web: *Addictions*

1. Carlos Castaneda, *The Teachings of Don Juan: A Yaqui Way of Knowledge* (Berkeley: University of California Press, 1968). See Wikipedia, at: <en.wikipedia.org/wiki/The_Teachings_of_Don_Juan>. This and later works by Castaneda, derived from participant observation in the same Aboriginal culture in the Sonoran Desert of northern Mexico, reportedly document the events that took place during an apprenticeship he claimed to have served with a self-proclaimed Yaqui Indian sorcerer between 1960 and 1965. The authenticity of the book, along with the rest of Castaneda's series, has been the subject of debate since they first appeared in print. What is not debatable is that Castaneda's writing has been used for the last 40 years as a rationalization for hallucinogenic drug use as a means to supposedly "superior" forms of spiritual understanding.
2. *A Beautiful Mind*, 2001, directed by Ron Howard, with Russell Crowe, Jennifer Connolly, Ed Harris, and Paul Bettany, adapted by Akiva Goldsman from a book of the same name by Sylvia Nasar.
3. This situation was recapitulated in Daniel's most recent interval of psychosis when it seemed to him that he had several brains and several bodies: a drug body, a medicated body, the memory of a body he felt guilty about, the memory of a body he felt hopeful about, a warrior body to fight temptations that became specialized to fight particular temptations. Sometimes he felt he had as many as 11 disparate personal dimensions; on good days, only three or so.
4. Craig Lambert, "Deep cravings: New research on the brain and behavior clarifies the mysteries of addiction," *Harvard Magazine*, 102, 4 (Mar.–Apr. 2000), at: <harvard-magazine.com/2000/03/deep-cravings.html>. Accessed 25 July 2008.
5. Nora D. Volkow, "The Addicted Brain: Beyond Willpower," in John Hoffman and Susan Froemke, eds, *Why Can't They Just Stop? Addictions: New Knowledge, New Treatments, New Hope* (New York: Rodale, 2007), 56.
6. The study, of which both I and our general practitioner took note, is no longer available on-line. A related literature review, done later, was reported by Reuters, Medical News, *Medscape*, 2001, "Lower dose, longer titration recommended for Risperidone," based on a literature review by Dr. Richard Williams of the Royal Jubilee Hospital, Victoria, British Columbia, *Journal of Clinical Psychiatry* 62 (Apr. 2001): 282–9. The Reuters report reads: "In patients not previously exposed to the drug [Risperdal], elderly patients, first-episode younger patients and patients with indications other than schizophrenia, he recommends an initial target dose as low as 1 to 2 mg/day, commencing with a dose of 0.25 mg/day in the elderly." From the *Journal of Clinical Psychiatry* abstract:

    > Background: Drug dosages utilized during controlled clinical trials are not always optimal for patients encountered in day-to-day practice. The original trials of risperidone, a novel antipsychotic, suggested that an initial target dose of 6 mg/day was appropriate, but these trials were necessarily conducted among patients who were chronically impaired, hospitalized, and often partly drug resistant.
    >
    > . . . Conclusion: On the basis of naturalistic studies, clinical audit, phase 4 trials, positron emission tomography data, and 5 years of clinical experience, the currently recommended target dose of risperidone is 4 mg/day for most patients, with less-rapid titration than previously recommended Moreover, a lower dose than this and slower titration may be appropriate for elderly patients, young patients, and first-episode patients.

A number of studies were being carried out in Vancouver in the early 2000s, some of them funded by Janssen Pharmaceuticals to learn more about Risperdal. The following study, from the Department of Psychiatry at the University of British Columbia, found that Risperdal affects metabolism in the left hemisphere of the brain, which is the desirable hemisphere for dominance: E.T.C. Ngan et al., "Immediate and delayed effects of risperidone on cerebral metabolism in neuroleptic naïve schizophrenic patients: correlations with symptom change," *Journal of Neurological and Neurosurgical Psychiatry* 72, 1 (Jan. 2002): 106–10. Accessed 18 July 2008 at: <jnnp.bmj.com/cgi/content/abstract/72/1/106>. "The evidence suggests that the reduction in medial-frontal activity after treatment with risperidone is a direct effect of risperidone and not a consequence of symptom improvement. Reduction of medial frontal metabolism may be one of the physiological mechanisms by which risperidone alleviates symptoms of psychosis in schizophrenia."

Or, as I found in Daniel's case, also causes it and predisposes the brain to further psychosis.

7. Norman Cousins, *Anatomy of an Illness As Perceived by the Patient* (New York: Norton, 1979).

Chapter 9
Long Muddy Road: *Addictions and Mental Illness*

1. Art Garfunkel, *Angel Clare* (New York: Columbia Records, 1973). "Woyaya," Mac Tontoh, Teddy Osei, Robert Bailey, Wendell Richardson, Loughty Lasisi Amao, R. Bedeau, and Sol Amarfio.
2. See Chapter 8, note 6.
3. John E. Upledger and Jon D. Vredevoogd, *Craniosacral Therapy* (Seattle: Eastland Press, 1982), 14, 19.
4. Ibid., 41.
5. Ibid., 260–4.
6. *The String Quartet Tribute to R.E.M.* (Los Angeles: Vitamin Records, 1999), Cat. No. CD-8517.
7. *The Last Emperor*, 1987, directed by Bernardo Bertolucci, with John Lone, Joan Chen, Peter O'Toole.
8. This dog, Garm, was killed along our country road in the late winter of 2007 when he went after a pickup truck pulling a trailer. The dog didn't see the trailer and died almost instantly. Dan was devastated. That summer he and Eric went to the nearest city one day and returned with a golden retriever puppy, variously called Lox (by Richard and me) and Minister (by the boys).

Chapter 10
Listening for the Light: *Tomatis Reclaimed*

1. Madaule, *When Listening Comes Alive*, 61–2.

Chapter 11
Listening to Psychosis: *Cerebral Non-Dominance*

1. Campbell, *The Mozart Effect*, 155.
2. *Shine*, 1996, directed by Luke Hicks, with Geoffrey Rush as David Helfgott.
3. Wheatley, *My Sad Is All Gone*, 253 ff., for her descriptions of Dr. Joseph E. Huggins's work with Risperdal. Also <www.bitemarks.com/dtmodels/vol18. html> and <www.netrover.com/~southgve/articles/htm>. These sites were accessed by the author in 2001, but seem not to have been maintained as of July 2008. Copies of the work of

Dr. Huggins et al. were submitted to the National Library of Canada, Ottawa: Dr. Joseph E. Huggins, Dr. Soula Homatidis, and Jason G. Huggins, "The 100% 'Evidence-Based' Universal Diagnostic & Treatment Model© Version 5.00-2000: A new clinical approach & methodology utilizing neurotransmitter - cytochrome P450 isoenzyme affinity & binding contro." As they write: "The new 100% 'evidence-based' D&T Model methodology evolved from work done with hundreds of management-refractory cases, the majority from Ontario, spanning the entire DSM-IV and ICD-9-CM diagnostic spectrum, ~ Autism/PDD/ADHD/DD to Fibromyalgia & Chronic Fatigue Syndromes. We have also had innumerable opportunities over the years to witness how quickly symptom and behaviour control could be lost once the decision making process reverts back to the traditional 'principal diagnosis-target symptom' basis for determining drug choice. It has now become very apparent that the 'experience-based' and the 'evidence-based' approaches to patient management are completely incompatible and must not be blended. The precision of drug selection afforded by a knowledge of the specific properties of a drug (neurotransmitter & cytochrome P450 isoenzyme affinity & binding potential) in any specific case, will far exceed any drug choices based on 'pragmatic and other empirical considerations' that depend largely on art, savvy and previous drug usage experience. In light of the scientific data currently available, the traditional methodology and approach are inadequate and require a major overhaul in methodology, attitude and thinking." These doctors are advocating treatment for patients based on how the individual patient responds to medication under the supervision and regulation of his/her physician ("evidence-based") rather than the standardized protocols recommended by the pharmaceutical companies on the basis of their experiments ("experience-based").

4. *Physicians' Desk Reference*, 317, 1301–5; <www.risperdal.com/risperdal/>. Bloomberg.com Science at: <www.bloomberg.com/apps/news?sid=aaw9D 5Ty8E ho&pid =20601 124>. "Current medicines [for schizophrenia] include Eli Lilly & Co.'s Zyprexa and Johnson & Johnson's Risperdal, the 10th and 11th best-selling drugs in the world in 2007. Zyprexa had $4.76 billion in sales last year while Risperdal generated $4.8 billion."
5. Wheatley, *My Sad Is All Gone*, 253.
6. *Bach for Meditation* (Naxos Canada, 2004), Cat. No. 8.557650.
7. Tomatis, *The Ear and the Voice*, 28. Madaule, *When Listening Comes Alive*, 34, writes: "Upon studying the auditory control of the voice further, Tomatis concluded that it is the ear, in conjunction with the brain, that controls voice production. The larynx simply does what it is told to do. This was summarized in a phrase, 'We sing with our ear,' and was later refined as 'the voice only contains harmonics the ear is likely to hear.' This phenomenon was demonstrated in an experiment before the Académie Nationale de Médecine, and later before the Académie des Sciences, both in Paris, and given the name The Tomatis Effect."
8. Henry Gray, *Anatomy, Descriptive and Surgical, 1901 edn* (Philadelphia: Running Press, 1974), 740.
9. Ibid., 858.
10. Tomatis, *The Ear and the Voice*, 52.
11. Gregory, ed., *Oxford Companion to the Mind*, 304–5.
12. Dr. Harold Levinson, at: <www.dyslexiaonline.com/information/brain/reverb_circuits.html>:
    "Reverberating circuits link the cerebellum . . . to the sensory nerves which connect tactile, visual, proprioceptive and auditory sense organs to the cerebrum. While part of the messages from these organs goes to the cerebrum, part detours through the cerebellum, then 'reverberates' through the cerebrum to the cerebellum. It is thought that these cir-

cuits serve a feedback function. Proprioceptive impulses from muscles . . . may reach more than one cerebral center." (Reprinted from *Scientific American*, Aug. 1958).

13. Madaule, *When Listening Comes Alive*, 60.
14. Tomatis, *The Ear and the Voice*, 52–3.
15. Madaule, *When Listening Comes Alive*, 68–70.
16. Campbell, *The Mozart Effect*, 53 passim.
17. Tomatis, *The Ear and the Voice*, 53.
18. Ibid., 67–74.
19. Gregory, ed., *Oxford Companion to the Mind*, 305.
20. Madaule, *When Listening Comes Alive*, 64–6. A similar method of sound stimulation is Auditory Integration Training (AIT). "Essentially, AIT is a mechanical means of achieving an improvement in auditory discrimination through an intensive form of music therapy." Developed by Guy Berard, M.D., "the French otolaryngologist who developed the therapy in France over the last four decades," and represented by Annabel Stehli, who wrote *The Sound of a Miracle* and founded the Georgiana Institute, it posits the same concept of middle-ear stimulation: "At the mechanical level, it seems likely that AIT exercises the muscles controlling the three ossicles, the small bones found in the middle ear. The treatment apparently strengthens the muscles and improves the reaction involved in the prevention of sensory overload." Georgiana Institute at: <www.georgianainstitute.org/index.htm>. Accessed 13 January 2009. Tomatis, it might be noted, decried his imitators: *The Conscious Ear*, 178–80.
21. John P.J. Pinel, "Brain damage from the other side of the knife : a biopsychological perspective," *Higher Brain Function Research,* 27 (2007): 125–32.
22. An interesting personal account of the pros and cons of cochlear implant for one Canadian woman can be found at: <www.deafened.org>.
23. Stefan Koelsch, Matthias Wittforth, Angelika Wolf, Joachim Müller, and Anja Hahne, "Music perception in cochlear implant users: An event-related potential study," *Clinical Neurophysiology* 115, 4 (2004): 966–72. At: <www.journals.elsevierhealth.com/periodicals/clinph/issues/contents?issue_key=S1388-2457(00)X0070-3>. Abstract conclusions: ERAN and N5 (as well as timbre-MMN and P3) can be elicited in CI users. Although amplitudes of effects were considerably smaller in the CI group, the presence of MMN and ERAN indicates that neural mechanisms of both physical and musicsyntactic irregularity-detection were active in this group.
24. Tomatis, *The Ear and the Voice*, 56–8; Campbell, *The Mozart Effect*, 192.
25. Tomatis, *The Ear and the Voice*, 29, 61–2.
26. University of Nebraska Medical Centre, on-line neurology text at: <www.unmc.edu/Physiology/Mann/mann17.html>: "However, there are some cortical functions that are not served equally by both hemispheres. In one cerebral hemisphere, there is a region, called Broca's area, that is involved in speech . . . . Broca's area . . . is located in the frontal cortex just anterior to the face area of the precentral gyrus. Actually, there are two additional areas of the same hemisphere that are involved with speech. These are the inferior parietal lobule, consisting of the angular and supramarginal gyri . . ., and Wernicke's area . . ., located on the posterior aspect of the superior temporal gyrus just behind the auditory cortex. . . . These areas are found in only one hemisphere, and their presence defines the dominant hemisphere. The dominant hemisphere is usually the one opposite the favored hand (in most individuals, the left hemisphere), but, in a few cases, it is the hemisphere on the same side as the favored hand (92% of right-handed people are left hemisphere dominant, 50 to 60% of left-handed people are right hemisphere dominant)."

27. Wikipedia, the Free Encyclopedia, at: <en.wikipedia.org/wiki/Laterality>. "Laterality of motor and sensory control has been the subject of a recent intense study and review. These may be viewed at mimickingman.com (review of papers). It turns out that the hemisphere of speech is the hemisphere of action in general and that the command hemisphere is located either in the right or the left hemisphere (never in both)."
28. Madaule, *When Listening Comes Alive*, ch. 8.
29. Tomatis, *The Conscious Ear*, ch. 10: "Laterality—It Must Be Right", 140–69.
30. Ibid., chs 9, 10.
31. Ibid., 132.
32. Wikipedia, the Free Encyclopedia, at: <en.wikipedia.org/wiki/Laterality>.
33. Jill Bolte Taylor, *My Stroke of Insight* (New York: Penguin Group, 2006). Readers of Dr. Taylor's work will immediately see the consonances between her experience of losing the function of her left brain and aspects of Daniel's experience related to his right brain functioning with inadequate communication with the left hemisphere.
34. Mark Bruzon at: <www.geocities.com/player2000gi/hemispheric_specialization.htm>. Teri James Bellis, *When the Brain Can't Hear: Unravelling the Mystery of Auditory Processing Disorder* (New York: Simon & Schuster, 2002), makes frequent reference to the specialized functions of the cerebral hemispheres in her descriptions of clients with "auditory processing disorder" but without Tomatis's understanding of laterality or, apparently, of the role of the functional capacity of the right ear in left-hemisphere dominance that I develop in this book. She does recommend art and activities such as sports for hemispheric integration, although music doesn't make her list for that purpose. Most of her clients fall in the range of ear-related brain dysfunction I recognize as dyslexic syndrome and chronic fatigue syndrome, although some clients verge on autism. Oddly, the one avenue for intervention largely neglected is that of the ear, although that organ usually is acknowledged as the source of the client's difficulties. Her extensive descriptions are helpful for those parents wanting to locate their children on a spectrum of auditory processing deficits; however, the therapies she recommends primarily remediate and support a brain that continues to be sound-deprived. The Tomatis Method, or focused listening, "opens" the ear by sound stimulation and hence the brain so that the left hemisphere can access the higher frequencies of sound via a more functional ear.
35. Campbell, *The Mozart Effect*, 192.
36. Tomatis, *The Ear and the* Voice, 22.
37. Sollier, *Listening for Wellness*, 227.
38. Ibid., 274.
39. Gregory, ed., *Oxford Companion to the Mind*, 740.
40. *Physicians' Desk Reference*, 317, 1301–5.
41. Tomatis, *The Ear and the Voice*, 55–63. See also Dr. Harold Levinson, at: <www.dyslexiaonline.com>.
42. Dr. Daniel G. Amen, *Change Your Brain, Change Your Life* (New York: Three Rivers Press, Random House, 1998), ch. 9.
43. University of Nebraska Medical Centre, on-line neurology text, at: <www.unmc.edu/Physiology/Mann/mann17.htm>.
44. Madaule, *When Listening Comes Alive*, 75.
45. Tomatis, *The Conscious Ear*, 128.
46. Personal communication by e-mail 9 Jan. 2009.
47. Sollier, *Listening for Wellness*, 274.
48. Ibid., 276.
49. Ibid., 277.

50. Petrofski e-mail, 9 Jan. 2009.
51. See also note 20, this chapter; Georgiana Institute at: <www.georgianainstitute .org/index.htm>. Accessed 13 Jan. 2009.
52. Ibid.
53. These phrases recapitulate the explanation kindly confirmed for me by The Listening Centre Co-Director, Morana Petrofski. Personal communication by e-mail 9 Jan. 2009.
54. Upledger and Vredevoogd, *Craniosacral Therapy.*
55. Ibid., Appendix C, 282–90.
56. Ibid., 23, 176–84, 210–12, 269.
57. Ibid., 85. "As the temporoparietal sutures disengage, the side-to-side swinging of the temporal squama continues as the temporal bones move caudad [to the posterior, or back, of the body]. This change in placement lowers the leaves of the tentorium cerebelli and imposes traction first upon the straight sinus and then upon the falx cerebri.

    "The supratentorial intracranial compartment is enlarged, and new fluid is sucked into the space by negative pressures. All of this can be perceived if you will listen carefully to your hands. As the fluid redistributes itself, a new horizontal balance of the mechanisms is achieved and the side-to-side swinging of the mandible and temporals will stop. The whole mechanism seems then to settle into a new and balanced pattern of reduced membrane tension."
58. Ibid. Upledger and Vredevoogd stress the gentleness of touch required and the non-invasive nature of successful craniosacral therapy. The techniques are far less forceful than interventions of the gentlest chiropractic manoeuvres. It shouldn't hurt.
59. Campbell, *The Mozart Effect*, 155.

Chapter 12

Tomatis's Golden Key: *The Neurology of Integration*

1. Madaule, *When Listening Comes Alive*, 1–15. In the first two chapters Paul Madaule describes his own mixed dominance and how it caused dysfunction and handicapped him physically and socially until he had been treated by Dr. Tomatis.
2. Jill Bolte Taylor describes this right-brained cognitive state eloquently in her first chapter of *My Stroke of Insight.*
3. Tomatis, *The Ear and the Voice*, 22.
4. Ibid., 22–3.
5. Norman Doidge, *The Brain That Changes Itself: Stories of Personal Triumph from the Frontiers of Brain Science* (New York: Penguin Books, 2007), 75. Doidge points out the links between autism and epilepsy. Dr. Oliver Sacks in *Musicophilia* devotes an entire chapter to epilepsy and seizures pertaining to the sounds of music as a stimulus, sometimes referring specifically to differences between the brain's hemispheres. In several instances recounted by Dr. Sacks, an abrupt shift in the patient from the dominance of one hemisphere to the other will seem the likely explanation for some of these behaviours. Sacks, *Musicophilia*, ch. 2 passim.
6. Tomatis, *The Ear and the Voice*, 84–5.
7. Wikipedia, the Free Encyclopedia, at: <www.wikinfo.org/index php/Corpus_ callo sum>; although some research data are disputed.
8. Valerie Dejean, at: < http://216.194.201.205/SI/valeriedejean.org/>. Based in Bethesda, Md, Valerie Dejean specializes in application of the Tomatis Method to creating the neurological developmental stages in autistic children that define their syndrome. The Spectrum Center, at: <http://216.194.201.205/SI/ tomatis.net/>. Dejean's Spectrum Centre in New York is a Tomatis Method facility that treats learning disorders,

autism, and other problems related to the function of the middle and inner ear.

9. Gregory, ed., *Oxford Companion to the Mind*, 77.
10. Ibid., 307.
11. *Science Magazine*, American Association for the Advancement of Science, at: <www.sciencemag.org/cgi/content/abstract/216/4552/1331>. J.F. Willot and S.M. Lu, "Noise-induced hearing loss can alter neural coding and increase excitability in the central nervous system of mice," *Science* 216, 4552 (1982): 1331–4. Abstract accessed on-line, 16 May 2008. Copyright © 1982 by American Association for the Advancement of Science.
12. United States Drug Agency, at: <www.ars.usda.gov/is/AR/archive/aug07/aging0807.htm>. This study suggests the importance of a diet rich in oxygen radical absorbing nutrients (anti-oxidants) for those persons with neuronal problems because they affect the level of dopamine production in the brain.

    "The oxygen radical absorbance capacity (ORAC) scores of fruits vary. In this mix of fruit, the ORAC score of blueberries is highest, followed by (in order) the scores of black plum, blackberries, raspberries, strawberries, sweet cherries, avocado, navel orange, and red grapes."

    "One of the first of Joseph's studies, published in the *Journal of Neuroscience*, showed a protective effect of consuming antioxidants. Study rats were fed—from adulthood to middle age—vitamin E, strawberry extracts, or spinach extracts, all with similar ORAC [the oxygen radical absorbance capacity] values. Animals receiving the high-antioxidant diets did not experience the age-related cognitive performance losses seen in control rats fed standard chow. A later study, also published in the *Journal of Neuroscience*, showed a reversal of functional loss among rats on special diets. . . . The rats fed the spinach, strawberry, or blueberry extracts effectively reversed age-related deficits in neuronal and cognitive function. In addition, the blueberry-fed group far outperformed their peers while traversing a rotating rod to test balance and coordination . . . . [and] showed remarkable stamina on neuromotor function tests," says psychologist and co-author Barbara Shukitt-Hale, also with the Neuroscience Laboratory.

    Examination of the brain tissue of those blueberry-fed rats showed much higher levels of dopamine than were found in the other groups. Dopamine has many functions within the brain. In particular, it can affect the way the brain controls movements. "We suspected that the combined antioxidant potency of compounds in blueberry extract may have reduced inflammatory compounds in the brains of these older animals," says Joseph. "Inflammation ordinarily contributes to neuronal and behavioral shortfalls during aging."

    Other relevant studies, on dopamine production in psychosis, at: <lansbury.bwh.harvard.edu/da_and_agonists.htm>, include T.R. Kosten, T.P. George, et al., "The potential of dopamine agonists in drug addiction," *Expert Opin Investig Drugs* 11, 4 (2002): 491–9; and an alternative to dopamine agonists: F.P. Bymaster and C.C. Felder, "Role of the cholinergic muscarinic system in bipolar disorder and related mechanism of action of antipsychotic agents," *Mol Psychiatry* 7, Suppl. 1 (2002): S57–63. A study of dopamine within the cochlear system itself is G. Halmos, B. Lendvai, et al., "Simultaneous measurement of glutamate and dopamine release from isolated guinea pig cochlea," *Neurochem Int* 40, 3 (2002): 243–8.
13. Richard S.E. Keefe and Philip D. Harvey, *Understanding Schizophrenia* (New York: Free Press, 1994), 22: "Schizophrenics are addicted to smoking 3:1 as compared with the general population." Lacking control in the left hemisphere, which is the seat of rational self-control in most persons, would explain that statistic. Persons suffering de-

pression also find it very difficult to quit smoking.

14. Gregory, ed., *Oxford Companion to the Mind*, 305.
15. Tomatis, *The Ear and the Voice*, 23. “The same test applied to the left ear has no effect and may facilitate the flow of sounds by getting it out of the way of the right ear.”
16. Matthew C. Holley, “Application of new biological approaches to stimulate sensory repair and protection,” *British Medical Bulletin* 63 (2002): 16–19. Holley discusses a number of agents, including antibiotics, that attack the cilia and notes several strategies being developed for stimulating cell regrowth. He suggests agents that may protect the cilia prior to damaging drug interventions (it is not clear whether sound damage might thus also be deflected), including aspirin.
17. Levinson, at: <www.dyslexiaonline.com>. Campbell, *The Mozart Effect*, 39, writes: “Tomatis . . . recommends a more traditional diet, high in whole grains and fresh foods, he warns against eating acid-producing foods, yogurt, and highly processed foods, which he believes interfere with normal hearing.” His emphasis is reflected by Madaule in advice related to avoiding infections of the middle and inner ear by avoiding milk and milk products and acidic beverages such as canned orange juice. Madaule, *When Listening Comes Alive*, 102.
18. I am referring here to the *Earobics* Series #1 and #2 tapes. The program of breathing and sound production augments and supports the Electronic Ear™ and listening retraining.
19. 23 Apr. 2008 around 1:45 p.m., after listening to *The String Quartet Tribute to R.E.M.* for about an hour.
20. Willot and Lu, “Noise-induced hearing loss,” 1331.
21. Some of the meanings that emerged (after many months of our hearing them referred to in their purely symbolic form) from Daniel’s “naming of demons” included “Podamy” from the brand of cigarettes he was smoking, “Podium”; “Legular,” as the demon of legibility; and “Alphabetica,” as “the demon who makes children put letters in a certain order.” I was particularly familiar with symbolic naming of emotionally painful elements of experience as a “distancing strategy” through a survivor of child sexual abuse my husband had counselled and written about.
22. For example, Freed and Parsons, *Right-Brained Children in a Left-Brained World*. However, some of these authors recognize symptoms, feed one another’s desire for highly emotional interactions (in which they have the credentials to establish personal domination), and tell emotional stories of partial success without achieving results close to what is accomplished even in focused music listening, let alone in the precision intervention of Tomatis’s Electronic Ear. Ratey’s incomplete diagram of the neurological route of hearing stops where Tomatis begins. Ratey’s description of the brain’s maps is essentially a projection of his own private organizational system of “theatres” on aspects of brain function. Edward M. Hallowell and John J. Ratey, *Driven to Distraction: Recognizing and Coping with Attention Deficit Disorder from Childhood through Adulthood* (New York: Simon & Schuster, 1994); John J. Ratey, *A User’s Guide to the Brain: Perception, Attention, and the Four Theaters of the Brain* (New York: Simon & Schuster, 1996).
23. Griffiths, *Teaching the Dyslexic Child*, 18. This classic in remedial pedagogy has been reproduced by The Ann Arbor Company UK, at: <www.annarbor. co.uk/acatalog/Teaching_the_Dyslexic_Child.html>.
24. Personal communication, Ruth Welham-Umphrey; Maureen Jennings, personal communication, 7 Jan. 2009. William L. White, *Slaying the Dragon: The History of Addiction Treatment and Recovery in America* (Normal, Ill.: Lighthouse Institute Publications, 1998), 38. The number of situations in which music has been “tried” with

mental patients is prodigious. The unfortunate fact is that such listening opportunities have been ambient rather than focused, intermittent and widely spread out in time (typically once a week rather than every day), full-range instead of high-frequency sound, and viewed as peripheral rather than as essential to the patients' recovery. Of course, no one has been expecting the patients to recover fully in any case. And of recent years the prescription of large doses of medication may have impeded the full effects that are potential in the music. Oliver Sacks's use of L-dopa for his *Awakenings* subjects who already were showing significant responses to music (primarily piano) may provide a particularly poignant instance of drugs trumping music and providing only temporary healing. Sacks, *Musicophilia*, 272, 283.

25. Tomatis, *The Concious Ear*, 132, 133. But for Dolto and other such associates, I have no doubt Tomatis would have come to understand the nature of autism better. Overlays of psychologizing clouded his observations of the results of his own listening interventions with autistic children. He might have learned that the "breaks" the psychiatrists took to contemplate and ruminate about their subjects would have been better spent reinforcing the children's sound treatments, because the psychotic individual needs consistent and persistent exposure to high-frequency sound and will relapse without it, a circumstance he describes, but without understanding it (ibid., 136). He, too, might have recorded the phases of recovery through which the psychotic individual attains the left-hemispheric dominance we consider normal. In that way, he might have seen the effect on the left hemisphere, not only in regard to "laterality," as he understood the role of the dominance of the left brain for language (Tomatis, *The Ear and Language* [Norval, Ont.: Moulin, 1996], 146–7), but more profoundly in regard to establishing left-hemisphere dominance in the brain for recovery from a range of mental illnesses.

    The psychiatric notion Tomatis embraced with reference to autism, namely, that a non-receptive mother caused the malfunction of her infant's ear–brain connection, is particularly unfortunate. Undoubtedly, a mother faced with an unresponsive or abnormally responsive infant may not know what to do to help her child and may be afraid to show the usual affection or to use the teaching strategies familiar to her when the child repels or seems harmed by her advances. Tomatis may have been too ready to read into others' lives the atypical misfortunes of his own birth and early life experiences (*The Conscious Ear*, 1-11), a vulnerability easily exploited by the followers of Freudian thought he had approached "with pilgrim's staff in hand." Tomatis recapitulates the unhelpful explanations for human behaviour he received from these self-proclaimed experts for vulnerable mothers of autistic infants in the uncharacteristically unscientific "explanation" he invents for autism (ibid., 110–13, 142–3). For a glimpse of some of the extraordinary efforts of mothers to find solutions for their children's autism, see <www.autism.com>.

26. Wikipedia, the Free Encyclopedia, at: <en.wikipedia.org/wiki/Dual_brain_theory>; emphasis mine.27. Timothy J. Crow, "Did Homo sapiens speciate on the Y chromosome," *Psycholoquy*, 11, 001, American Psychological Association, at: <www.cogsci.ecs.soton.ac.uk/cgi/psyc/newpsy?11.001>. Accessed 12 July 2008.

28. It is fascinating to notice how various researchers are prone to associate their specialty with the mate recognition system of our ancestors. The musician imagines the capacity for making music to be the precursor to language that led to mate selection (Daniel J. Levitin); the lateralist is sure right-handedness evolved through selection against lefties and promoted language development (Chris McManus); the researcher preoccupied with language names language alone for selection (Tim Crow). The pairings and couplings of my high school friends (better representative of *Homo sapiens* than my university classes) suggest motivations for mate recognition lie elsewhere. One study of

mates suggests the way the ear lobe joins the head is a strong contender for selection bias.

29. Rachel L.C. Mitchell and Tim J. Crow, "Right hemisphere language functions and schizophrenia: the forgotten hemisphere?" *Brain: A Journal of Neurology* (2005), at: <brain.oxfordjournals.org/cgi/content/full/128/5/963>. Accessed 25 June 2008.
30. Regush, "The listening cure," 5–6.
31. Relevant, too, may be the two ultrasound examinations to which Dan was subjected in utero. Do we know how ultrasound impacts the bone-conducted sound reception of the fetus? The new learning about the impact of sound on the fetus should be carried forward into further research on the possible effects of ultrasound on the developing organs that carry out the mechanisms discussed here, especially in view of the rise in the incidence of autism and dyslexic syndrome.
32. Wikipedia, the Free Encyclopedia, I. Derakhshan, MD, Neurologist, at: <en.wikipedia.org/wiki/Laterality>.
33. Gregory, ed., *Oxford Companion to the Mind*, 746.
34. Wikipedia, the Free Encyclopedia, at: <http://en.wikipedia.org/wiki/Vincent_van_Gogh#Early_life_.281853.E2.80.931869.29>. Among the diagnoses suggested by dozens of psychiatrists are porphyria, epilepsy, bipolar disorder, schizophrenia, poisoning from paints and from absinthe, alcoholism (absinthe), smoking, and possibly digitalis.

    I think a case might also be made for dyslexic syndrome in Vincent's youth. If, as proposed here, dyslexic syndrome, (some kinds of) depression, bipolarity, schizophrenia, and epilepsy can be considered to lie on a continuum of right-ear-related hemispheric instability, all of those syndromes could have existed in Van Gogh sequentially and become exacerbated towards that increasing instability not only by psychological issues and poisoning by various substances, but also by his deteriorating physical condition due to neglect and illness. See also Martin Bailey, "Drama at Arles: new light on Van Gogh's self-mutilation," *Apollo*, Sept. 2005, note 72, at: <findarticles.com/p/articles/mi_m0PAL/is_523_162/ai_n15880259>. Accessed 1 Feb. 2009. Bailey confirms Van Gogh's left-ear mutilation pictured in mirror image in his self-portrait, giving the appearance of a right-ear mutilation.
35. University of Nebraska Medical Center, on-line version of text, at: <www.unmc.edu/Physiology/Mann/mann17.html>; emphasis mine. <www.unmc.edu/Physiology/Mann/index.html> Michael D Mann, Ph.D. *The Nervous System in Action* © 1997-2008 by Michael D. Mann, Ph.D. All rights reserved.
36. See, for example, Sollier, *Listening for Wellness*, 105, 227.
37. Dr. Lee Schultz, personal communication, 12 September 2008. As I had done, Dr. Schultz leapt from his recognition of the differing functions of the hemispheres to a reassessment of his understanding of how the analysis of English language and literature might proceed. "I wish I had had access to [Dr. Jill Bolte Taylor's account of her recovery from stroke in *My Stroke of Insight*] while I was still teaching. All I can think about when I see it is [William Carlos?] Williams' ideas of the dynamic 'Local' and 'Universal.' . . . . Taylor's presentation mimics the sub-atomic structure of language and life as we know it. I keep wondering what access to this would have done for Witkenstein [*sic*], Hayakawa, Chomsky, the psycholinguists, and even [the] French deconstructionists."
38. Clayton L. Thomas, ed., *Taber's Cyclopedic Medical Dictionary*, 15th edn. (Philadelphia: F.A Davis, 1985), strabismus, 1635; nystagmus, 1147. See also Dr. Harold Levinson, at: <www.dyslexiaonline.com>; Upledger and Vredevoogd, *Craniosacral Therapy*, 176, 269.

39. The Listening Centre Co-Director, Morana Petrofski, personal communication by e-mail, 9 Jan. 2009.
40. Tomatis's conception of "La Nuit Utérine" is nicely elaborated by Sollier, *Listening for Wellness*, 276–7.
41. Madaule, *When Listening Comes Alive*, 8–10.
42. Tomatis, *The Ear and the Voice*, 75. "This asymmetry [in the right and left branches of the recurrent nerves] has enormous influence over language and over gesture."
43. Dr. Jerome Goldstein, "The Neurobiology of Sexual Orientation—The Gay Brain," San Francisco Neurological Society Meeting, 11 Nov. 2008; personal communication. Dr. Goldstein's website is at: <www.sfcrc.com/html/aboutsfcrc.htm>. An external electronic device to be attached to the skull to cure headache was demonstrated on television by Dr. Goldstein. Video clip aired on *Good Morning America*, 7/8/2006: <abcnews.go.com/ Video/playerIndex?id=21678362006>.
44. Mark Bruzon, at: <www.geocities.com/player2000gi/hemispheric_specialization.htm>. A layman's synthesis of what is generally known about the differential function of the hemispheres.
45. Ibid.; see also <www.unmc.edu/Physiology/Mann/mann17.html>; emphasis mine.
46. University of Nebraska Medical Centre, on-line neurology text, at: <www.unmc.edu/Physiology/Mann/mann17.html>.
47. Ibid.
48. Ibid.
49. Such as that of the singer for whom ear stimulation restored his voice, even his singing voice, after several surgeries that successively butchered his soft palate, tonsils, and nose without repairing nodular vocal cords. Tomatis, *The Ear and the Voice*, "Crazy about Singing," 33.
50. Professor Doctor Lars Heslet, at: <www.musicahumana.dk/documents/00015.pdf>, 2.
51. Ibid., 3.
52. Ibid., 4.
53. Ibid., 3.
54. Wheatley, *My Sad Is All Gone*, 253.
55. Alan Loveday, Iona Brown, Carmel Kaine, Roy Gillard, solo violin, Academy of St. Martin-in-the-Fields, Sir Neville Marriner, conductor, *Antonio Vivaldi Violin Concertos* (London: Decca, 1973), Cat. No. 4674322.
56. Campbell, *The Mozart Effect*, 125. See also "entrainment" [the effect on the brain of listening to music] with long-term effects from listening programs that help stroke victims to walk better (p. 273).
57. Tomatis, *The Conscious Ear*, chs 9, 10.
58. Campbell, *The Mozart Effect*, 19–22; Madaule, *When Listening Comes Alive*, 72–3.

Chapter 13
The Language of Psychosis: *Speech and Integration Disorder*

1. The lattice with emphasized points of intersection likely is the cognitive precursor to a representation on paper of a three-dimensional object. In this instance, Daniel was producing both forms of "box," almost simultaneously. The Gestalt (hemispheric) shift most commonly experienced in observing the faces of a Necker cube (described by Benjafield et al.) must be anchored somehow to prevent that effect in drawing a box; dotting the intersections of lines may be the first step in preventing the shift while rendering a 3-D object on a 2-D surface. John G. Benjafield, Daniel Smilek, and Alan Kingstone, *Cognition*, 4th edn (Toronto: Oxford University Press, 2010), ch. 2.

2. Tomatis, *The Ear and the Voice*, 75.
3. Tomatis corrects poor posture to maximize vestibular function. Ibid., ch. 12.
4. Ibid., 53, 62–3.
5. Ibid., 26.
6. A layman's summary of the point of view of the San Diego, California, neurologist Dr. Ramachandran on belief systems (Mark Bruzon, at: <www.geocities.com/player2000gi/hemispheric_specialization.htm>: "According to Ramachandran, our beliefs about ourselves and the world develop in the left side of our brain. The function of the right hemisphere is to detect anomalies, forcing the left to revise the established belief structure. Through this process, we generate a unified model of reality, a consistent storyline that allows us to live without confusion and indecision."

   I question, however, whether belief systems per se eliminate confusion. They provide the foundation for decision-making, but may vary in their relative degrees of success in reducing confusion. I also question whether belief systems originate *only* in the left hemisphere. The rationalizing aspects of the religion likely do, but I think belief systems draw also on the holistic thinking and metaphorical sorts of associations offered by the right brain. The "mystic" experiences scanned by Beauregard and others indicate some religious experiences affect the brain bi-hemispherically. See Mario Beauregard and Denyse O'Leary, *The Spiritual Brain: A Neuroscientist's Case for the Existence of the Soul* (New York: HarperCollins, 2007), 266–77. Furthermore, brain function may be only part of the picture in the human being's perception of things unseen, an issue Richard L. Gregory pursues in *Mind in Science*.
7. Tomatis glimpsed some of these implications of hemispheric instability in his observations of left-handedness: "Such a left-hander remains slow, absent, given to day-dreaming, inhibited when faced with movements that are too fast, often drawn into rhythms *that are no doubt imposed on him by uncontrolled and instinctive acts*." Tomatis, *The Ear and the Voice*, 147; emphasis mine.
8. Personal communication, Nancy Joan Gall.
9. Tomatis, *The Ear and the Voice*, 28.
10. Humans are fascinated with technology, whether or not it may be harmful; patients, particularly children and people with mental problems, are vulnerable to the curiosity of those with powerful scanners. I had read two years ago that the technicians who conduct MRIs will not lightly submit to them because they are aware of the significantly increased risk of cancer for those who receive them. Today, laws against MRIs are being considered in Europe because of the effects of the high-powered magnetism and radio waves on those technicians who operate them. "But now the legislation—already approved by Euro-MPs and EU Governments—is being postponed until at least April 30 2012 while the MRI health risks are studied again." The British medical profession was nstrumental in halting that process. Virgin Media, at: <latestnews.virginmedia.com/news/health/2008/04/06/eu_law_on_mri_scans_delayed>. Accessed 11 Aug. 2008. In the opinion of some doctors, all radiation is potentially hazardous to the human body. See Burton Goldberg. M.D., at: <www.burtongoldberg.com/health-articles/medical-x-rays-a-cause-of-cancer-and-heart-disease.htm>.
11. For example, the author of a popular self-help book, Dr. Daniel G. Amen, out of sheer curiosity, scanned a friend/patient who had fallen in love to see the changes that emotional condition made in the man's brain (he had an earlier SPECT scan of the individual on file). Fortuitously, his technician had a "shot" of radioactive dye left over at the end of the day; the doctor suggested the scan to his compliant patient. See Amen, *Change Your Brain, Change Your Life*, 86. Dr. Amen's book, while useful for some kinds of

learning, lacks longitudinal information on case studies, lacks sufficient detail on patient responses to medications including side effects, and gives shallow consideration to most other interventions by comparison with his recommendations for drugs. In his account of a case of prefrontal cortex damage he overlooks the differences in function between the brain's hemispheres that would explain the patient's personality changes. Ibid., 129.

12. Such as Dr. Robert Fischel's implant technology, an invasive surgical procedure to place a "medical device" into the brain to alter its electrical signals to "cure depression." *Fortune*, 13 June 2005, 27. We are reminded of an example of massive surgical interventions on persons with mental problems: "In the 1940s and 1950s, prefrontal lobotomies were frequently performed in mental institutions and prisons on patients whose behavior was considered undesirable and intractable. The procedure made them docile and carefree, but the alterations in their personalities were severe. At one point, it became so fashionable to have a lobotomy performed [that is, at the request of someone having some measure of control over the unfortunate patient] that it actually became an office procedure. Only a small incision in the orbit above the eye is required. Some physicians did as many as 1000 leukotomies per month during one period." University of Nebraska Medical Center, at: <www.unmc.edu/Physiology/Mann/mann17.html>.
13. Mitchell and Crow, "Right hemisphere language functions and schizophrenia," 969.
14. Ibid.; emphasis mine. See note 6, this chapter, for summary.
15. Ibid., 973.
16. Ibid., 972.
17. Gregory, ed., *Oxford Companion to the Mind*, 809.
18. Ibid.; emphasis mine.
19. Ibid., 34.
20. Ibid., 809.
21. Mitchell and Crow, "Right hemisphere language functions and schizophrenia," 972.
22. Ibid.
23. Ibid.
24. Beauregard and O'Leary, *The Spiritual Brain*, 268, 274. See also Benjafield et al., *Cognition*, ch. 5, who refer to the research of "Roediger, Gallo, and Geraci (2002) [who] observed that one of the implications of the levels of processing approach is that there is no particular place in the brain where memories are stored. Rather, the same parts of the brain used to comprehend an event will be activated when the event is recalled." This position is founded on "an MRI study done by Wheeler, Petersen, and Buckner (2002)" and supported by Nyberg (2002).
25. Beauregard and O'Leary, *The Spiritual Brain*, 263.
26. Gregory, ed., *Oxford Companion to the Mind*, 525.
27. Ibid. For the possibilities for "plasticity" in the brain due to these neural interconnections, see Doidge, *The Brain That Changes Itself.*
28. Mitchell and Crow, "Right hemisphere language functions and schizophrenia," 972.
29. Gregory, ed., *Oxford Companion to the Mind*, 426, 525.
30. Ibid., 526.
31. Ibid., 304.
32. Mitchell and Crow, "Right hemisphere language functions and schizophrenia," 973.
33. Timothy J. Crow, "The two-syndrome concept," *Schizophrenia Bulletin* 11, 3 (1983): 479, at: <brain.oxfordjournals.org>. Accessed 14 Apr. 2009.
34. Gregory, ed., *Oxford Companion to the Mind*, 809.
35. Mitchell and Crow, "Right hemisphere language functions and schizophrenia," 970.
36. Ibid.
37. Ibid., 973.

38. Ibid., 963. See Chapter 14, note 2, for summary.
39. Ibid., 972. The "scratchpad" referred to here presumably reflects Alan Baddeley and Graham Hitch's model for the concept of "working memory" or "temporary memory" that is commonly called "short-term memory." "The original model of Baddeley and Hitch was composed of three main components; the *central executive system*, which acts as supervisory system and controls the flow of information from and to its *slave systems*: the *phonological loop* and the *visuo-spatial sketchpad*. The slave systems are short-term storage systems dedicated to a content domain (verbal and visuo-spatial, respectively). In 2000 Baddeley added a third slave system to his model; the *episodic buffer*." Wikipedia, the Free Encyclopedia, at: <en.wikipedia.org/wiki/Baddeley%27s_model_of_working_memory>. Accessed 13 Mar. 2009. See Alan D. Baddeley and Graham J. Hitch, "Working Memory," in G. Bower, ed., *Recent Advances in Learning and Motivation* (New York: Academic Press, 1974), 47–89; Baddeley, "The episodic buffer: A new component of working memory?" *Trends in Cognitive Sciences* 11 (2000): 417–23.

In several points, this model fails to account for hemispheric integration. First, all episodic memory, whether short- or long-term, entails the activity of both hemispheres because all knowledge is obtained in a setting of concrete objects: parents' acculturating, teachers' instructing, books' describing with illustrations and verbal imagery, audio-visual media employing an integration of sound and pictures to convey information, and so on. To separate phonological knowledge from the images attached to phonological learning produces a false dichotomy and introduces a "ghost"—the episodic buffer—into the way the hemispheres communicate. The need for an "episodic buffer" is obviated in a system where the right and left hemispheres are engaged throughout the life course in constructing a unified logical structure under the supervision of the left brain. The information supplied by the right brain has been referred to by Baddeley as "general knowledge" (Benjafield et al., *Cognition*, chs 3, 5, 8.) In the right brain, visual images, which function as templates in anticipatory ways, provide patterns of organization for the left brain. For example, some images can function like chess pieces and be manipulated in consciousness (as Daniel would remark in his comment about being like Bobby Fischer, which also, of course, subsumed Fischer's paranoid schizophrenia). The King, Queen, knights, bishops, rooks, and pawns have other lives on different chessboards—they allow one to think in terms of CEOs and general managers and various other members of the corporate structure, for example. They might stand for the interactions of staff members in hierarchical settings, such as hospitals or universities. They can stand in for families involved in intergenerational conflict as easily as Freud's co-opting of the ancient Greek's dramatic heroes. By the means of such template images, the right brain also offers the units for comparisons and for metaphors, which the left brain assesses while assigning names and values. Geometric structures provide other sorts of templates, such as pyramids and Venn diagrams and other categorizing structures. Musical structures, which employ both hemispheres, also provide templates for other kinds of thought where flow and rhythm and other antiphonal notions are applicable to problem-solving. "General knowledge" seems an inadequate term for the holistic concepts operating through the ceaseless activity of the right brain; "holistic knowledge" might be a preferable term.

An alternate model that to me seems closer to my observations of Daniel's gradual recovering of his short-term memory and sudden resumption of normal storage of and normal access to long-term memories is that of Nelson Cowan, who regards working memory as an aspect of long-term memory. See, e.g., N. Cowan, "What are the differences between long-term, short-term, and working memory," *Progress in Brain Re-*

search 169 (2008): 323–38. Also pertinent is the work of M.J. Kane and R.W. Engel in reference to working memory (WM) that brings the "executive" capacity of the prefrontal cortex to attend (to pay attention) into the model. Unfortunately, among human subjects, they do not consider patients with neuropsychiatric illness, but only consider studies of patients with various types of damage to the dorsolateral prefrontal cortex (dPFC) including left, right, both, and extra-dPFC, which are compared with human controls under brain-imaging techniques. While they do not identify the ear as the source of attention-producing capacity, they touch on laterality to some extent. However, they do infer that the capacity for WM and attention is related to the range of general fluid intelligence (*Gf*) or psychometric intelligence (measured with IQ tests) in normal persons, which tends to support my concept of the spectrum of ear-related mental conditions producing the variety of attentional conditions from the range of normal through the ranges of abnormality. M.J. Kane and R.W. Engel, "The role of prefrontal cortex in working-memory capacity, executive attention, and general fluid intelligence: An individual-differences perspective," *Psychonomic Bulletin and Review* 9, 4 (2002): 637–71. See also Wikipedia, the Free Encyclopedia at: <en.wikipedia.org/wiki/Working_memory>. Accessed 13 Mar. 2009. It should be noted that some of the kinds of studies to which Kane and Engel refer are confounded in peculiar ways; for example, the *n*-Back tests of hypothetically left-brain activity that present letters under both upper- and lower-case forms thereby eliciting first from the subject a right-brained comparison of the shape of "b" with "B" before the numerical order of the b/Bs in the sequential display of letters can be noted and retained in WM. Seldom are aural cues distinguished from visual cues; visual cues dominate; subvocal activity often is assumed rather than ascertained; only occasionally is the laterality of the subject noted; rarely is the gender of the subject noted; age ranges appear to be limited. Leigh E. Nystrom, Todd S. Braver, Fred W. Sabb, Mauricio R. Delgado, Douglas C. Noll, and Jonathan D. Cohen, "Working memory for letters, shapes, and locations: fMRI evidence against stimulus based regional organization in human prefrontal cortex," *Neuroimage* 11 (2000): 424–46. Accessed 14 Mar. 2009, at: <www.csbmb.princeton.edu /ncc/publications/2000/NystromBraverSabb2000NI.pdf>.

40. *Physicians' Desk Reference*, 2596.
41. Bonnie J. Kaplan, J. Steven A. Simpson, Richard C. Ferre, Chris P. Gorman, David M. McMullen, and Susan G. Crawford, "Effective mood stabilization with a chelated mineral supplement: an open-label trial in bipolar disorder," *Journal of Clinical Psychiatry* 62 (2001): 936–44. Note the small sample size, but remarkable efficacy of the intervention. Of eight patients who completed the "minimum 6-month open trial, symptom reduction ranged from 55% to 66% on the outcome measures; need for psychotropic medications decreased by more than 50%. . . . In some cases, the supplement replaced psychotropic medications and the patients remained well. The only reported side effect (i.e., nausea) was infrequent, minor, and transitory."
42. Tomatis, *The Ear and the Voice*, 65.
43. Ibid., 61–2.
44. Ibid, 62.
45. Jay A. Goldstein, M.D., "Treatment protocol for physicians," *The Forum*. At: <www.ncf-net.org/forum/jay.htm>. Accessed 14 Apr. 2009.
46. Madaule, *When Listening Comes Alive*, 61–2. Madaule does not mention schizophrenic and schizo-affective disorders, but his description of depression induced by a "downward spiral" of sound deprivation is relevant, including how this behavioural syndrome relates to some sufferers of chronic fatigue syndrome. See also Chapter 11, note 3: Dr. Joseph Huggins "places on the DSM-IV and ICD-9-CM diagnostic spectrum, ~

Autism/PDD/ADHD/DD . . . Fibromyalgia & Chronic Fatigue Syndromes."
47. See Chapter 11, note 3, re Dr. Huggins.
48. Mitchell and Crow, "Right hemisphere language functions and schizophrenia," 974.
49. Doidge, *The Brain That Changes Itself*, 47, 75.
50. The reader can compare my observations with the disabling conditions also purported to respond to sound stimulation at Annabel Stehli's website: "Distortions in auditory discrimination are present in many common learning disabilities such as dyslexia and attention deficit disorder and in many more complex problems such as autism, pervasive developmental delay and depression. In addition, it has been found that AIT has sometimes proved useful in arresting progressive deafness." At: <www.georgianainstitute.org/index.htm>. Accessed 13 Jan. 2009.
51. M. Hoke, G. Kauffmann, and E. Bappert, "Preface," in Hoke, Kauffmann, and Bappert, eds, *Cochlear and Brainstem Evoked Response Audiometry and Electrical Stimulation of the VIIIth Nerve* (Stockholm: Almqvist & Wilsell Periodical Co., 1980).
52. *Taber's Cyclopedic Medical Dictionary*, 1141.
53. Ibid.
54. Tomatis, *The Ear and the Voice*, 55–8.
55. Evelyne Veuillet, Annie Magnan, Jean Ecalle, Hung Thai-Van, and Lionel Collet, "Auditory processing disorder in children with reading disabilities: effect of audiovisual training," *Brain*, at: <brain.oxfordjournals.org/cgi/content/abstract/130/11/2915>. Accessed 31 Jan. 2009.
56. Ibid.
57. Ibid.
58. Ibid.
59. S. Khalfa, N. Bruneau, B. Rogé, N. Georgieff, E. Veuillet, J.-L. Adrien, C. Barthélémy and L. Collet, "Peripheral auditory asymmetry in infantile autism,". *European Journal of Neuroscience* 13: 628-32, 2001. At: <skhalfa.com/ doc/Khalfa_autisme.pdf>. Accessed 18 July 2009.
60. Veuillet, "Auditory processing disorder."
61. Ibid.
62. Doidge, *The Brain That Changes Itself*, 81.
63. May, *Addiction and Grace*.

Chapter 14
The Cerebral Society: *Cerebral Dominance Institutionalized*

1. Memory for auditory word sequences may reside in the left brain but will not make much impression on long-term memory without hemispheric integration. During periods of right-ear damage causing right-brain psychosis, it may be that the electrical alteration in the left hemisphere as the right surrenders its dominance prevents the retention of acts and perceptions of the right-brain phase of cognition in the long-term memory. This may be the same mechanism that explains the loss of memory during episodes of epileptic seizure, bipolar rages, emotional shocks, and some physical traumas. At a much lower level of impairment, the switching of hemispheres in dyslexic syndrome children also can interfere with memory and learning. Bellis describes a case of a learning disabled child who seemed nonetheless so intelligent that she questioned his diagnosis as mentally retarded: "In Randy's case, his rapidly cycling seizure activity was so regular and subtle that the clinician who had administered all three of his cognitive assessments had simply missed it." Bellis, *When the Brain Cannot Hear*, 189–93.

   Other family members have noticed that Daniel does not remember much about how he has behaved or what has happened to those around him during an episode of prevalent psychosis. When he recovered from his recent episode, his memories of the previous 11 months retreated like a dream from which he had awakened. Sometimes, with reminders, he could reclaim aspects of his psychotic episode.

See also <www.unmc.edu/Physiology/Mann/mann17.html>: "Animals given electroconvulsive shocks or other treatments to interrupt brain activity within about 5 minutes after a training session do not remember the task they were trained to perform when tested some time later. When no treatment is given, the same animals remember the task 5 minutes and even weeks after the training sessions. If the same treatment is given to the animal 4 hours after training sessions, it has no effect at all on their ability to remember the task even weeks later. It thus appears that there is a period shortly after an event when the memory is vulnerable to disruption; then something happens to make the memory trace resistant to erasure. It is this event that constitutes the conversion from short-term to long-term memory."

2. Wikipedia, the Free Encyclopedia, at: <en.wikipedia.org/wiki/Schizophrenia> and <en.wikipedia.org/wiki/Integration_disorder>. Accessed 30 Sept. 2008: See Mitsumoto Sato, "Renaming schizophrenia: a Japanese perspective," *World Psychiatry* 5, 1 (2006): 53–5, at: <www.ncbi.nlm.nih.gov/ pubmed/16757998?dopt=AbstractPlus>. Accessed 30 Sept. 2008, 30 Apr. 2009. I have adopted this more useful term for the syndromes not only of schizophrenia but of the other disorders I place on the spectrum of ear-related brain dysfunction.
Summary:
In order to contribute to reduce the stigma related to schizophrenia and to improve clinical practice in the management of the disorder, the Japanese Society of Psychiatry and Neurology changed in 2002 the old term for the disorder, "Seishin Bunretsu Byo" ("mind-split-disease"), into the new term of "Togo Shitcho Sho" ("integration disorder"). The renaming was triggered by the request of a patients' families group. The main reasons for the renaming were the ambiguity of the old term, the recent advances in schizophrenia research, and the deep-rooted negative image of schizophrenia, in part related to the long-term inhumane treatment of most people with the disorder in the past. The renaming was associated with the shift from the Kraepelinian disease concept to the vulnerability-stress model. A survey carried out seven months after renaming in all prefectures of Japan found that the old term had been replaced by the new one in about 78% of cases. The renaming increased the percentage of cases in which patients were informed of the diagnosis from 36.7% to 69.7% in three years. Eighty-six percent of psychiatrists in the Miyagi prefecture found the new term more suitable to inform patients of the diagnosis as well as to explain the modern concept of the disorder. The Japanese treatment guideline for "Togo Shitcho Sho" was developed in 2004 under the framework of the vulnerability-stress model.
3. Wikipedia, the Free Encyclopedia, at: <en.wikipedia.org/wiki/Schizophrenia> and <en.wikipedia.org/wiki/Integration_disorder>.
4. Gregory, *Mind in Science*, 8–9.
5. Julian Jaynes, *The Origins of Consciousness in the Breakdown of the Bicameral Mind*
6. Maureen Cech, *Interventions with Children and Youth,* (Don Mills, Ont.: Oxford University Press, 2010 [forthcoming]), ch. 3.
7. Gregory, *Mind in Science*, 8–9.
8. Keefe and Harvey, *Understanding Schizophrenia*, 22.
9. Doidge, *The Brain That Changes Itself*, 47, 75.
10. Erving Goffman, *Stigma: Notes on the Management of Spoiled Identity* (Englewood

Cliffs, NJ: Prentice-Hall, 1963), 40.
11. Hoffman and Froemke, eds, *Why Can't They Just Stop?*, 56.
12. Gregory, *Mind in Science*, 547.

Chapter 15
The Nuclear Family: *Microcosm and Macrocosm*

1. Anthony Thomson, *The Making of Social Theory: Order, Reason, and Desire* (Toronto: Oxford University Press, 2006), 379–86, 403–4.
2. Ibid.
3. Georg Simmel, "The Problem of Sociology," in Simmel, *On Individuality and Social Forms*, ed. Donald N. Levine (Chicago: University of Chicago Press, 1971).
4. C.S. Lewis, "A Confession," *Poems* (London: Geoffey Bles, 1964), 1.
5. Samuel Johnson, *A Dictionary of the English Language* (London, 1755).
6. Samuel Johnson (1709–1784), *The Lives of the most Eminent English Poets with Critical Comments on their Works, a new edition, corrected, in Four Volumes* (London: printed for C. Bathurst, J Ruckland, W. Strahan, J. Rivington and sons, et al., 1783).
7. Diary entry, William A. Mohun, 1930.
Scarlet Fever
Took sick: March 27 (Thurs) 1930.
I had a cold which seemed to be getting better. Woke up in morning with a very sore throat, but went to school. Felt very miserable all day and was stiff, especially my neck. Went out with the Camera Club to the Canadian Kodak Co. after school. Come home and went straight to bed.
Next morning temperature 101° and we couldn't get it down by sweating.
Doctor came Saturday 29th and said it was Scarlet Fever. The rash just came out this afternoon. . . .
Placard put up April 1st
Feeling much better April 3rd
Stiff neck started April 4th
Can only lie on one side and my back. Cannot turn my head at all.
Had to keep hot linseed poutices on my neck; they were changed about every twenty minutes which was very tiring.
Neck lanced - Sat. April 8th by Dr. J.S. Hart. Local Anaesthetic (ethyl Chloride) Incision ¾" long x ¾" deep. Very little pus. Has to be dressed with gauze pad dipped in a warm 1% solution of Lysol every 3½–4 hrs.
April 16 - Neck easier but still some stiff.
April 24 - Neck got quite sore and stiff again this evening.
April 25 - Very sore. Infection coming to a head.
April 26th, Lanced Again—only surface anaesthetic (spray of Ethyl Chloride) Incision ¾" x ¾"
Great quantity of pus. Hole w[h]ere pus pocket was makes the cavity over an inch deep.
(26th) — Mother pulled out the plug this evening that the doctor put in. Pus flowed freely for 2 hrs.
May 3rd — Neck still draining slightly. Mother has to push a plug in every night to keep it open and also probe it with a bodkin each time to open it up when it has started to heal.
May 5th – Sat up on edge of bed
May 9 – Sat up in chair
May 10th – Placard off.

May 12th – Room fumigated
My eyes have been so weak all along that I had to have the room darkened but they are a bit better to-day (14th)
May 18 (Sunday) Got up, dressed and went downstairs for supper
May 19 [no entry, diary ends]

8. John Stuart Mill, *System of Logic,* 1843, Book III Of the Ground of Induction, Ch. 3, para.1 Axiom of the uniformity of the course of nature. Quoted in Gregory, *Mind in Science*, 242.
9. *The Madness of King George*, 1994, directed by Nicholas Hytner, with Nigel Hawthorne, Helen Mirren, Ian Holme.

Bibliography

Books, Articles, Websites, Personal Communication

Adrian, E.D. 1928. *The Basis of Sensation* (quoted in Gregory, *Mind in Science*).

Amen, Daniel G. 1998. *Change YouBrain, Change Your Life*. New York: Three Rivers Press, Random House.

Baddeley, Alan D. 2000. "The episodic buffer: A new component of working memory?" *Trends in Cognitive Sciences* 11: 417–23.

——— and Graham J. Hitch. 1974. "Working Memory," in G. Bower, ed., *Recent Advances in Learning and Motivation*. New York: Academic Press, 47–89

Bailey, Martin. 2005. *Apollo* (Sept.). At: <findarticles.com/p/articles/mi_m0PAL/is_523_162/ai_n15880259>. Accessed 1 Feb. 2009.

Beauregard, Mario, and Denyse O'Leary. 2007. *The Spiritual Brain: A Neuroscientist's Case for the Existence of the Soul*. New York: HarperCollins.

Bellis, Teri James. 2002. *When the Brain Can't Hear: Unravelling the Mystery of Auditory Processing Disorder*. New York: Simon & Schuster.

Benjafield, John G., Daniel Smilek, and Alan Kingstone. 2010. *Cognition*, 4th edn. Toronto: Oxford University Press.

Bergler, Edmund. 1962. *Homosexuality: Disease or Way of Life?* New York: Hill and Wang.

Blackadar, Bruce. 1996. "American dream goes sour, and we're not far behind," *Toronto Star*, 21 Jan., F6.

Bloomberg.com Science, at: <www.bloomberg.com/apps/news?sid=aaw9D5Ty8 Eho&pid =20601124>.

Bruzon, Mark. At: <www.geocities.com/player2000gi/hemispheric_specialization.htm>.

Bymaster, F.P., and C.C. Felder. 2002. "Role of the cholinergic muscarinic system in bipolar disorder and related mechanism of action of antipsychotic agents," *Molecular Psychiatry* 7, Suppl. 1: S57–63. At: <lansbury.bwh.harvard. edu/da_and_agonists.htm>.

Campbell, Don. 1997. *The Mozart Effect: Tapping the Power of Music to Heal the Body, Strengthen the Mind, and Unlock the Creative Spirit*. New York: Avon Books.

Castaneda, Carlos. 1968. *The Teachings of Don Juan: A Yaqui Way of Knowledge*. Berkeley: University of California Press.

Clarke, Juanne Nancarrow. 1996–2009. *Health, Illness, and Medicine in Canada*. Toronto: Oxford University Press, various editions.

Cousins, Norman. 1979. *Anatomy of an Illness As Perceived by the Patient*. New York: Norton.

Cowan, Nelson. 2008. "What are the differences between long-term, short-term, and working memory," *Progress in Brain Research* 169: 323–38.

Crow, Timothy J. 1983. "The two-syndrome concept," *Schizophrenia Bulletin* 11, 3: 479, at: <brain.oxfordjournals.org>. Accessed 14 Apr. 2009.

———. 2000. "Did Homo sapiens speciate on the Y chromosome?" *Psychology* 11, 001. American Psychological Association, at: <www.cogsci.ecs.soton.ac.uk/ cgi/psyc/newpsy?11.001>. Accessed 25 June 2008.

Deafened. At: <www.deafened.org> (on cochlear implant).

Dejean, Valerie. At: < http://216.194.201.205/SI/valeriedejean.org/> (treating children with autism).

Doidge, Norman. 2007. *The Brain that Changes Itself: Stories of Personal Triumph from the Frontiers of Brain Science*. New York: Penguin Books.

Elliott, G.B., and K.A. [Elliott?]. 1964. "Some pathological, radiological, and clinical implications of precocious development of the human ear," *Laryngyscope* 79: 1160–71 (quoted in Madaule, *When Listening Comes Alive*).

Forcese, Dennis. 1997. *The Canadian Class Structure*, 4th edn. Toronto: McGraw-Hill Ryerson (quoted in Wotherspoon, *The Sociology of Education in Canada: Critical Perspectives*).

Fischel, Robert. 2005. "Implant technology," *Fortune*, 13 June 27.

Freed, Jeffery, and Laurie Parsons, *Right-Brained Children in a Left-Brained World: Unlocking the Potential of Your ADD Child*. New York: Simon & Schuster.

Gall, Nancy Joan. 2008. Kingston, Ontario. Personal communication, 3 Sept.

Gilmor, Tim, Paul Madaule, and Thompson. 1987. *About the Tomatis Method*, trans. from the French by Dr. Paula Kacher. Toronto: Listening Centre Press.

Goffman, Erving. 1961. *Asylums: Essays on the Social Situation of Mental Patients and Other Inmates*. Garden City, NY: Anchor Books, Doubleday.

———. 1963. *Stigma: Notes on the Management of Spoiled Identity*. Englewood Cliffs, NJ: Prentice-Hall.

Goldberg, Burton, M.D. At: <www.burtongoldberg.com/health-articles/medical-x-rays-a-cause-of-cancer-and-heart-disease.htm>. Accessed 11 Aug. 2008.

Goldstein, Jay A., M.D. n.d. "Treatment protocol for physicians," *The Forum*. At: <www.ncf-net.org/forum/jay.htm>. Accessed 14 Apr. 2009.

———. n.d. "Looking in the Mirror: an M.E./CFIDS Symptom checklist by Jay A. Goldstein, M.D., Anaheim Hills, California."

Goldstein, Jerome, M.D. 2008. "The Neurobiology of Sexual Orientation—The Gay Brain," San Francisco Neurological Society Meeting, 11 Nov. Personal communication. Dr. Goldstein's website is at: <www.sfcrc.com/html/aboutsfcrc.htm>.

Gray, Henry. 1974. *Anatomy, Descriptive and Surgical*, 1901 edn. Philadelphia: Running Press.

Greenberg, Joanne. 1964. *I Never Promised You a Rose Garden*. New York: New American Library.

Gregory, Richard L. 1981. *Mind in Science*. Harmondsworth, Middlesex: Penguin Books.

———, ed. 1987. *The Oxford Companion to the Mind*. London: Oxford University Press.

Griffiths, Anita. 1978. *Teaching the Dyslexic Child*. Novato, Calif.: Academic Therapy Publications, 1978. Reproduced by The Ann Arbor Company UK at: <www.annarbor.co.uk/acatalog/Teaching_the_Dyslexic_Child.html>.

Hallowell, Edward M., and John J. Ratey. 1994. *Driven to Distraction: Recognizing and Coping with Attention Deficit Disorder from Childhood through Adulthood*. New York: Simon & Schuster.

Halmos, G., B. Lendvai, et al. 2002. "Simultaneous measurement of glutamate and dopamine release from isolated guinea pig cochlea," *Neurochem Int* 40, 3: 243–8. At: <lansbury .bwh.harvard.edu/da_and_agonists.htm>.

Heslet, Lars, Professor Doctor. At: <www.musicahumana.dk/ documents/ 00015.pdf>.

Hoffman, John, and Susan Froemke, eds. 2007. *Why Can't They Just Stop? Addictions: New Knowledge, New Treatments, New Hope*. New York: Rodale.

Hoke, M., G. Kauffmann, and E. Bappert, eds. 1980. *Cochlear and Brainstem Evoked Response Audiometry and Electrical Stimulation of the VIIIth Nerve*. Proceedings from the Workshop Held at Münster, Federal Republic of Germany, March 1979. *Scandinavian Audiology* Supplementum II. Stockholm: Almqvist & Wilsell Periodical Co.

Holley, Matthew C. 2002. "Application of new biological approaches to stimulate sensory repair and protection," *British Medical Bulltetin* 63: 16–19.

Huggins, Joseph E. (quoted in Wheatley, *My Sad Is All Gone*).

———, Dr. Soula Homatidis, and Jason G. Huggins. "The 100% 'Evidence-Based' Universal

Diagnostic & Treatment Model© Version 5.00-2000: A new clinical approach & methodology utilizing neurotransmitter - cytochrome P450 isoenzyme affinity & binding contro." Copies of work submitted to the National Library of Canada, Ottawa.

Hyde, Byron, Jay Goldstein, and Paul Levine. 1992. *A Symposium: The Clinical and Scientific Basis of Myalgic Encephalitis/Chronic Fatigue Syndrome*. Ottawa: Nightingale Foundation.

Jefferson, Lara. 1975. *These Are My Sisters: A Journal from the Inside of Insanity*. Garden City, NY: Anchor Press/Doubleday.

Johnson, Hillary. 1996. "The forgotten epidemic," *Rolling Stone*, 21 Mar.

———. 1996. *Osler's Web: Inside the Labyrinth of the Chronic Fatigue Syndrome Epidemic*. New York: Crown Publishers, Random House.

Johnson, Samuel. 1755. *A Dictionary of the English Language*. London.

——— (1709–1784). 1783. *The Lives of the most Eminent English Poets with Critical Comments on their Works, a new edition, corrected, in Four Volumes*. London: printed for C. Bathurst, J Ruckland, W. Strahan, J. Rivington and sons, et al.

Kane, M.J., and R.W. Engel. 2002. "The role of prefrontal cortex in working-memory capacity, executive attention, and general fluid intelligence: An individual-differences perspective," *Psychonomic Bulletin and Review* 9, 4: 637–71.

Kaplan, Bonnie J., Steven A. Simpson, Richard C. Ferre, Chris P. Gorman, David M. McMullen, and Susan G. Crawford. 2001. "Effective mood stabilization with a chelated mineral supplement: An open-label trial in bipolar disorder," *Journal of Clinical Psychiatry* 62: 936–44.

Keefe, Richard S.E., and Philip D. Harvey. 1994. *Understanding Schizophrenia*. New York: Free Press.

Koelsch, Stefan, Matthias Wittforth, Angelika Wolf, Joachim Müller, and Anja Hahne. 2004. "Music perception in cochlear implant users: An event-related potential study," *Clinical Neurophysiology* 115, 4: 966–72. At: <www.journals.elsevierhealth.com/periodicals/clinph/issues/contents?issue_key=S1388-2457(00)X0070-3>.

Kosten, T.R., T.P. George, et al. 2002. "The potential of dopamine agonists in drug addiction," *Expert Opin Investig Drugs* 11, 4: 491–9. At: <lansbury.bwh.harvard.edu/da_and_agonists.htm>.

Lambert, Craig. 2000. "Deep cravings: New research on the brain and behavior clarifies the mysteries of addiction," *Harvard Magazine* 102, 4 (Mar.–Apr.). At: <harvard-magazine.com/2000/03/deep-cravings.html>.

Lewis, C.S. 1964. "A Confession," *Poems*. London: Geoffey Bles.

Lehtinen, Laura E. c. mid-1980s. "Have you ever known a perceptually handicapped student?" From the New Jersey Association for Brain Injured Children.

Levinson, Harold, M.D. 1980. *A Solution to the Riddle—Dyslexia*. New York: Springer-Verlag.

———. 1994. *Smart but Feeling Dumb: The Challenging New Research on Dyslexia—And How It May Help You*. New York: Warner Publications.

———. 2000. *The Discovery of Cerebellar-Vestibular Syndromes and Therapies: A Solution to the Riddle—Dyslexia*. Stonebridge Publications.

———. at: <www.dyslexiaonline.com/information/treatment.html>.

Levitin, Daniel J. 2008. *The World in Six Songs: How the Musical Brain Created Human Nature*. New York: Dutton.

Madaule, Paul. 1987. "The Dyslexified World," in Gilmor, Madaule, and Thompson, *About the Tomatis Method*, trans. from the French by Dr. Paula Kacher. Toronto: The Listening Centre Press. "L'Univers Dyslexie" was presented at the Fifth International Congress of Audio-Psycho-Phonology (Tomatis Method) in Toronto in 1978.

———. 1993. *When Listening Comes Alive*. Norval, Ont.: Moulin Publishing.

Mann, Michael D. 1997–2008. *The Nervous System in Action*. © 1997–2008 by Michael D. Mann, Ph.D. At: University of Nebraska Medical Center, on-line version of text: <www.unmc.edu/Physiology/Mann/mann17.html>; <www.unmc.edu/Physiology/Mann/index.html>.

May, Gerald. 1988. *Addiction and Grace*. San Francisco: Harper Row.

McManus, Chris. 2002. *Right Hand, Left Hand: The Origins of Asymmetry in Brains, Bodies, Atoms and Cultures*. Cambridge, Mass.: Harvard University Press.

Mental Help.net at: <www.mentalhelp.net/poc/view_doc.php?i=439&type =book&cn=7>.

Mitchell, Rachel L.C., and Tim J. Crow. 2005. "Right hemisphere language functions and schizophrenia: The forgotten hemisphere?" *Brain: A Journal of Neurology*. At: <brain.oxfordjournals.org/cgi/content/full/128/5/963>. Accessed 25 June, 12 July 2008.

Mohun, William A. 1930. "Diary."

National Longitudinal Survey of Children and Youth. 1996. Ottawa.

Ngan, E.T.C., C.J. Lane, T.J. Ruth, and P.F. Liddle. 2002. "Immediate and delayed effects of risperidone on cerebral metabolism in neuroleptic naïve schizophrenic patients: Correlations with symptom change," *Journal of Neurological and Neurosurgical Psychiatry* 72, 1: 106–110. At: <jnnp.bmj.com/cgi/content/abstract/72/1/106>. Accessed 18 July 2008.

Nystrom, Leigh E., Todd S. Braver, Fred W. Sabb, Mauricio R. Delgado, Douglas C. Noll, and Jonathan D. Cohen. 2000. "Working memory for letters, shapes, and locations: fMRI evidence against stimulus based regional organization in human prefrontal cortex," *Neuroimage* 11: 424–46. At: <www.csbmb.princeton.edu/ncc/publications/2000/NystromBraverSabb2000NI.pdf>. Accessed 14 Mar. 2009.

Payer, Lynn. 1988. *Medicine and Culture: Varieties of Treatment in the United States, England, West Germany, and France*. New York: Holt.

Petrofski, Morana, Co-Director, The Listening Centre, Toronto, 2009. Personal communication, 9 Jan.

The Physicians' Desk Reference, 50th edn. 1996. Montvale, NJ: Medical Economics Company.

Rabinowicz, Theodore, Dorothy E. Dean, Jean McDonald-Comber Petetot, and Gabrielle M. de Courten-Myers. 1999. "Gender differences in the human cerebral cortex: More neurons in males; more processes in females," *Journal of Child Neurology* 14, 2: 98–107. At: <jcn.sagepub.com/cgi/content/abstract/14/2/98>.

Ratey, John J. 1996. *A User's Guide to the Brain: Perception, Attention, and the Four Theaters of the Brain*. New York: Simon & Schuster.

Reaume, Geoffrey. 2000. *Remembrance of Patients Past: Patient Life at the Toronto Hospital for the Insane, 1870–1940*. Toronto: Oxford University Press.

Regush, Nicholas. 1997. "The listening cure," *Equinox Magazine* (May): 5–6.

Risperdal. At: <www.risperdal.com/risperdal/>.

Ruben, Sharon. 2004. *Awakening Ashley: Mozart Knocks Autism on Its Ear*. Bloomington, Ind.: iUniverse. See "Awakening Ashley" at: <www.awakeningashley.com>.

Sacks, Oliver. 2008. *Musicophilia: Tales of Music and the Brain*. New York: Vantage Books.

Sato, Mitsumoto. 2006. "Renaming schizophrenia: A Japanese perspective," *World Psychiatry* 5, 1: 53–5. At: <www.ncbi.nlm.nih.gov/pubmed/16757998? dopt=AbstractPlus>. Accessed 30 Sept. 2008, 30 Apr. 2009.

Shilts, Randy. 1987. *And the Band Played On: Politics, People, and the AIDS Epidemic*. New

York: St. Martin's Press.

Schultz, Lee. 2008. Personal communication, 12 Sept.

Simmel, Georg. 1971 [1908]. "The Problem of Sociology," in Simmel, *On Individuality and Social Forms*, ed. Donald N. Levine. Chicago: University of Chicago Press.

Slinger, Joey. c. 1996. *Toronto Star* column.

Smith, Edith Joyce. 1985. "Does it make you happy that I tell you of a kindness?" *Toronto Star*, 27 Aug.

Sollier, Pierre. 2005. *Listening for Wellness: An Introduction to the Tomatis Method*. Walnut Creek, Calif.: Mozart Centre Press.

Spectrum Centre, The. At: <216.194.201.205/SI/tomatis.net/> (Dejean's Spectrum Centre in New York).

Stehli, Annabel. 1991. *The Sound of a Miracle: A Child's Triumph over Autism*. New York: Doubleday.

Taylor, Jill Bolte. 2006. *My Stroke of Insight*. New York: Penguin Group.

Thomas, Clayton L., ed. 1985. *Taber's Cyclopedic Medical Dictionary*, 15th edn. Philadelphia: F.A. Davis.

Thomson, Anthony. 2006. *The Making of Social Theory: Order, Reason, and Desire*. Toronto: Oxford University Press.

Tomatis, Alfred A. 1991. *The Conscious Ear: My Life of Transformation through Listening*. Barrytown, NY: Station Hill Press.

———. 1996. *The Ear and Language*. Norval, Ont.: Moulin.

———. 2005. *The Ear and the Voice,* trans. Roberta Prada. Lanham, Md.: Scarecrow Press.

Tomatis Method. At: <www.tomatis.com/English/Articles/Biography.html>.

Toronto Star. 1991. "Mom says 'miracle' treatment healed her daughter's autism," 1 Feb., B1.

———. 1996. 22 Sept., B1 (on death of James Lonnee).

Upledger, John, and Jon D. Vredevoogd. 1983. *Craniosacral Therapy*. Seattle: Eastland Press.

United States Drug Agency. At: <www.ars.usda.gov/is/AR/archive/aug07/ aging0807.htm>.

Vacca, John, ed. 2005. *The World's 20 Greatest Unsolved Problems*. Upper Saddle River, NJ: Pearson Prentice-Hall.

Veuillet, Evelyne, Annie Magnan, Jean Ecalle, Hung Thai-Van, and Lionel Collet. 2007. "Auditory processing disorder in children with reading disabilities: Effect of audiovisual training," *Brain* 130, 11: 2915–28. At: <brain.oxfordjournals.org/ cgi/content/abstract/130/11/2915>

Virgin Media. At: <latestnews.virginmedia.com/news/health/2008/04/06/eu_law _on_mri_scans_delayed>. Accessed 11 Aug. 2008.

Volkow, Nora D. 2007. "The addicted brain: Beyond willpower," in Hoffman and Froemke, eds., *Why Can't They Just Stop?*

Welham-Umphrey, Ruth, Personal communication, *passim*.

Wheatley, Thelma. 2005. *My Sad Is All Gone*. Lancaster, Ohio: Lucky Press.

White, William L. 1998. *Slaying the Dragon: The History of Addiction Treatment and Recovery in America*. Normal, Ill.: Lighthouse Institute Publications.

Wikipedia, The Free Encyclopedia. At: < www.wikinfo.org/index.php/>, on the following topics: corpus callosum; dual brain theory; Geschwind-Galaburda hypothesis; integration disorder; laterality (Dr. L. Durkashan); schizophrenia; Vincent Van Gogh; working memory.

Williams, Richard. 2001. "Lower dose, longer titration recommended for Risperidone," *Journal of Clinical Psychiatry* 62 (Apr): 282–9 (also reported in Reuters, Medical News, Medscape, 2001; literature review by Dr. Richard Williams, Royal Jubilee Hospital, Victoria, British Columbia, Canada).

Willot, J.F., and S.M. Lu. 1982. "Noise-induced hearing loss can alter neural coding and increase excitability in the central nervous system of mice," *Science* 216, 4552 (1982): 1331. At: <www.sciencemag.org/cgi/content/abstract/216/4552/ 1331>.

Wotherspoon, Terry. 2008. "Education," in Lorne Tepperman, James Curtis, and Patrizia Albanese, eds, *Sociology*, 2nd edn. Toronto: Oxford University Press, 244–75.

———. 2009. *The Sociology of Education in Canada*, 3rd edn. Toronto: Oxford University Press; 2nd edn, 2004.

Discography

Angel Clare. Art Garfunkel. New York: Columbia Records, 1973.

Antonio Vivaldi Violin Concertos. Alan Loveday, Iona Brown, Carmel Kaine, Roy Gillard, solo violin, Academy of St. Martin-in-the-Fields, Sir Neville Marriner, conductor. London: Decca, 1973. Cat. No. 4674322.

Bach for Meditation. Naxos Canada, 2004. Cat. No. 8.557650.

Earobics® Series #1 and #2, The Listening Centre, Toronto (audiocasette tape program of breathing and sound production that augments and supports the Electronic Ear™ and listening retraining).

In Utero. Nirvana. DGC Records, 1993.

The String Quartet Tribute to R.E.M. Los Angeles: Vitamin Records, 1999. Cat. No. CD-8517.

Filmography (chronological)

The Snake Pit, 1948, directed by Anatole Litvak, with Olivia de Havilland, Mark Stephens, Leo Genn.

David and Lisa, 1963, directed by Frank Perry, with Keir Dulea, Janet Margolin, Howard Da Silva.

One Flew Over the Cuckoo's Nest, 1975, directed by Milos Forman, with Jack Nicholson, Louise Fletcher.

The Gods Must Be Crazy, 1980, directed by James Uys, with Marius Weyess, Sandra Pensloo, N!xau.

The Last Emperor, 1987, directed by Bernardo Bertolucci, with John Lone, Joan Chen, Peter O'Toole.

Awakenings, 1990, directed by Penny Marshall, with Robert de Niro, Robin Williams, Julie Kavner. Based on Oliver Sacks's memoir; see www.oliversacks.com.

Lorenzo's Oil, 1992, directed by George Miller, with Susan Sarandon, Nick Nolte, Zach O'-Malley Greenburg.

And the Band Played On, 1993, directed by Roger Spottiswoode, with Matthew Modine, Alan Alda, Richard Gere, Angelica Huston.

The Madness of King George, 1994, directed by Nicholas Hytner, with Nigel Hawthorne, Helen Mirren, Ian Holm.

Shine, 1996, directed by Luke Hicks, with Geoffrey Rush as David Helfgott.

The Empire Strikes Back, 1997, Special Edition theatre re-release of original 1980 film, directed by George Lucas, with Harrison Ford, Mark Hamill, Carrie Fisher, Alec Guinness; voiceover by James Earl Jones.

A Beautiful Mind, 2001, directed by Ron Howard, with Russell Crowe, Jennifer Connolly, Ed Harris, Paul Bettany.

Index

The Notes begin on page 398; page references appear in italic.

To order copies of

Listening for the Light:

A New Perspective on Integration Disorder

in

Dyslexic Syndrome, Schizophrenia, Bipolarity,
Chronic Fatigue Syndrome, and Substance Abuse

by

Laurna Tallman

visit

www.northernlightbooks.ca

email

laurnatallman@kos.net

or mail your order to

Northern Light Books

Box 494

Marmora, Ontario, Canada

K0K2M0